DEAREST BARB

"DEAREST BARB"
FROM KARACHI, 1943–1945

Letters and Photographs in the World War II Papers
of a Naval Intelligence Officer
Lieutenant Albert Zimmermann, USNR

By

George J. Hill, M.D., M.A., D.Litt.
Captain, Medical Corps, USNR (ret)

HERITAGE BOOKS
2018

HERITAGE BOOKS
AN IMPRINT OF HERITAGE BOOKS, INC.

Books, CDs, and more—Worldwide

For our listing of thousands of titles see our website
at
www.HeritageBooks.com

Published 2018 by
HERITAGE BOOKS, INC.
Publishing Division
5810 Ruatan Street
Berwyn Heights, Md. 20740

Copyright © 2018 George J. Hill, M.D., M.A., D.Litt.

Heritage Books by the author:

*American Dreams: Ancestors and Descendants of John Zimmermann and
Eva Katherine Kellenbenz, Who Were Married in Philadelphia in 1885*

*"Dearest Barb": From Karachi, 1943–1945, Letters and Photographs in the World War II Papers
of a Naval Intelligence Officer, Lieutenant Albert Zimmermann, USNR*

Edison's Environment: The Great Inventor Was Also a Great Polluter

Four Families: A Tetralogy Reader's Guide to Western Pilgrims, Quakers and Puritans, Fundy to Chesapeake, *and* American Dreams;
*Synopsis of 481 Immigrants and First Known Ancestors in America from Northern Europe in the Families of George J. Hill
and Jessie F. Stockwell, William T. Shoemaker and Mabel Warren, William H. Thompson and Sarah D. Rundall,
John Zimmermann and Eva K. Kellenbenz, with Outlines of Their Descent from the Immigrants.*

*Fundy to Chesapeake; The Thompson, Rundall and Allied Families: Ancestors and Descendants of
William Henry Thompson and Sarah D. Rundall, Who Were Married in Linn County, Iowa, in 1889*

*Hill: The Ferry Keeper's Family, Luke Hill and Mary Hout,
Who were Married in Windsor, Connecticut, in 1651 and Fourteen Generations of Their Known and Possible Descendants*

John Saxe, Loyalist (1732–1808) and His Descendants for Five Generations

*Quakers and Puritans: The Shoemaker, Warren and Allied Families; Ancestors
and Descendants of William Toy Shoemaker and Mabel Warren, Who Were Married in Philadelphia in 1895*

*Western Pilgrims: The Hill, Stockwell and Allied Families;
Ancestors and Descendants of George J. Hill and Jessie Fidelia Stockwell, Who Were Married in Wright County, Iowa, in 1882*

Illustrations
The photos dated 1942-45 are from the Wartime Papers of Albert Zimmermann, of which thirty-one were previously
published by the U.S. Naval Institute Press. All are now in Special Collections at the U.S. Naval Institute.
Uncredited photos are from the personal collection of the author. Permission requested for
reproductions of paintings by April Swayne-Thomas, *Indian Summer*.

Cover designed by Debbie Riley

All rights reserved. No part of this book may be reproduced or transmitted in any form or by any means,
electronic or mechanical, including photocopying, recording or by any information storage and retrieval system
without written permission from the author, except for the inclusion of brief quotations in a review.

International Standard Book Numbers
Paperbound: 978-0-7884-5824-8

Lieutenant Albert W. Zimmermann, USNR
Commanding Officer, Naval Liaison Office, Karachi, India – 1944-1945

U.S. Naval Liaison Office, Karachi – 1944
243 Engle Road

Thanks to my father-in-law, Albert Zimmermann, for his service,
and to his wife, Barbara, for preserving the records of it

Albert Walter Zimmermann   Barbara Shoemaker

c. 1924

Contents

Dedication	vi
Contents	vii
Preface	ix
Chapter 1 – Albert Zimmermann and Barbara Shoemaker, 1902-1942	1
Chapter 2 – Zimmermann Joins the Navy, 1942-43	9
Chapter 3 – "Dearest Barb": Zimmermann's Letters from Karachi, 1943-45	11
1943	17
1944	68
1945	150
Chapter 4 – The North-West Frontier Trip, November-December 1943	179
Chapter 5 – Aftermath: Zimmermann Returns Home, and His Post-War Life	261
Chapter 6 – Zimmermann's Correspondence with Others	267
Major Gordon Enders	268
Major Sir Benjamin Bromhead	296
Lieutenant Curtin Winsor and Others in the Office of Naval Intelligence	307
Other Correspondence and Invitations	320
Notes	323
Appendices	335
Appendix A: AWZ Wartime Papers at U.S. Naval Institute	335
Appendix B: Zimmermann's Biographical Sketch	336
Appendix C: Maps Used by AWZ on his North-West Frontier Trip	337
Appendix D: Report from NLO Karachi to JICA-FE, Declassified	342
Appendix E: Conversation with Col. H. R. A. "Tony" Streather, OBE	352
Glossary and Abbreviations	355
Acknowledgements	358
Bibliography	361
List of Illustrations	378
Index	382
Other Books by the Author	398
About the Author	399

Preface

Until 2007, all that was known about the contents of this book, *Dearest Barb*, were two letters, written in pen and ink by Lieutenant Albert Zimmermann, a U.S. Naval Intelligence Officer, to his wife, Barbara, in November and December 1943. The letters were transcribed and typed by her after she received them, and they were shared with her children and friends.

The first letter was written when he was in Peshawar – near the Khyber Pass – then in India, and now in Pakistan, on 25 November 1943. It mentioned Zimmermann's trip by train from Karachi to Peshawar, where he met two other men who were to accompany him on a trip along the border of Afghanistan. Zimmermann was met at the Peshawar train station by Major Sir Benjamin Bromhead, O.B.E., a British baronet, who would be the guide for the trip; and he was joined two days later by Major Gordon B. Enders, a U.S. Army Intelligence Officer. Zimmermann's letter told of meetings with British officials in Peshawar, and with Maliks of the Tribal Territories, the Wali of Swat, the Nawab of Dir, and the Mehtar of Chitral. It told of their unique journey by jeep beside high terraced rice fields into the highest mountains of the Hindu Kush range; how they became the first to travel in any type of motor vehicle over the hazardous Lowari Pass; of Zimmermann's trip to the historic Khyber Pass on the Silk Road, perhaps one of the first twenty or so Americans ever to cross this point; and of the Governor's garden party in Peshawar where the Viceroy, Lord Wavell, came to hear of their trip. Zimmermann was one of the first Americans to view the highest mountain in the Hindu Kush – Tirich Mir – which was at that time still unclimbed, and he was probably the first ever to capture the south face of it in a movie. Zimmermann and Enders were to become the first Americans to travel along the southern border of Afghanistan, on the Durand Line, from its highest accessible point, in Chitral, to Quetta, in Baluchistan. They may be the only Americans who have ever made this trip, openly, under arms, and in uniform.

A copy of Zimmermann's typed letter was preserved by Zimmermann's daughter, Helene. I wrote a brief essay about this letter in 2003, annotating it with comments about events that had occurred in those places over the ensuing 60 years. My essay was accepted by *Naval History*, but the journal eventually declined to publish it and returned the publishing rights to me. Zimmermann's second letter to his wife about this trip, which was sent after he returned to Karachi, had apparently been lost.

On 25 February 2007, the second letter that Zimmermann wrote to his wife about this trip was discovered, along with many other documents and photographs related to the trip and to the rest of Zimmermann's career in the Navy. And with it, other items of family interest, dating back to the generation of his grandparents and the grandparents of his wife. These items were in the attic of his daughter, Barbara, who had taken them to her house after her mother died. When the items were gathered together, I inventoried them and put them into two footlockers, which they filled completely. I then studied carefully the items related to Zimmermann's career in the Navy. I copied and transcribed many of the letters and other documents, and I also copied many of the photos and movies that he took while he was in India. I called these documents and photos the AWZ-Wartime Papers (or AWZ-WP).[1] I used the AWZ Wartime Papers to prepare a paper that was published in the journal *Appalachia* about his trip along the border of India and Afghanistan. I then used the AWZ Wartime Papers as the basis for a book, *Proceed to Peshawar*, which was published in 2013 by the Naval Institute Press. My original submission was edited by the Naval Institute Press, and the final version was shorter than the original. For this book, *Dearest Barb*, I include the full text of all of Zimmermann's letters to his wife, and also all of the photos that he took on the trip along the Afghan-Indian border. I have also added some additional images from the AWZ Wartime Papers, including other photographs and still shots from his 16mm movies.

In the five years since *Proceed to Peshawar* was published by the Naval Institute Press, the book has been the subject of many reviews and lectures. It won a Finalist's Medal in the Military History category at the 2015 Indie Book Awards in New York City. My lecture on March 13, 2014, to the Naval War College Foundation has had 520 views on YouTube. I was invited to speak about the book as a participant in the Asia Seminar of the Virginia Book Fair in 2015; and it was featured at the Annual Book

---

[1] See Appendix A for a summary of the AWZ-Wartime Papers.

Fair of the Harvard Club of New York City. I have spoken about *Proceed to Peshawar* at Great Decisions, Fearrington Village, N.C.; at Pantops Mountain, Charlottesville, N.C.; in the evening lecture series of the Harvard Club of New York City; and to the Annual Meeting of the Descendants of American Colonists at the University Club, Washington, D.C. Six reviews of *Proceed to Peshawar* have been published, in addition to readers' comments on Amazon.com.[2]

Zimmermann and Enders traveled openly along the southern border of Afghanistan in 1943. The Trip itself was not secret, but its true purpose was not disclosed in Zimmermann's orders, and no copies of Zimmermann's official written report have survived. Zimmermann and Enders gave verbal reports to their superiors; Zimmermann traveled to Ceylon to give his verbal report, and Enders gave his verbal report to General Patrick Hurley in Kabul. However, Zimmermann realized that this Trip was historic, for he carefully preserved a copy of the Trip report, and he retained copies of more than 140 photographs that he took. Nevertheless, a reader must read between the lines to see that the travelers were watching for evidence of Russian intrigue in Afghanistan, and penetration by Russian agents into northern India.

The intelligence mission by Zimmermann and Enders is a long-forgotten piece of the history of the Great Game. The "Great Game" is the term used to define the centuries-old conflict between Russia and England for control of the highlands of Central Asia. Russia wanted to have a warm-water port on the Indian Ocean, and Britain wanted to keep Russia out of India. The term, "Great Game," was popularized in Rudyard Kipling's book, *Kim*, although the words first appeared in a letter written by the doomed Captain Arthur Connolly, when he was in a dungeon in Bukara, in what is now Uzbekistan. The Czar's Foreign Minister, Karl Nesselrode, called the contest the "Tournament of Shadows." Lenin continued the Great Game. He planned "to set the East ablaze," by sponsoring the Communist Part of India. And then Stalin, too, who called it *Bol'shaia igra*, meaning "Great Game" in Russian. I believe the Great Game was the rationale for the intelligence mission taken by Bromhead, Enders, and Zimmermann in 1943. The Trip that the three men took has relevance today, because a New Great Game is now underway for control of Central Asia. The U.S. has replaced Britain and is now a participant in the New Great Game – as was predicted by the British officials in 1943 who encouraged and authorized Zimmermann and Enders to take a guided tour along the Afghan border: to take the Trip that is the focus of *Proceed to Peshawar* and which is also the major theme in *Dearest Barb*.[3]

For the past two years, I have been working with several people from Hollywood on a movie, to be based on the Trip along the Afghan border that is the focus of *Proceed to Peshawar*. The working title of the movie is *Khyber Pass*. For information about the movie, see www.khyberpassmovie.com.

The AWZ Wartime Papers are now in the Special Collections at the U.S. Naval Institute, Annapolis, Maryland. A brief summary of the contents of the papers is in the Appendix to this book. The AWZ Wartime Papers are the only complete set of papers known to exist of any field Naval Intelligence Officer in World War II. The only comparable documents are those based on the lives and work of three other men: one was a World War I secret operative; one was a Japanese language code-breaker in World War II; and one was the Deputy Chief of Naval Intelligence in World War II.[4]

---

[2] YouTube 8 Bells Lecture / George Hill: Proceed to Peshawar. See Bibliography for full citations of reviews by C. Naseer Ahmad, Robert Baumann, James Cox, Roger Cunningham, Charles Kolb, Emma Reid, Ronald Rosner, and Michael Swisher.

[3] For publications about the Great Game, see full citations in the Bibliography for Rudyard Kipling, *Kim*; Peter Hopkirk, *The Great Game*; Karl Ernest Meyer and Shareen Blair Brysac, *Tournament of Shadows*; and Lutz Kleveman, *The New Great Game*.

[4] See full citations in Bibliography for Charles H. Harris, III, and Louis R. Sadler, *The Archaeologist Was a Spy*; Elliot Carlson, *Joe Rochefort's War*; and Ellis M. Zacharias, *Secret Missions*. Several hundred intelligence officers were trained by the Office of Naval Intelligence in World War II. Many microfilmed records of those who served as Naval Attachés have been preserved in NARA II, Record Group RG38.4.3, but no comparable collection of records of Naval Liaison Officers has survived. Naval Attaches served at U.S. embassies, at the capital of countries; Naval Liaison Officers, such as Zimmermann, served at consulates.

2015 Next Generation Indie Book Awards

Finalist

Military

Awarded to

Proceed to Peshawar: The Story of a U.S. Navy Intelligence Mission on the Afghan Border, 1943

Presented by the Independent Book Publishing Professionals Group, May, 2015

Albert Walter Zimmermann

Barbara Shoemaker

1

INTRODUCTION

Albert W. Zimmermann and Barbara Shoemaker[1]
1902 - 1942

Albert Walter Zimmermann was born in Philadelphia, Pennsylvania, on 11 June 1902, the youngest of the four daughters and three sons of John (Johannes) Zimmermann and his wife, Eva Katherine (Evakaterina) Kellenbenz. He died at Haverford, Pa., 24 July 1961. His father had come to America in 1874 from Gussenstadt, a village in the district of Heidenheim, near Stuttgart, in Baden-Württemberg (now Germany). He was a nearly penniless man of 18 who had distant relatives in rural New York state, but no close friends in America. His mother came from the nearby town of Klein-Eislingen, near Gross Süssen, also in Baden-Württemberg. She worked as a chambermaid when she arrived in America. John and Eva met in Philadelphia and were married there in 1885.

John Zimmermann was a member of the hard-working and close-knit German community in Philadelphia, and he ultimately achieved the American Dream. As weaver, John Zimmermann started by selling his own carpets from a push cart on the streets of Philadelphia, where he was spotted by the owners of the Philadelphia Tapestry Mills. They took him in as an employee, and as an inventor and skillful mill manager, he soon became their partner. His patents made it possible to weave enormous carpets, and their company, renamed Artloom Corporation, became one of the largest of its type in America. He became a very wealthy and charitable person, and he later became Bishop of the Reorganized Church of Latter Day Saints in the eastern United States. His children each received a gift of $1 million dollars when they married.

John Zimmerman's youngest son, Albert Walter Zimmermann, who I will often refer to as AZ or AWZ, received the degree of B.S in Mechanical Engineering when he graduated from the University of Pennsylvania in 1923. He was a member of Sigma Tau (Honorary Engineering Fraternity), Sphinx Honorary Senior Society, and President of the Glee Club. He entered the carpet and fabric business which his father established, John Zimmermann Sons, and he eventually became a vice president of the company. In 1926, he married Barbara Shoemaker, the daughter of a prominent Philadelphia ophthalmologist, in what some said was the society wedding of the season. In contrast to Zimmermann's immigrant roots, her ancestors went back to the fleets of William Penn and John Winthrop. After a grand honeymoon, they settled in a new home that they built in the suburbs, "Cotswold Corners," in Haverford. He joined the usual Main Line golf and country clubs: Philadelphia Country Club, Merion Cricket Club, and Fourth Street Club; and a men's singing group, the Orpheus Club. With his knowledge of textiles, he formed a wool brokerage in partnership with another Club member. He was very comfortable, but not ostentatious; and he was a shrewd observer, but he kept his mouth shut about what he saw. When he was in the Navy, he wrote to his wife that she needn't worry about spending money. He wrote to remind her that they were "rich" and "wealthy," although he would never have said it aloud.[2]

One of Al Zimmermann's closest friends was a Naval Reservist named Jack Kane, who was the Deputy District Intelligence Officer (DDIO) for the Third Naval District, headquartered at the Philadelphia Navy Yard. Zimmermann's wife Barbara's best friend was Kane's wife, Amelie, née Sexias. She was a distant relative of the famous tennis player, Vic Sexias. Zimmermann was also a friend of Captain Thomas Thornton, who was DIO for the Third Naval District. Tom Thornton was responsible for recommending Zimmermann for Naval Intelligence. It was probably either Kane or Thornton who made the connection with the FBI for Zimmermann to take pictures of the German Bund in Philadelphia in 1937, when the Bund turned out in uniform, with swastikas on their armbands, for the wedding of one of the Zimmermanns' household employees. These photos were Zimmermann's first known contact with the federal intelligence service, although also in 1937, he had also surreptitiously taken movies of Nazi troops marching through Stuttgart. These movies had likely been seen by Thornton and Kane, and they showed Zimmermann's ability to do good photo intelligence work. Indeed, Thornton and Kane may have asked Zimmermann to take movies in Germany when he and his wife sailed on an extended summer trip to Europe.[3]

After war broke out in Europe in September 1939, Zimmermann's friends were making plans for what they would do when the U.S. entered the war. Some, like AZ's father-in-law, who had been a major, were too old to serve again, but they encouraged others to prepare. Some decided to join the reserves, or to reactivate their previous commissions. AZ was a too old to learn to be a front-line fighter, and he was also color-blind, but intelligence work appeared to be right for him. Army Intelligence (G-2) was a possibility, but he had no connections with that branch of service. Kane and Thornton encouraged him to apply for the Office of Naval Intelligence, or ONI. War with Japan was believed to be inevitable, and an attack was expected by many Americans. The surprise on December 7, 1941, was the location: it was at Pearl Harbor, not the Philippines. After the U.S. declared war on Japan, Germany declared war on the U.S. Roosevelt was then officially authorized to support England in its war against Germany, and the U.S. mobilized to fight a war in both the Atlantic and the Pacific. If AZ had waited a bit longer, he might have joined the Office of Strategic Services (OSS). His good friend Freeman Lincoln and his next-door neighbor Clarence Lewis joined the OSS. But the OSS didn't even exist until 13 June 1942 (it had previously been known as the Coordinator of Information or COI), and AZ was already talking with the Navy by that time.[4]

Zimmermann also had two other door-openers for the ONI. The head of Naval Intelligence in the First Naval District (New York) was Commander (later Captain) Vincent Astor, who led "The Room," Franklin Delano Roosevelt's secret intelligence operation in New York City. Astor had been secretly named by FDR to vet candidates on the East Coast of the U.S. for the ONI, and AZ had at least two connections with this group. His good friend John "Jack" Thayer, Jr., had survived the sinking of *Titanic* as a young man; Jack's father and Vincent Astor's father had gone down together on the *Titanic*. And AZ's friend Malcolm Aldrich, who he called "Mac," had been a member of Skull and Bones at Yale; Mac Aldrich was a cousin of the banker, Winston Aldrich, who was also a member of "The Room."[5]

FDR with Vincent Astor, his secret intelligence agent. Astor headed "The Room" in New York City

Museum of Rhinebeck History

John Zimmermann and Eva Kellenbenz

Their Family and His Business – Artloom in Philadelphia

John and Eva and their five surviving children in 1907
Albert is the youngest, on the lap of his sister Lillian

The Zimmermann family roadster with Albert in the right front seat, in front of the family home (L)
Art Loom factory, Philadelphia, Pa. – John Zimmermann's carpet mill (R)

Barbara Shoemaker

Barbara Shoemaker was the second daughter and fourth child of Dr. William Toy and Mabel (Warren) Shoemaker. She was born at Philadelphia, Pa., 13 February 1902, and died at Haverford, Pa., 24 July 1985. She was active throughout her life as a leader in Philadelphia society. She belonged to the Acorn Club and the National Society of Colonial Dames, and she was a member of the Church of the Redeemer, Bryn Mawr, Pa.

Barbara Shoemaker grew up in the city of Philadelphia near Rittenhouse Square, where her father, an ophthalmologist, had his office. Her older sister, Dorothy, recalled that although the family was prominent in society, they were not wealthy, and never had their own horse and carriage. Barbara was educated at private schools and did not attend college – a college education then rarely being undertaken by young women in her social group. An attractive woman with blond hair and fair skin, she had many suitors and was very popular at the parties and balls where young people of her age gathered. Her hand was won by a bright and handsome young graduate of the University of Pennsylvania, and she married him. Her husband was Albert Walter Zimmermann, youngest son of a wealthy immigrant German engineer who was co-owner of Artloom, one of the nation's largest carpet manufacturing companies. He had also established a family textile manufacturing firm, John Zimmermann & Sons.

The wedding of Barbara Shoemaker and Al Zimmermann was a major event in the social season of 1926. The wedding and their honeymoon cruise to the Caribbean was recorded on the first of more than 50 8-inch rolls of 16 mm film that Al Zimmermann took over the next three and a half decades. Their new house, "Cotswold Corner," in which they lived for the rest of their lives, was constructed at 400 North Rose Lane in Haverford, Pa., in the style of an English manor house. It had ivy covered walls and flower and vegetable gardens, above a stream on a steep wooded hillside. After spending summer vacations in Chatham, Mass., and elsewhere in New England, they purchased a large summer home, "Full House," in East Hampton, on Long Island, N.Y. While in East Hampton they were members of the Maidstone Club, with a cabana near the clubhouse which they bought from Jack ("Black Jack") Bouvier – whose daughter, Jacqueline, later became Mrs. John Kennedy.

In Philadelphia, Barbara (Shoemaker) Zimmermann was active in support of many charitable organizations, most notably the Church Farm School in Exton, Pa., and the Graduate Hospital, on whose board she represented the Volunteers' organization, which she had founded. Barbara and Al were members of both the Merion Cricket Club and the Gulph Mills Golf Club. They were avid bridge players and golfers.

Residence of Dr. William Toy Shoemaker, 1903-1927
109 South 20th St., Philadelphia, Pennsylvania

The Zimmermann and Shoemaker Families

Albert Walter Zimmermann's Diploma on Parchment – 1923
Bachelor of Mechanical Engineering, University of Pennsylvania

Albert and Barbara (Shoemaker) Zimmermann
Wedding party in 1926

Albert and Barbara (Shoemaker) Zimmermann
Their Family Home and Their Children

Home of Albert and Barbara Zimmermann
"Cotswold Corners" – 400 North Rose Lane, Haverford, Pa.

Their four children (L to R):
Albert Jr. (1937-), Barbara (1927-2011),
Helene (1929-), and Warren (1934-2004)

Albert Zimmermann's First Photos for Intelligence Agencies – in about 1937

German Bund guard at St. Paul's Church, Philadelphia
Taken for the FBI, and returned to AWZ in 1949

Nazi troops marching in Stuttgart, Germany.
From Zimmermann's 16mm movie, taken from his cousin's second floor window

Albert Walter Zimmermann's Commission as a Lieutenant, USNR, signed by Frank Knox, Secretary of the Navy on 8 September 1942, with a date of rank of 1 August 1942.

2
ZIMMERMANN JOINS THE NAVY

By September 1942, Albert Zimmermann had decided on Navy Intelligence. He tidied up his personal affairs, got a waiver for color-blindness, and on 21 September, he took the Oath of Office as a Lieutenant, I-V(P), USNR – the "P" standing for Probationary. His appointment was back dated to 1 August 1942, and his appointment was effective on 8 September. He went to basic training as an officer at the Washington Navy Yard from 18 October – 14 November. He passed the course; the "P" was dropped, and he became an Intelligence Officer-in-training.

LT Albert W. Zimmermann, USNR

Albert Zimmermann's Basic Orientation Class, Washington Navy Yard, October 1942

Zimmermann then reported to the Intelligence Indoctrination Course at Dartmouth College for ten weeks, from 18 November – 28 January 1943. While he was at Dartmouth, the new Naval Intelligence officers developed a sense of camaraderie that persisted for the rest of their time in service, and the wives, too, often joined in this vicariously. It was like college, in many ways, and the students "roasted" their instructors at graduation. They did not roast, but instead treated with suitable respect two men who spoke at their graduation: Director of Naval Intelligence (DNI), Rear Admiral Harold C. Train, and his deputy, Captain Ellis Zacharias – who later became known as "The Man Who Wanted to be DNI." How much the new Intelligence Officers knew about the conflict between these two men – Train and Zacharias – is unknown, but it was probably a good introduction to the arcane and back-biting world of intelligence.[6]

RADM Harry Train (L)

CAPT (later RADM) Ellis Zacharias (R)

Google Images   Google Images

Navy Department Building, "Main Navy," 1941

A classmate at Dartmouth was another Philadelphia socialite, Curtin Winsor, who was the brother of his friend in the Orpheus Club, Jim Winsor. LT Curtin Winsor, younger than Zimmermann, went to Washington in the Far East desk of ONI, at "Main Navy," in Washington. Curt Winsor later became AZ's desk officer, or "handler." Zimmermann thought, but could never prove, that he was being manipulated by Curt Winsor, and he believed that Winsor cleverly escaped whenever he was caught in a web of lies. Winsor was married to the divorced wife of James Roosevelt and was thus the step-father of a grandson of President Roosevelt.

Zimmermann began French Language School in Washington, D.C, in preparation to go to Dakar, French West Africa. These orders were cancelled after two months, and he was instead sent to the Advanced Intelligence Course in New York City. It is not clear why he went there, instead of going to Dakar. It may be that his German connections were too close for comfort, or perhaps Winsor, now head of the Far East desk at ONI, wanted to form a Far East team from men that he knew.

Zimmermann began the Advanced Operational Intelligence School at the Henry Hudson Hotel on 19 April and completed it on 28 May. He spent most of the month of June at the ONI in Washington in preparation for his next job. His assignment was to the Naval Liaison Office, Karachi, India (now Pakistan). He would spend nearly two years there. It was a challenging and sometimes worrisome experience, but it was the most interesting adventure of his life.

3

"Dearest Barb . . . With Love"

**Letters of Lieutenant Albert W. Zimmermann, USNR
Naval Intelligence Officer, China-Burma-India Theatre
Commanding Officer, Naval Liaison Office, Karachi**

To His Wife, Barbara

Karachi to Haverford, 1943-1945

Can't wait till you're in my arms again (AWZ, 20 October 1943)

Here's a big hug and a kiss sweetheart for the best wife ever
(AWZ, 29 January 1944)

I want to be with you – <u>alone</u>
(AWZ, 5 January 1945)

Background for the Letters

Albert Zimmermann's trip to Karachi would be familiar to anyone who lived through World War II, but it seems exotic to us. After saying good bye for what he thought would be at least a year, and perhaps forever, on 23 June he boarded a commercial airline flight for New York. He changed there to one of the world's great seaplanes, one of only two in existence, a four-engine American Export Airlines Sikorsky VS-44. He flew to Botwood, Newfoundland, and on to Foynes, near Limerick, in the Irish Free State. He began to send letters home, numbering each as he was taught to number his Intelligence Reports. All passed the censor, and in two years, only one word – the name of one of the places on his route to Karachi – was ever redacted by the censor. He touched down in Lyautey and Casablanca, Morocco; Oran and Algiers, Algeria; Constantine and Sousse, Tunisia; Tripoli and Benghasi, Lybia, and he arrived in Cairo on 1 July to get in the long line to continue on to the east.

He stayed in relative comfort at the famous Shepheard's Hotel, and – as was the custom with expats and intelligence officers – he made good use of the time. He saw the Sphinx and the pyramids, and visited the Philadelphia physicians who were at Army General Hospital 38. They began to receive casualties from the invasion of Sicily, which began while he was in Cairo. While he was cooling his heels in Cairo, he met Edgar Snow and Tom Treanor (in his letter, he misspelled the correspondent's name as Trainer), who were also staying at Shepheard's. They were war correspondents who had been in India and China and were on their way to cover the invasion of Sicily. Treanor had been in Cairo before; he arrived on 1 July 1942, just after Rommel pulled up at El-Alemain, and left for India just after Rommel began his retreat. In his war stories, Treanor showed a droll sense of humor, even when describing close calls under fire. But his humor largely disappeared when he was in India, and he may have given AZ an earful of the struggle that was going on there. He had passed through Karachi on his way to New Delhi. Treanor wrote that Ghandi and Nehru wanted the British to leave India, so the Hindus could rule it; and on the other hand, there was Jinnah, who refused to bargain with them, and intended to put three-fourths of the Muslims of India into a new country, Pakistan. All three men postured for their political constituencies, and none were willing to compromise. (AZ would see that this was also the British perception of these three men, and would transmit the details in an Intelligence Report in 1944.) And "despite protestations and promises of the British that India one day is to get her independence, no apparent step is being made in this direction." Treanor never completely lost his sense of humor, though. He wrote that, "If you want to empty a hall, all you have to do is announce the next speaker will discuss India."[7]

On July 4th Zimmermann took movies (which were lost in transit home) of a baseball game with "Quentin Reynolds at bat and General Strong umpiring." General Strong was doubtless the Army Chief of Intelligence (G-2) who famously detested General Donovan of the OSS. His host on several days was a Naval Reservist from Philadelphia, Captain Thomas Anthony "Tom" Thornton, who was now with the U.S. Forces-Cairo. He had seen LCDR Jack Kane only the previous week and reported that he was well, although by now they both knew that Kane had a fatal case of cancer of the stomach – he would soon die. Thornton planned to take AZ and Commander Gene Markey to Alexandria to pass the time while they were waiting for a flight out of Cairo, but AZ couldn't make the trip because he was placed on standby for his flight to the east. On one of his last nights in Cairo, he dined with Henry Hotchkiss, the Assistant Naval Attaché and several other old friends from Philadelphia. Cairo was a place where many paths crossed. He finally flew out on an RAF B-24 "Liberator" bomber on 16 July via Habanniah, Iraq, and another unnamed remote airfield. He commented to his wife that "I didn't need the parachute so I didn't find out whether it worked or not. They say you can always get another if it doesn't." He arrived in Karachi at 0500, Monday 19 July 1943.[8]

The Naval Liaison Office in Karachi was a spacious stucco-on-stone residence at 254 Ingle Road, on a frontage of about 100 yards, and equally deep – a typical American square block. AZ called it "palatial." It was surrounded on three sides by a 5-foot wall, and it was enclosed at the rear by the houses of the twenty or so household servants and their families. In the real estate market in India, it was called a "bungalow," but this hardly expresses its size and beauty. It was a rental property owned by a well-to-do Hindu family who lived next door. They sold most of the furniture to the Navy, retaining ownership of a few objects that had personal value, such as hunting trophies, which they left behind. The house was surrounded by trees, which insured privacy, and there was a one-way driveway marked for entrance and exit. The first floor of the house had 25-foot ceilings, and the table in the dining room seated 18 people with ease; it could be expanded to hold more. The first floor was for offices – the leading petty officer, the code room, the commanding officer's – and spaces for guests, including a lounge and library, and the dining room, with its fireplace and George Washington's portrait over the mantle. The second floor was the residence; the commanding officer's suite, the executive officer's suite, a small library, and dormitories for the junior officers and a half-dozen enlisted men. The executive officer's suite was used by VIPs. RHIP – Rank hath its privilege, especially in the Navy – and there were so many VIPs (very important persons) who passed through Karachi that it was simpler to leave the suite alone than to vacate it every time it was needed. There were social events there on many evenings, and grand parties, too, but there was an ever-present reminder of the war, too; a locked gun closet was on the stairway with a dozen M-1 rifles, and their slings and ammunition.

When AZ arrived, there were four officers at the NLO Karachi: the Commanding Officer, LCDR F. Howard Smith, USN; and LTs Burns, Callahan, and Browning. Except for Smith, all were junior in years to AZ. Browning developed eye trouble and soon headed for home. The rest worked long days, taking their turns on "duty" call. It was secret work, and AZ said almost nothing about what they did during the days. We know that the Navy's long distance radio network provided service for the Army and the State Department, so the code room must have been very busy. Karachi was also a busy seaport, located near the mouth of the great Indus River. It was the entry point for everything that passed on land into India from the west, and to the north. The State Department cables show that Afghanistan was interested in selling and transshipping wool, of which AZ had special knowledge; and Karachi was used as the exit port for Axis diplomats who traveled under safe conduct to and from Kabul.

AZ was introduced to the strange ways of the Naval Liaison Office almost immediately on arrival, when he received an invitation to dinner with a Madame Dubash. He soon learned that she was a "good-looking" Russian woman of about 50 who had been married to an Indian, and that she was the "girl friend" of the Commanding Officer. That was something to keep quiet about. AZ later learned that she and Indian man named Sheikh were employed in counter-espionage activities.[9]

He also believed that the NLO Karachi was spending U.S. government funds for lavish entertainment, that they were living in a "fool's paradise." All of this was contrary to the instructions to Naval Intelligence officers, and eventually he was proved right. Sex and money have proved to be the downfall of many intelligence officers.[10]

In the evenings and on weekends, AZ played bridge with Brigadier Hind and his American-born wife (Hind was the commander of the British force in the area), and with Major and Mrs. Smyth,[11] he being the Military Secretary to the Governor of the Sind Province. AZ played tennis and partied at the Sind Club, where the Governor and Lady Dow held forth, and at the Boat Club. He was a frequent guest in private homes, where his ability to sing was much appreciated. He was also invited to the Karachi Club, which was for Indians (Parsi, Muslim, and Hindus), which was rare for an American. He was invited to dinner by his prewar Indian wool supplier, and learned (with difficulty) to eat the food, and to appreciate their customs (apologizing for his occasional faux pas). On one occasion, he got a pass to go into the out-of-bounds area to a movie with them. On weekends he often went to the private offshore beach known as Sandspit with members of the American commercial community. He was there on 12 September with Arlo Bond, who was with Standard Oil; Bond had sent his wife and children back to America. They would attend the same summer camp in Vermont where one of the Zimmermann girls had gone to the previous summer. The war rarely intruded; it appeared that Japan would not attempt an invasion of India, and the Germans were slowly retreating. In his first three months, he mentioned only the fall of Mussolini and the surrender of Italy on 10 September in letters to his wife.

The visitors who passed through the NLO were too numerous to write home about. Some such as four groups of "Freeman's friends" (sent out by J. Freeman Lincoln, a major in the OSS in London) were so secret that he couldn't mention their names, but he was able to identify some of them to BSZ because she knew them, too. Others who were named included Senator Richard Russell on 23 August, Ambassador Gauss "and his attachés" on his way to China on 15 September, and a Mr. Preston, the Consul General at Lorenzo Marques on 9 November. The new American commander, General Julian B. Haddon, arrived in town on 15 September, but AZ was bedridden with sand fly fever and didn't meet him until later.[12]

Other diseases were always threatening. The insect-borne diseases included yellow fever (hopefully protected by vaccination), malaria (although you could take atabrine to prevent it, most people decided to use mosquito netting at night), and dengue. Almost everyone got dengue if they stayed in Karachi for a year or so; it was called "break-bone" fever, a wretched, unpreventable illness. And there was diarrheal disease, known as dysentery, which could be from amoebae (lingering and bad) or bacterial (even worse).

Through all of this, AZ was being observed by Clarence Macy, the hardboiled but wise American Consul in Karachi, who AZ met for the first time on 15 August and who he saw many times after that. Macy was in frequent contact with J. R. Harris, the British Intelligence officer who received a message from IB Quetta [translated as Intelligence Bureau Quetta], on 26 October. IB Quetta asked for an American at the Naval Liaison Office to go to the North-West Frontier Province. He would accompany Sir Benjamin Bromhead and Major Gordon Enders on this trip. IB Quetta had asked Bromhead if he could also take one of the Americans from the Naval Liaison Office in Karachi, to be added as a third traveler. Bromhead agreed, subject to the Governor's sanction "which he said he thought would certainly be forthcoming." The man who was picked for this by J. R. Harris, with Macy's support, was Albert W. Zimmermann, U.S.N.R. Zimmermann would appear to have been picked at random, because he was available, and the Commanding Officer of the Naval Liaison Office in Karachi was not. However, there are good reasons to believe he was a specific choice, and the message was carefully written to ensure that he, and only he, would be sent.[13]

When he arrived, Zimmermann was the second ranking officer at the NLO Karachi. He would have been the Executive Officer, but he was not named to that position; he would later become the Commanding Officer. His position as a Naval Liaison Officer, abbreviated as ALUSLO, was analogous to a Naval Attaché. Like all Naval Attachés, Naval Observers, and Naval Liaison Officers, he was trained as an intelligence officer. And, like most of them at this point in the war, he was a reservist.[14]

AZ had made a lot of new friends, and he hadn't made any enemies. Rumor had it that he also had someone in Kabul who wanted him on the trip – the Minister's Secretary and Chargé d'Affaires, a man named Charles Thayer.

AZ was sent to Colombo, Ceylon (now Sri Lanka), on a courier mission on 29 October – only three days after the message was received by Harris, so the decision to send him had been made quickly. His courier mission was surely just a cover; he must have been there to be briefed by Navy intelligence on what to look for and how to comport himself on the North West Frontier. He stayed in Colombo for two nights and a day with a Lieutenant Commander B. W. Goldsborough. He wrote to his wife that Goldsborough was from Baltimore, and "knew the right people." Indeed he did; Brice Worthington Goldsborough II was the son of Phillips Lee Goldsborough, who was Governor of Maryland in 1911, and later a U.S. Senator. Ceylon had just been designated as the headquarters of the new South East Asia Command, known as SEAC, and the Supreme Commander of SEAC, Lord Louis Mountbatten, was moving his staff from Delhi to Kandy, in the interior of Ceylon. AZ would visit Kandy later, but this trip was just to visit Navy intelligence, in Colombo. He left on Friday 29 October via Ahmedabad, Bombay, and Hyderabad, to Colombo; and he retraced his route, going back to Bombay and then to Karachi, where he arrived on Tuesday, 2 November. He saw another ALUSLO, Lieutenant Al Payne, presumably passing on information verbally to him, on both trips through Bombay.[15]

The purpose of the trip on the North West Frontier was stated in the message from IB Quetta, to make "it clear to the American Legation in Kabul what are our frontier problems and our ideas and policy in dealing with them and the Afghans." This was restated with little change by AZ when he sent his typed report to his family: "the trip had been already instigated by the Military Attaché to Kabul (Maj. Enders) to give him an opportunity to see what was on the other side of the fence."[16] But there were many more things that AZ and Enders were looking for. Some of these would be details that provided depth to the main goal, but others were quite different. They may not have been told about some things to watch for, but like good intelligence officers, they would be on the lookout for whatever they could see. Some of these items will appear in the trip reports, and others can best be understood from other sources.

Zimmermann was added to balance Enders, who was expected to show his creative imagination. Those who have heard Enders, and those who have read his books, see his genius for telling a story, but his story is sometimes very selective. He leaves out facts that don't suit his narrative, and he often tells the story differently each time he says it. It is, in short, hard to know if Enders is telling the truth, or not. Zimmermann, on the other hand, though not always interesting to read, would be inclined to tell it as he saw it, not as he wished it would be.

* * * * * *

Letters from Lieut. Albert W. Zimmermann (AWZ)
1943-1945

All are to Mrs. Albert W. Zimmermann (Barbara [née Shoemaker] Zimmermann; BSZ),
400 North Rose Lane, Haverford, Penna., U.S.A., except as noted below
Return addresses on the envelopes included:

Lt. Albert W. Zimmermann, U.S.N.R. [AWZ], **or**
AWZ / U.S. Naval Mess / 254, Ingle Road / Karachi [USNM]**, or**
AWZ, U.S.N.R./ Mail & Dispatches / Office of Naval Intelligence / Wash D.C., [ONI] **or**
AWZ / Office of the United States Naval Liaison Officer / Karachi, India [NLO] **or**
AWZ / APO 886 Postmaster N.Y. [APO]

All envelopes were stamped "Passed by Naval Censor" and initialed by the censor, except where otherwise noted below. Most of the postmarks were "U.S. Army Postal Service / A.P.O. 886" [APO], with some exceptions noted. Letters are shown in the order that they were assembled, tied, and boxed by BSZ, which was principally in the order of the postmarks.

The letters were in a cardboard box, on which was written by BSZ:
Letters from India – '43-'44 from AWZ

The letters were in two bundles, each of which was tied with a string, and the box was tied with a string. The letters were probably placed in the box in about 1945, and it is unlikely that it had ever been opened thereafter until January 2011, when I first opened it.

Except for the first set of letters, most of the letters are not numbered. Most of them are in individual envelopes, although a few are loose, and their envelopes are missing. There is more than one letter in some envelopes. The first envelopes were numbered on the outside by BSZ, but she discontinued that practice after #14. For the sake of clarity, I have added numbers to follow #14. There are about 149 envelopes plus a few loose letters, viz.:

23 June-31 December 1943   53 (#1-53)
1 January-31 December 1944   83 (#54-137)
1 January-2 April 1945   11 (#138-149)

Family members included the four children of AWZ and BSZ, identified as follows:
    Barbara Warren "Babs" Zimmermann (BWZ), b. 1927
    Helene "Nanie" or "Lanie" Zimmermann (HZ), b. 1929
    Warren "Warr" Zimmermann (WZ), b. 1934
    Albert Walter "Albie" Zimmermann, Jr. (AWZJr.), b. 1937

Other correspondence from AZ to his family and correspondence to and from him to others is in the AWZ Wartime Scrapbook, and in other folders, such as the Enders Folder, Bromhead Folder, and Winsor Folder. Many of the photos that are referred to in his correspondence are in the AWZ Photo Album and Scrapbook and in the envelope and folder for Loose Photos, and some of these are referenced in my transcriptions (*q.v.*).

       Zimmermann's handwriting was often difficult to decipher, and he used abbreviations that were familiar to his wife. His letters must have been a challenge for the censor to read, and he knew that. However, his handwriting was never good, even when he was not trying to evade the military censor. Look, for example, at the comment that he wrote on Tuesday, December 30:

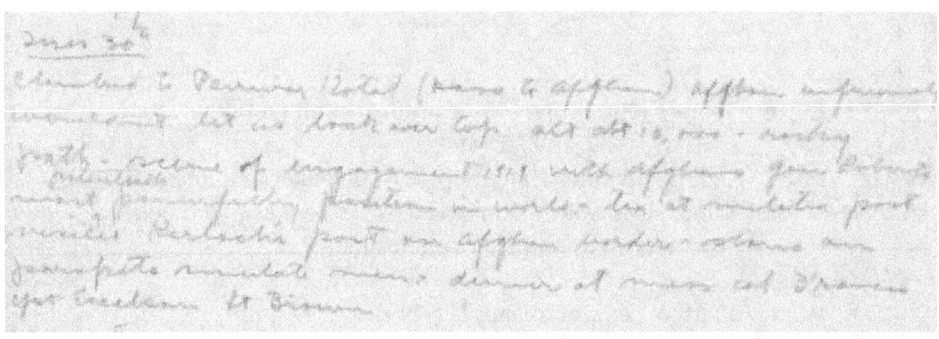

"Climbed to Peiwar Kotal (Pass to Afghanistan) Afgans unfriendly, wouldn't let us look over top. Abt. alt. 10,000 – rocky path – scene of engagement 1919 with Afghans. Gen. Roberts. Most potentially powerful position in world. Tea at militia post. Visited Karlachi post on Afghan border. Stairs on parapets simulate men. Dinner at mess. Col. Francis [illegible] Lt. Brown."

1943

23 Jun 1943, postmarked "Botwood" From AWZ[1] / British Overseas Airways Corporation / Caledonia Camp / Botwood, Newfoundland / marked by BSZ #1 (on reverse, marked "Opened by Examiner / DC/21"), to BSZ[2] ALS-P[3]

[No heading]

I [Botwood, Newfoundland, 24 June 1943][4]

Dearest,

So far so good, although feeling about the way I felt at the cow camp at Rapid Creek[5].

By the way, I forgot to tell you, Daddy Jo and Bitz looked very well the last time I saw them. Bitz's garden doesn't look nearly as nice as last year but I suppose *C'est en guerre.*[6]

Expect to be leaving here any minute now so must close. Lots & lots of love to everybody and especially you.

Al

Postmarked 28 Jun 1943 To BSZ Rec'd 13 Jul ALS-I[7]
from AWZ / Royal George Hotel, Limerick   marked by BSZ #2 (on reverse and edge, "Examiner" with symbol of crown)

ROYAL GEORGE HOTEL.
LIMERICK.

II Saturday [26 June 1943]

Dearest Barb,

We are staying here an extra day waiting for a part which evidently turned up today as we just received notice we're leaving this evening. I hope the office had sense enough to let you know of my arrival as soon as they got the cable. I though it was the surest way of getting thru. When I sent it I had

---

[1] AWZ = Lt. Albert W. Zimmermann, USNR.
[2] BSZ = Barbara (née Shoemaker) Zimmermann (Mrs. Albert W. Zimmermann).
[3] ALS-P = Autograph Letter Signed-Pencil (i.e., handwritten in pencil)
[4] AWZ's endorsed travel orders, endorsed on 20 July 1943 after arrival in Karachi, are on sheet 8 (verso), of his Wartime Scrapbook (AWZ-SB). These orders show his stops as follows: Departed NY 0815, 24 June 1943; arrived Botwood (N.F.) 0430; departed Botwood 0600; arrived Foynes (I.F.S.) 0600, 25 June; departed Foynes 2100; arrived Iyautey [should be Lyautey] (Meroc) [Morocco] 0610, 27 June; departed 0750, 08 June; arrived Casablanca 0830; departed 0850; arrived Oran 1115; departed 1145; arrived Algiers 0110; departed 0905, 30 June; arrived Constantine 1010; departed 1050; arrived Sousse 1230; departed 1255; arrived Tripoli 1455; departed 0720 1 July; arrived Benghasi 1055; departed 1135; arrived Cairo 1530; departed 0815, 16 July; arrived Habanniah 1215; departed 0815, 19 July; arrived Karachi 1330, 19 July 1943.
[5] The dude ranch where they stayed on vacation in Wyoming before the war.
[6] Daddy Joe = Joseph Crosby Lincoln (1870-1944) was a friend of the Zimmermanns, who before World War II summered near Lincoln's country home in Chatham, Mass. He was an American author of novels, poems, and short stories, many set in a fictionalized Cape Cod. His work appeared in the *Saturday Evening Post* and *The Delineator*. Two of his books were adapted for movies. In Chatham, he lived in a shingle-style house named "Crosstrees" that was located on a bluff overlooking the Atlantic Ocean.

Bitz = "Bitz" Durand, a woman, whose given name is unknown. She was a friend of the Zimmermanns in Chatham, Mass., and her name appears several times in AWZ's letters. Her daughter Virginia, nicknamed "Peachy," came to India in 1944. "Peachy" was mentioned in AWZ's correspondence as one of the "12 beautiful girls" who worked for the OSS in Kandy, Ceylon. She is seen relaxing in a photo of "Peachy's Place" with two U.S. Navy Lieutenant Commanders, seen later in AWZ's letters.
[7] ALS-I = Autograph (i.e. handwritten) letter, signed; in ink.

no idea of doing anything in wool but I spent the morning talking to two wool men (that I wrote to Jack[8] about).

/p.2/ The trip over was very exciting, nothing unusual happened, but it certainly is thrilling to float along with a crystal clear sky overhead and an infinite blanket of cotton clouds underneath. One of the sights is to see your shadow on the clouds below encircled by a perfect rainbow.

Four of us go on from here (Henson making another connection) and we probably will pick up a couple more. The Irish are very nice, happy to do all they can to make our stay pleasant & comfortable. Ames and I had a room together so I've seen quite a lot of him.

Yesterday we took a walk /p.3/ around town and bought some lace at a convent – real Limerick. We came back to the hotel for dinner and after rushing thru we were told we'd be here another day. After dinner we went to a movie (San Francisco – they are all American) and then went to a dine & dance place and looked over the local talent and finally broke down & danced with a couple of colleens. Two dances were enough so we came back to the hotel.

The girl at the desk invited /p.4/ us out to the kitchen to have a drink (the bar was closed) so we ended up with two girl hotel clerks, the bar maid and a very nice Irish lad (the men's furnishing buyer of a department store here) singing Irish & American songs till 1 A.M. We know the Irish songs better than they did & they knew the American songs better than we. Same old stuff.

The man has just come to say we're off so have to close.

Lots & lots of love,
Al

Postmarked 7 Jul from AWZ to BSZ   U.S. Navy / airmail stamp, BSZ marked #3

ALS-I

[No header]

Sunday [Casablanca,[9] 27 June 1943]

Dearest Barb,

Here today – gone tomorrow. I'm another step further along. I'm down to Ames now from the original group. We leave him tomorrow along the way. We arrived here early this morning and rather than stay here for the day we took a sight seeing tour via the mail truck. I had my picture taken and am enclosing same.

I had an RHIP[10] pulled on me on the trip by friend Ames. He saw me all dressed up and asked where I was heading and after he found out he decided it was a good idea and wanted to go along. The

---

[8] John "Jack" Ott was his partner in the wool brokerage firm, Ott and Zimmermann.

[9] AWZ could not say where he was when he wrote this letter, but correspondence with BSZ after he reached Karachi shows that his route included landings at Casablanca, Algiers, Tripoli, and Cairo in North Africa. This being his first stop, it was Casablanca, where AWZ later said that the enclosed photo was taken. The Anglo-American invasion of North Africa (Operation Torch) began 8 November 1942. Algiers surrendered on that date, following a coup and a short battle. Casablanca surrendered on 10 November. The Casablanca Conference of FDR and Churchill took place at the Anfa Hotel, 14-24 January 1943. Heavy fighting continued to the east, however, until the German surrender in Tunisia on 13 May 1943.

When BSZ and her friends saw AWZ's photo, they must surely have thought of the movie "Casablanca," which premiered on 26 November 1942 and was released on 23 January 1943. Starring Humphrey Bogart, Ingrid Bergman, Paul Henreid, and Peter Lorre, it is now regarded as iconic. In the enclosed snapshot, one can visualize AWZ in the role of Bogart, standing beside a French officer, played in the movie by Henreid.

At about the same time that Operation Torch was launched, the battle of El-Alamein was raging near Cairo. It began on 23 October, with British and New Zealand forces under Montgomery and the Nazis under Rommel. After see-saws, Allied armor broke the German and Italian lines through on 7 November and by 11 November the Germans and Italians had been driven out of Egypt.

[10] RHIP = The old saying, "Rank hath its privilege." It is not clear if AWZ was joking, or if "friend Ames" (about whom nothing more is known) actually outranked him.

cab of the mail truck holds three and I was to be the third but he very neatly got in the cab and left me to ride on the mail bags and today's mail must have been filled with ingots. I never /p.2/ felt anything so hard and my how the roads are and how the truck bounced. If vibrations have anything to do with reducing I should be the thin man. We were gone practically the whole day. The country here of course is very interesting. Got my first glimpse of a camel in its native haunts and the famous veiled women. They didn't fool me however. I didn't want to see their faces anyway. I saw enough of those that were unveiled.

I'm getting to be a hardened traveller now. I hardly feel any qualms at all. In fact I rather enjoy that part of it. I do wish you were here. I miss you terribly. I'm afraid you wouldn't enjoy the inconveniences but you certainly would the sights and the people.

I guess I'd better close as I have to be up early in the morning.
Loads of love,
Al

[enclosed with this letter, a 3 ½ by 2 ¼ in. black and white print of AWZ between two men in Eastern attire. On the left, a uniformed man wearing a fez, and on the right, a man in a long gown, white turban, and sandals. AWZ is in service dress khaki with combination cover and shoulder boards, long trousers. On the back, he wrote: "Sunday June 27th. I'm the one in the center (Keep this out of strong light. It will fade.)" Although neither the letter nor the photo specifies the location, in later correspondence AWZ reveals that it was Casablanca, although I originally thought it was Algiers.]

Postmark 31 Jul 1943 to BSZ   Received 7 Aug
from AWZ / Navy 1925, postmarked U.S. Navy, airmail stamp, marked #4 by BSZ
ALS-P

[no header]
Monday [doubtless Algiers, 28 June 1943]

Dearest,

Another stop nearer destination. Arrived here a little after noon after getting up at six to leave at seven. Had lunch & dinner at four which was five at the time here. Got my reservation for leaving tomorrow and then took a look about town. This sure is a busy /p.2/ place. I wish I could tell you more about it. It's interesting to see a picture of Roosevelt with "notre victorie est votre liberte" underneath and the words of Giraud, "un seul bout – victoire." All stores are closed so I couldn't get writing paper so I'm writing this on some Irish scot tissue which forgot to send you since you were interested in everything.
Love,
Al

[The paper is a piece of tan toilet tissue, perforated edges at top and bottom, 6in. by 5in.]

No postmark to BSZ   Received 7 Aug   from AWZ / Navy 1925, no postmark, airmail, passed by Navy censor, BSZ marked #5   ALS-I
[no header]
Tuesday, June 29, '43 [French North Africa, probably Algiers]

#5
Dearest,

Wrote from here last night but as there is nothing much to do thought I'd write again this morning. I can't do much right now as I picked up a touch of A.F.[11] and can't do much walking. Looked forward to taking a nice tub last night in this delightful villa but found the water turned off and it's been off ever since so you can imagine what that means. For breakfast I had three glasses of canned fruit juice. They couldn't even wash the dishes from last night's dinner. No shave, no teeth brush or face wash this morning. I have no canteen and everybody else was caught short. I'm still limited to 55# so I don't intend to get any even if I could which is very doubtful.

This really is a very nice place. Probably very pleasant living here under normal conditions. It's warm during the day especially in the sun but the nights are cool. Last night a blanket felt very comfortable. I've had a chance to work out on /p.2/ a little of my French. I successfully found out what bus to take back last night from town and had quite a conversation with an employee of the hotel about where I could go to play roulette. As there was only bacharat & chemin de fer I gave up the idea of gambling in favor of going home to take that nice bath I spoke of. I got a good night's sleep anyway.

In town I managed finally to elbow my way up to an American Bar in the main hotel and drank two glasses of some god awfull stuff they call champagne. It was wet and cold anyway. They served it in sort of a mug with ice. There are all sorts & kinds of people here and plenty of them. I'm going to try & get some pictures before I leave so will stop. My news is about exhausted in the things I can say.

Wish you were here. You certainly could make a wonderful story of these experiences. I'm afraid I'm going to forget an awful lot of it. That's really not the only reason. I miss you terribly and it's going to be such a long time.

    Loads of love,
    Al

Postmarked 23 May 1943 Received by BSZ 15 July from AWZ / American Legation / Office of the Naval Attaché (location redacted), passed by Naval Censor / postmarked again in Washington, DC 13 July, marked on envelope by BSZ: #6, #7

[no header]

ALS-P
Wednesday the 30th [June 1943]

#6
Dearest,

I'm writing this 3500 ft up. The scenery is very dull. Once in a while we go up to 8-10 000 ft to clear some mountains. When we do this it gets quite cool. This is an Army plane, the first I've been on. The seating arrangement is quite unique. There are two rows on either side facing each other. You sit in bucket seats – concave aluminum. We have three women (American) with us. Quite an unusual occurrence in these parts. They are probably diplomatic wives returning to their posts. Now we see a bit of sea. We are going to ride over it for a while. I couldn't buy any cholocat bars and I couldn't get any sandwiches so I guess I'll go without lunch today. I won't mind much. I don't feel very hungry.

Last night, as there was no water at our place at least what there was so chlorinated it was awful to drink, I went in town to try some of their van blanc. I met up with four RAF guys and after four or five (with ice mind you) we had dinner together. They were very nice and it was a lot of fun, in fact so much I missed the last bus up the hill. I really /p.2/ didn't realize they stopped so early (9:30) I got a lift part way and then had to walk the rest. I was lucky to find home in that strange city but I finally did by following the bus wires. It was pitch black. Well I'm sleepy and I think I'll stretch out on the mail.
Later 2:30

---

[11] AZ appears to have use the initials A.F. to mean a digestive disorder.

We arrived at our destination[12] a little before I thought. Believe me this is some place. 105° in the shade. They say it cools off at night and Oh boy I can't wait. I've got to spend the night at the airport here at the Gedink Hotel,[13] which consists of a number of huts with cots – no mattresses or sheets, just blankets and they certainly seem superfluous. One of the nice features is when the planes come in blowing clouds of dirt all over the place. The drinking water is just about as hot as the temperature. I hope to goodness I'll be able to get out in the morning. Even though it might be hot in the plane at least it's not dirty.

Well, I guess enough for now.
                Loads of love,
                   Al

I have just arrived at [REDACTED][14] and this will go back by fast mail. Love, Al

----

                                                                                                      ALS-P
                          Thursday, July 1st [1943, somewhere over North Africa]

#7
Dearest,

Up again. We will be landing in a little while and I thought I might as well get off another letter. I couldn't get air mail stamps at the place we stopped last night so I'll be mailing this and #6 at the same place.

I'm glad I made this plane as our quarters last night were rather primitive. We had to walk about a mile & a half to supper and then again this morning for breakfast. Not too easy walking in the sand. Last night my bunk mates broke out a can of orange juice and believe me it tasted pretty good after the hot chlorine cocktails. We cooled the can first by soaking a towel and letting the water evaporate. I say cooled but I really mean cooler than it was. During the night the temperature must have fallen to about 45° because you needed two blankets to keep warm. One of my bunkies was from Portland Oregon and the other from Texas. Very nice guys. We talked till about 10 PM on our cots outside the bunk and then decided to sleep there as it was cooler, no mosquitos and the flies had gone to sleep at sun down. Somehow or other the place reminded me of Bill Hirst[15] and his favorite song.

We had lunch a little while ago at an airport[16] where we stopped to fuel and /p.2/ drop off the mail & some passengers and take on some new ones. Lunch consisted of a spam sandwich, a chlorine cocktail and a can of peaches. I've talked quite a lot on the plane with a boy from Mississippi who has been over here a year in the Air Corps and knows all about this country. He's been acting as guide, pointing out the many places of interest.

This about brings me up to date. I hope the next place will afford more comfort than last night. Perhaps Mrs. Reilly[17] was right.
                Loads of love,
                   Al

----

9 July 1943 U.S. Navy postmark. Received 20 July by BSZ #8 from AWZ, passed by Naval Censor / postmarked again Washington, DC 19 July

                                                                               ALS-P

---

[12] Probably Tripoli, Libya.

[13] The name of the hotel is probably a joke. There is no hotel now in existence in Tripoli with a name that is anything like this. "Geedunk" or "gedunk" is the standard expression in the Navy and Marine Corps for whatever is standard GI, snack-bar quality stuff. Wikipedia says that the term first appeared in *Leatherneck* magazine in 1931.

[14] He had arrived in Tripoli. AWZ wrote, after he reached Karachi, that he had stopped in Casablanca, Algiers, Tripoli, Cairo, and Habanniah, Iraq. One other stop between Cairo and Habanniah has not been identified.

[15] Bill Hirst was apparently a friend of the Zimmermanns, probably in the Orpheus Club with AWZ.

[16] The airport between Tripoli and Cairo where they made a brief stop has not been identified.

[17] "Mrs. Reilly" appears to have been a joke between AZ and BSZ

[no header]   Friday, July 2nd [1943, Cairo[18]]

#8
Dearest,

    I hope you got the letters I sent yesterday by Navy fast mail. You should get them quicker than previous ones. Let me know if you did. It would be interesting to see how much quicker they get home compared to regular air mail.

    I had quite a time getting a decent room here. The assistant naval attaché[19] got me a room at the National[20] where I spent last night. It had one small window that overlooked the elevator shaft. It was a quarter the size of our room in New York and pretty grim. This morning after talking to one of the naval officers who was staying at Shepard's [sic][21] I went over and talked to the /p.2/ manager and ended up with a swell room. I might have to get out Monday but it's pleasant being here for a while. This is one of the most famous hotels in the world and everyone I've talked to marvels at the fact I'm here.

    Most of today I spent trying to get transportation to my destination. It looks as though it's going to be pretty difficult. One way of going there is not possibility [sic] for a month. The weather here is quite nice but warm. They say it's usually cool for this time of the year.

    This evening I went out on the terrace to have dinner – a really beautiful setting and very colorful – the waiters in their night gowns and fezes – the bus boys in elaborate red uniforms. An American colonel at the next table looked /p.3/ very familiar and kept looking over at me and finally came over and spoke. It turned out to be Wally (?) McClanahan from Chestnut Hill. He was with a Major Mahare[?] and asked me to join them which I did with alacrity for I was feeling pretty lonely. We talked till midnight mostly about Chestnut Hillites. He invited me to go to a ball game on Sunday if I'm still here and it certainly looks as though I will be. It's a service game to celebrate the Fourth and will afford an opportunity to meet some Americans who are here. It will be pretty grim if I have to stay here a long time and not know anybody. I'm getting pretty sick of my own company.

    Well, that's about all. I certainly /p.3/ do miss you and I'm afraid it's going to [get] worse & worse.

        Loads of love,
        Al

The envelope postmarked U.S. Navy, 9 July 1943, was passed by Naval Censor and postmarked again in Washington, DC, on 19 July, received by BSZ on 20 July, was marked by her #9. It was empty. The letter was, however, in the second envelope marked #11 (*infra*), and I transferred it to the empty envelope.

        ALS-P

[no header]   Saturday July 3rd [1943, Cairo]

#9
Dearest Barb,

---

[18] AWZ did not put the location in his letter, or it would have been redacted by the censor. However, it is clear from the text of this letter and those that follow, that he is now in Cairo, Egypt.

[19] The Assistant Naval Attaché in Cairo in early 1943 was Col. William A. "Bill" Eddy, USMC (1896- 1962), who was already famous as a WWI war hero, and who rejoined the service in WWII as an OSS officer. He later was Ambassador to Saudi Arabia, where he translated the ship-board conversation between FDR and King Saud. The assistant naval attaché who found a room for AZ at the National Hotel was, however, probably LT Henry Hotchkiss (letter #11, 9 July).

[20] The National Hotel, built in 1905, was the largest hotel in Cairo. It no longer exists with that name.

[21] Shepheard's Hotel was the leading hotel in Cairo and one of the most celebrated hotels in the world between the middle of the 19th century and 1952. It was destroyed in an explosion in January 1952. It has since been rebuilt.

You probably will say something is wrong with me, writing to you every day like this. As a matter of fact there is – I miss you terribly and the nearest thing I can think of to bring you closer is to write to you (parse that sentence if you can).

I brought a trinket for you today which to my mind is one of the prettiest pieces of jewelry that I have ever seen. The stones are probably glass. Anyway it struck my fancy and I hope you like it. The stones are supposed to by amethyst – the large one being surrounded by seed pearls. It's "second-hand" as M. Hatoun explained to me – that's why he was selling it so cheap. It' supposed to be over 100 years old and I believe it is – for nobody in our day and age would take the trouble to carry /p.2/ out the design – lotus flowers etc – on the back and it must have taken some time to wear smooth the lower pendant rubbing against some princess's bosom. I have picked out something for the girls which I'll probably send along with this. You might have these things appraised as I would like to know how much I'm stuck.

This morning was spent by going around to several people trying to get transportation out of here.

The best I could get, so far, is next Thursday and even that isn't definite. Money goes pretty fast here – it's almost as expensive as New York. However, the per diem furnished by the Navy helps out a bit.

After coming back to the hotel here /p.3/ for lunch I had a snooze and then went out to see the prescribed tourist must – the sphinx and the pyramids. They were something to see – that's about all. They didn't inspire me with awe but I couldn't leave here without seeing them. I took some movies which you may sometimes get (along with some I took in Algiers. I am afraid they are going to be pretty bad as they are not tropical __?__ and are & will be subjected to pretty hot weather which is especially bad for color film.

I joined up with another lonesome soul at the bar this evening, a New Zealander who has been here for four years. Wally (still ?) McClanahan joined us later and the three of us had /p.4/ together. The New Zealander did most of the talking being pumped along by Mac & myself. He has been in practically every major engagement including Greece, Crete, etc., and my what tales he can tell. Even Mac, who has been here for over a year, was amazed with his stories.

Well, it's time to stop dear. I am propped up in bed writing on a tablet and it's now 2:30. So good by and lots & lots of love,

Al

The envelope postmarked U.S. Navy, 9 July 1943, was passed by Naval Censor and postmarked again in Washington, DC, on 19 July, received by BSZ on 20 July, was marked by her #10

ALS-I

[no header]

Tuesday, July 6th [1943, Cairo]

#10
Darling,

On Sunday I slept late wrote some post cards and then went out to the Gezira Club[22] to meet Mac McClanahan for lunch. Gezira Club is a tremendous sporting club with a number of cricket fields, a swimming pool, golf course, etc. It is a British club with a few Americans and others I guess. You certainly wouldn't know there was a war on. There were thousands out there driving in their own cars or in taxis. Most all the men and a lot of the women in uniform.

---

[22] From Gezira Club website, accessed 1/23/2011: "The Gezira Club has a long, chequered history, and so does golf in the club. The Club, according to some sources was founded in circa 1882 by the British forces that had just been victorious at Tel el Kebir. The first horse races were run in 1883, on grounds that were part of the botanical gardens of the Gezira Palace (now the Marriot Hotel) built by Khedeive Ismail for the opening of the Suez Canal. The golf course in Gezira is one of the oldest in Africa and probably the oldest in the Middle East."

After lunch we watched cricket for a while and then a base ball game between the American public relations /p.2/ group and the war correspondents. I took some pictures. Quentin Reynolds[23] at bat and General Strong[24] umpiring. That finished up the second film (about 10 feet on the end of the main street of Cairo). I'm sending the two films I've finished home. They will be developed at Eastman in Rochester under Naval censorship and then be sent to you if they don't get fouled up somewhere. The first film was taken at Algiers right near the place where I stayed.

Mac had been invited to a Fourth of July reception by Mr. Kirk[25] – the minister to Egypt so he left me at Shepard's and I had dinner alone as usual.

Yesterday, when I reported to the Naval Attaché's office they said there was a good chance of my /p.3/ going out this morning on a bomber so I spent the rest of the day leisurely getting ready to leave. One of the important things to accomplish was to get some chocolat bars. They probably will be the only thing I'll eat along the way, at least I sorely felt the lack of them coming this far. I sent a package home to my girls. I certainly hope nothing happens to it. I wrote you about your necklace and yesterday I succeeded to get the ones I looked at for the girls. If they are too long, they can be shortened and you can decide who can have what.

Toward the end of the day I received the news I was not going to go this morning so it looks as though I'll be here for two more days /p.4/ mostly sitting around doing nothing. I keep in touch with the office when I'm not there in case some other means of transportation comes up.

Yesterday I talked to Alan Parker[26] on the phone and I'm trying to get hold of him for dinner tonight now that I'm staying over. I also called the Am. University but all I got was a native who couldn't tell me any thing and I was told by the hotel operator that the university was probably closed for the summer.

It's been pretty hot here but probably a lot cooler than where I'm going. During the day it gets around 90-100° in the shade. The food is fair but I don't have much appetite in this heat.

Lots & lots & lots & lots of love,
Al

Postmarked Washington, DC, 19 July 1943. Received 20 July by BSZ, marked by her #11. (no original postmark by U.S. Navy)   ALS-I

[no header]

Wednesday, July 7th [1943, Cairo]

#11
Dearest,

I finally got in touch with Alan Parker yesterday afternoon and he insisted I should come out to where he was, which I did in late afternoon and spent the night there. We had a cocktail at their officers club then dinner and the movies. I was rather in an awkward position as Alan would ask me about a number of people in Phila that were friends of his but not mine. As a matter of fact I don't think I've ever met either Jane or Alan. I'm sorry I didn't see Janet before I left because Alan seemed to be thirsty for

---

[23] Quentin James Reynolds (1902-1965) was a journalist and World War II war correspondent. After the war, he won the largest libel suit ever filed up to that time, against Westbrook Pegler, who had called him "yellow" and an "absentee war correspondent."

[24] Probably Maj. Gen. George Veazey Strong, Chief of U.S. Army Intelligence (G-2).

[25] Alexander Comstock Kirk (1888-1979) was Ambassador to Cairo from 1941-1944. He hosted the Cairo Conferences in 1943 and was a forceful advocate for internationalism in the post-war period. Sen. Joseph McCarthy later called him a homosexual and said that he was a friend of Charles W. Thayer (see below).

[26] Alan Parker was apparently a physician, probably a surgeon, on duty with a Pennsylvania medical unit in Cairo. It is not clear how he learned about the arrival of AWZ in Cairo, but he and AWZ were not friends before the war. I suppose BSZ learned about Mrs. Janet Parker from a mutual friend and asked AWZ to look for her husband, Dr. Parker, in Cairo.

recent personal news of his friends and of his wife & children. /p.2/ It was quite an experience visiting a base hospital which we did this morning and also very pleasant to meet and talk to a lot of Philadelphia doctors although I didn't know any of them. They are quite comfortably set up but of course all are anxious to get this bloody mess over with and get home. [BSZ's transcribed letter adds two sentences, which I don't know how she found; I may have missed them: Alan is executive officer of the base, which is second in command. They are near enough so that they can come in town when they want and far enough out to get away from the noise and dirt of the city.][27]

Tell Jack Kane[28] that Capt. Thornton[29] (who is sitting at my elbow) sends his best regards. Evidently the Capt. thinks a lot of Jack though realizing his shortcomings.

Well, that about brings me up to date. I'm still up in the air about transportation out, but I expect to be on my way tomorrow.

Lots of love,
Al

The Capt saw Jack Kane on Sunday & reports all's well with him.

Postmarked 9 July 1943, U.S. Navy. Received 20 July by BSZ, marked by her #11 [sic, same as above, and the number #11 is also given by AWZ].

ALS-I

[no header]

Friday, July 9th [1943, Cairo]

#11 [sic: two letters are numbered 11]
Dearest,

Saw Alan [Parker] last night. Had him come in town for dinner. He brought this letter along to Janet[30] which I think will get to her quicker thru me.

Lt. Henry Hotchkiss, the asst Naval Attaché and Maj. Maliare[?] (whom I met thru Wally McClanahan) joined us at dinner. It was a very nice party.

At lunch yesterday Captain Thornton took me out to Gezira Club which was a pleasant interlude. If nothing comes thru on my transportation the Capt asked me to drive up to another big city near here.[31] Com. Chipman and Markey[32] are going along too.

---

[27] AWZ visited Parker at General Hospital 38, a 1000-bed hospital at Camp Russell B. Huckstep, a few miles from Heliopolis. This was the major hospital for the Mediterranean Theater. It bore the same number as Base Hospital 38 of the Jefferson Medical College of Philadelphia in World War I, which was probably not a coincidence. From "U.S. Army Medical Department / Office of Medical History / CHAPTER II / Army Medical Service in Africa and the Middle East" (accessed 1/24/2011).
[28] CDR (sel) John "Jack" Kane, USNR, was Director of Naval Intelligence in Philadelphia. He and his wife, Amelie, were close friends of the Zimmermanns. Kane probably recruited AWZ into ONI.
[29] Capt. Thornton = ONI, Cairo. In AWZ Wartime Scrapbook, page 11(R): Business card of "Thomas Anthony Thornton / Captain / United States Naval Reserve / USAF [redacted] / Cairo [redacted]." He and AZ met several times in Cairo, and he later came to Karachi (cf. AZ's letter of 29 January 1944), and his letter to AZ in the Scrapbook, page 24(V), 17 February 1945, written from JICAME Cairo. USAF = United States Armed Forces.

[30] The letter was apparently passed on to Janet Parker and is not in this envelope.
[31] They were apparently planning to go to Alexandria, which AWZ mentions in letter #12.
[32] CDR, later CAPT Eugene Markey, USNR; promoted to RADM on retirement. He was the Senior Naval Liaison Officer in the China-Burma-India Theatre. He was a well-known Hollywood producer, who had been in World War I as a young college dropout. He joined Naval Intelligence before Pearl Harbor, and he had been in the South Pacific on Admiral Halsey's staff before coming out to India. He had previously been married to the actress Hedy Lamarr, with whom he had a child (see fn, AZ to BSZ, 7 June 1944). He first married Joan Bennett, and after the war he married two other women: The actress Myrna Loy and Lucille Wright, widow of the owner of Calumet Farms. He was highly decorated, and wrote at least two books and a biography of Lord Mountbatten, who he knew well in India.

This is about all that's new and as I've got some scurrying around to do if I'm going on this jaunt I'd better close.

I miss you an awful lot and wish you were here. Every day it gets worse.

      Lots & lots of love,
       Al

Postmarked 17 July 1943, U.S. Navy, and again in Washington, DC, 22 July. Marked #12 by BSZ on arrival in Haverford, Pa. Return address: ~~American Legation / Office of Naval Attache / Cairo, Egypt.~~ AWZ, ONI, Washington, D.C.

                         ALS-I

AMERICAN LEGATION
OFFICE OF THE NAVAL ATTACHÉ
CAIRO, EGYPT.

                 Saturday July 10th [1943]

#12
My dearest,

Foiled again. I've just been told I'm not to go tomorrow. This is the third postponement. This time it was after I was packed and weighed in and just had an overnight bag á la New York. Well, such is life. It's hot as hell here especially during the day. I'm tired of hotel food especially when you can't order what you want and have to eat what they put before you – no choice whatever.

Yesterday I scoured town looking for something to send to the boys from here and finally found some sandalwood animals. When they arrive you might give Warren his choice for I don't think Albie would care /p.2/ much. They were about the only things I could find. I hope they like them. I thought of getting them Fez hats but not knowing their size and being very unhandy to pack I decided agin it.

The news came out today about the attack on Sicily.[33] I, of course, saw a lot of the preparation for this along the way – Algiers especially, but I couldn't say anything about it at the time. This is probably one of the reasons I'm not able to get out. There is so much of more important cargo to transport.

Met two war correspondents today, Snow[34] of the Sat Evening Post & Trainer [sic][35] of a paper in Los Angeles. They have been to Chunking & India and are going to follow this present push. You might see some of their stuff as other papers will probably run it. /p.3/

Well, dear, there isn't much more to say except I miss you and the children so much. Give my love to all of them. One of the worst things about staying over here is the fact I get no news from any of you and won't until I get to my destination. My thoughts are always with you and I'm dying to get some word of how you all are.

My trip to Alexandria went by the boards when they thought I might leave tomorrow. It's doubly disappointing to fell I could be up there seeing something new instead of this boredom here.

      Loads & loads of love,
       Al

---

[33] "Operation Husky," the Allied invasion of Sicily, began on 10 July 1943. It was the first step in the endeavor to recover Europe from the Nazis. It was concluded on 17 August.

[34] Edgar Snow (1905-1972), journalist who was best known for his articles on Communism in China and his book, *Red Star Over China* (1937). He was a supporter of Mao Zedong. Shortly before he died, he helped to arrange the visit of President Richard Nixon to China.

[35] Los Angeles *Times* correspondent Tom Treanor wrote *One Damn Thing After Another* (Garden City, N.Y.: Doubleday, Doran & Co.) 1944. He was killed in a jeep accident on 18 August 1944 after being injured while covering the liberation of France.

17 July 1943 postmarked by U.S. Navy, marked by BSZ #13 from AWZ, postmarked again in Washington, DC, 22 July 1943 ALS-I
[no header]
#13 Monday, July 12th [1943, Cairo]
Dearest,

Another delay. They told me Sat. afternoon I probably wouldn't go on the next plane but nevertheless they want me to stick around in case something unexpected happens. I'm awfully bored with it all.

Yesterday afternoon I decided to go out to General 38 again after sitting around the hotel most of the day. I'm glad I did. It was a pleasant interlude and very nice to get some good American food instead of this "Wog" stuff we had hamburgers and potatoe salad and canned peaches – gee, they tasted good. After sundown we had some movies – Tale of Two /p.2/ Cities and Bob Hope & Dorothy Lamour in something pretty bad. The Tale of Two Cities was wonderful. We didn't get to bed till about one – it was such a long program. I slept under a blanket for the first time in a long time. In fact yesterday & today have been quite pleasant. Even in the middle of the day if you keep out of the sun it's not so bad. After a good breakfast I came in town to report – but there's nothing new so I hung around for a while and then came to the hotel for lunch and am now sitting in the garden in back, in a small pavilion. It's really beautiful. Tall stately palm trees, a fountain in /p.3/ middle, thousands of pots but nothing much in bloom. Over in a corner there is an enclosure with a mamma & papa & baby gazelle, or at least that's what I think they are – about the size of a dog with thin spindly legs, thin neck, large ears with the male having two straight horns. They are awfully cute. Lots of birds around, beautiful ones and strange – the kind you'd find only in the bird house in the zoo – many colored with long beaks.

Everyone is pretty excited about the invasion. At the moment things look quite hopeful. I guess, though, it's going to make Alan pretty busy later on. That's about all for now.

Lots & lots of love, Al

----

17 July 1943 postmarked by U.S. Navy, marked by BSZ #14 from AWZ, postmarked again in Washington, DC, 22 July 1943 ALS-I
[no header]
#14 Wednesday the 14th [July1943, Cairo]
Dearest,

Well it looks as though I'm off tomorrow. They decided to run a special plane and I'm one to go along. Instead of a regular transport it's to be a Liberator[36] – much faster but I'm afraid not as comfortable.

I'm not sorry to leave although I've had a nice room (plumbing down the hall of course) in the nicest hotel here. I'd rather get to where I'm going even though this is supposed to be a paradise in comparison. People have been very nice but there's an awful lot of time to put in by yourself with nothing to do.

Had dinner last night with Capt. Thornton & Wally McClanahan. Wally had just returned from North /p.2/ Africa and had a thrilling experience visiting Italy on a bomber.[37] It's not what one would call a quite [sic] Sunday afternoon's jaunt through the park.

Cairo is filled with army mostly British a lot of them here on furlough. The hotel crowd seems to change every 2 or 3 days. A surprising amount of women around in uniform – RAF and nurses and wrens,[38] some American Red Cross & nurses. You get awfully tired of the Wogs (worthy oriental

---

[36] "Liberator" = B-24, one of the great 4-engine U.S.-built bombers of the early years World War II.
[37] This was but one of many reconnaissance and bombing missions over Italy that preceded the invasion at Reggio Calabria on 3 September 1943 by British forces under Bernard Montgomery. The surrender of Italy was announced on 8 September but war with the Germans continued in the Italian peninsula until hostilities in Europe ceased in May 1945.
[38] "wren" = commonly used pronunciation for members of the Women's Royal Navy Service, WRNS.

gentlemen) who keep pestering you to show you the sights of the town or to try to sell you swagger sticks with knives. One came all the way down from 2½ pounds ($10) to fifty piastres ($2) without my showing the least bit of interest.

The country as a whole is pretty /p.3/ poverty stricken. Eleven to thirteen million out of seventeen are said to live on 2 to 3 pounds a year ($8 to 12). They are badly in need of more ground to cultivate and more industry. There are projects, held up because of the way, to irrigate more of the Nile banks and to harness the river to generate electricity. Another fantastic idea is to bring the Mediterranean Sea into the Qatara Depression which is consistently below sea level using the flow of water to generate electricity and depending on transportation to still keep a difference in level. I think there will be no doubt of evaporation /p.4/ with the heat they have on the desert.

The war must have helped Egypt a lot for there are thousands of foreigners here spending money paying high prices but I suppose this is all going to those that had it before – not the poorer classes. It's easy to get around here – plenty of taxis even though most of them you would rate as broken down jalopies at home.

Now to get packed again and get exchange to pay my bill.

I miss you all so much and can't wait till I get where there will be some mail from you.

Loads & loads of love,
Al

No postmark by U.S. Navy, postmarked in Washington, DC, 3 August 1943   ALS-I
8 pages, numbered 1-8 on recto, written on both sides.

[no header]

#15   Wednesday the 15th [July 1943, Cairo]

Dearest,

Foiled again. After getting up at 3.30 this morning and driving out to the airport we were told at 7.00 the plane wouldn't leave because of bad weather further on. Tomorrow we will go thru the same procedure again. There are three of us as passengers. A Flight Lt. Com. in the RAF and a private. Instead of going back to Shepards I decided to stay at Ghiza. Lamboit (the FC) and I are sharing a room – a tremendous one with a balcony and bath that opens up to a full view of the pyramids only a couple hundred yards away. We are sitting on the balcony, I writing and he reading of our successes in Sicily. /p.2/ This evening we will see the sphinx & pyramids in full moon light – quite a sight they say.

There is a swell swimming pool here that I took advantage of this morning before lunch. Our pilot and co-pilot had come over so we had drinks and luncheon together. It's their first flight past Cairo which adds to the excitement of the venture. Besides which we are to wear parachutes and a "Mae West." The Mae West is sort of a life preserver in case you come down in the Persian Gulf. That first step is certainly going to be a hard one if the eventuality arises. The trip will be done in two hops and promises to be the most interesting so far to say the least. Instead of being /p.3/ [sheet 2] treated like other bags of mail or super cargo – this time we will be a lot more intimate with the crew and know a lot more what's going on. Might even get to do a bit of flying myself. Who knows?

Took a nap this afternoon to catch up for lost sleep of last night and didn't wake up till seven. Had a swell bath – probably will be the last good one I'll get for a long, long while – and will shortly have a drink or two (maybe scotch or bourbon although they are both scarce) then dinner at eight (the earliest you eat in these parts).

There's quite a crowd here at the hotel – Egyptians, British officers and quite a few of the feminine sex – none very attractive. Some of them are officers' wives. It feels quite strange /p.4/ to be the only American in fact I'm looked at as quite an oddity. Well it's about time to stop. I'll continue this in the plane if there's any place to write or if I'll be able to maneuver in all my accoutrements.

The next day – We got off this morning – a bit later however as we had to wait for a mist to clear. Instead of 0700 it was 0815 – without parachutes or Mae Wests as they couldn't find any at the airport. It was only a short hop to here[39] – four hours in our fast plane and we didn't need any chutes anyway. As a matter of fact, quite a routine trip. I spent an hour in the pilot's seat although you'd hardly call it piloting as the plane practically runs itself after you get up. Landing & taking off gives you much more of a thrill in this plane as the speed /p.5/ [sheet 3] has to be so much greater (about 140 m p h). They are not as comfortable as a transport (C 53 or DC 3 for Warren). I sat much of the time in the mid-machine gunner's seat – an 8" x 8" stool with a little padding. I had two machine guns – double trained – at my disposal – but I didn't find any opportunity of using them. The plane just bristles with guns – all loaded for bear – also small bombs – incendiaries. The large bomb bay is taken up with extra fuel tanks. You have to double up every time you move. The plane wasn't designed for your comfort – especially if you're a passenger.

We saw Jerusalem and the Dead Sea at a distance and many other biblical places. The country all looks pretty forlorn & desolate from the air. I long for the verdure of Haverford. It's hot as hell here. In the middle /p.6/ of the day it gets up to 120 in the shade – Bad news – just heard our second pilot has a temperature so we are not leaving in the morning. The pilot says he doesn't feel any too well himself. I hope it will not be more than one day as this is really a pretty grim place. I'll stop now as the boys are gathering for a drink (lime squash probably) and I'll continue this later on Saturday.

We spent more than 24 hours here and I've almost enjoyed it. The British know pretty well how to make things comfortable. Our sleeping quarters are quite nice – high ceiling with a slow speed fan in the center of the room. A cool shower a few paces away, a good mosquito netting /p.7/ [sheet 4] quite fine to keep out the sand fleas as well as mosquitoes. This makes it quite hot but with the fan stirring the air a bit it's quite bearable even though its 120° during the day and 90° at night. In the hottest part of the day 3 to 5 everyone retires to one's quarters and has a siesta with all windows closed. In fact, they have it worked out to a pretty neat system to bottle as much of the cool air as possible.

The mess here is excellent considering where we are – eggs (real) & cereal (from New Zealand) & melon (local) for breakfast, good soup, sausage or cold meat, cucumber & tomato salad for lunch and an equally good meal for supper.

I've become quite "chums" with /p.8/ the crew especially the officers – the pilot, the radio officer, the gunnery officer and my fellow passenger Paul Lamboit. We eat and drink together. I'm all ears listening to the tales they had to tell. The gunnery officer was interned in Spain for three months among many other exciting experiences. He was decorated for something but I haven't been able to find out exactly what it was for.

It looks as though we leave in the morning. Swanby the pilot feels better and the flight sargent [sic] is reported all right. We were all examined by the M.O. (medical officer) today. I've been surprisingly well for a hypocondriac [sic], nothing worse, so far, than snapping my cookies at Cairo which I'm sure was due to taking Atabrine[40] /p.9/ [sheet 5] which seems to have that effect on most everyone. I've decided to give it up. Practically nobody takes anything in these parts. West coast Africa seems to be the worst place for malaria. Got measured for my parachute today. They have some here and they seem to think it's a good idea on this last hop. One views these things rather stoically as one goes on. Don't think I'd like the experience but if you gotta go you gotta go. I've talked the thing over with myself and have convinced myself it's the thing to do should the occasion arises.

Well enough for now – the boys are gathering for a drink before supper (bourbon & soda – the Scotch /p.10/ are sending their wiskey to America they say here) so toodle-do. Bad luck again the M.O.

---

[39] After he reached Karachi, AWZ wrote that his stops along the way included Habanniah, which is in Iraq, near Baghdad. He refers here to the Garden of Eden, which is an oblique reference to Mesopotamia rather than to a place of beauty.

[40] Atabrine = a drug developed to prevent malaria during World War II. It reportedly had unpleasant side effects such as nausea and causing the skin to turn yellow. Many refused to take it, in the (false) belief that atabrine was worse than malaria. Others, such as AWZ, decided to depend on mosquito netting and to weigh the risk, taking atabrine only in areas of high risk such as West Africa.

says we can't go tomorrow as two other crew members are sick with a fever. There are seven in the crew altogether – one was left sick in Cairo.

<u>Sunday</u>

It hardly seems like Sunday. It's just like any other day in the Garden of Eden. The schedule today was just the same as yesterday. My how monotonous it gets to have the sun glaring all day long – How nice it would be to have a good old cloud burst. Speaking of weather, a boy just told me that it got down to 16° below. Some climate.

It seems that the crew are all right and we leave tomorrow /p.11/ [sheet 6] if nothing else happens. I'll kind of miss this place with its funny looking bugs & insects, lizards and vampires. Last night we saw a good movie Deanna Durbin in something about Eve.[41] It was very strange to be sitting in the open air with a full moon overhead in this far away place, knee deep in British looking at an American movie. They don't say movies here they say "the flicks." It was a tremendous place and must seat 2000.

We got to bed fairly late – We knew weren't leaving early so we sat around & talked. It was "wizard" fun. I'm getting now so I almost understand them when they talk.

At night or in fact any time – no matter how hot it is you always /p.12/ keep your stomach covered. It you don't you're very apt to get "gyppy tummy" very prevalent in Egypt. One theory is – it has some effect on the fly eggs that have been hatched on your food which give you a high fever and upset stomach for a day or two. Well enough of this for now – it's 6.30 time again. Also one must drink at least 10 pints of water a day that's 20 good sized glasses.

<u>Monday</u> the 19th July.

Karachi at last. Got up at three to leave at day break. Arrived here at 0130 but it was 0500 Karachi time. I didn't need the parachute so I didn't find out whether it worked or not. They say you can always get another if it doesn't. The trip was quite uneventful outside of seeing some "shocking" country. The contours /p.13/ [sheet 7] around the mouth of the Persian Gulf is some of the most awe-inspiring I have ever seen. Besides no vegetation whatever the mountains come up like knife edges out of the abbyssmal [sic] valleys. I'm sure no human has ever penetrated this scorching inferno. Nothing whatever to support life and everything to take it away. We flew at a great height for two reasons – the engines overheated at low altitude and if anything happened there would be a lot better chance to glide to a safer place where your choice is landing on a razor or the shark-infested Persian Gulf. This was the first time I'd had oxygen but I had no mask and had to cup it to my nose with /p.14/ my hand. I didn't need much fortunately – surprising when you think how groggy I felt at the crew camp 2 years ago.

I sure expected the worst and got the best. Outside of missing you terribly life should be quite pleasant and comfortable. With the monsoon season on there is always a cooling breeze from the Arabian sea. Our quarters are palatial, the food excellent (fruit juice, eggs, steaks, etc.) and even liquor (scotch at $18 but reasonable bourbon although rationed to one bottle a month).

I've been very busy this morning making protocol visits and I'll probably be very busy the next few days getting "oriented" (which as I write it has a more literal meaning than before).

Your first two letters were waiting for me when I arrived and also /p.15/ [sheet 8] a swell letter from Nanie and boy was I glad to get them. It's been almost 4 weeks now without a word and now I'm set up no end. I love you very much and miss you so. I just can't wait till I'm back again in your arms.

I'll get this letter off now as it has gotten to be a very long one and until I'm settled I'm not going to have an awful lot of time to write – but I'll do my best.

Loads & loads & loads of love,
Al

----

11 Aug 43 postmark Washington, DC   Received by BSZ 13 Aug   From AWZ at NLO

ALS-I

---

[41] *It Started with Eve* (1941), often said to be Deanna Durbin's best film. Its music score won an Academy Award.

OFFICE OF THE
UNITED STATES NAVAL LIAISON OFFICER
KARACHI, INDIA

CABLE ADDRESS:
"ALUSLO"   Sunday July 25th

My love,

      Tomorrow will be a week in Karachi. I'll have that under my belt at least and it's that much nearer to getting home. I do miss all of you so I just can't wait till I'll be homeward bound.

      So far I've received letters #1 & #2[42] and your cable of the 21st. It was a swell idea to send the cable. It did me a world of good to know everything was all right at home. Your letters were marvellous. Please keep up the good work. Also I have received a letter each from Babs & Nanie that were very sweet. I'm very proud of my family and have shown my pictures to everyone here.

      I'm sorry I haven't written you sooner after my arrival here but they've kept me awfully busy showing me office routine during the day and meeting /p.2/ people in the evening.

      The night of my arrival a Russian lady by the name of Dubash was here for dinner. She was married to a Parsi Indian who died about a year ago. She's quite attractive, I would say pushing fifty. The next night one of the boys here Harmon Burns a j.g. had the vice consul and a Standard Oil man here to dinner and afterwards we went to the movies (American – most of them 1-3 years old). Then the following night I took Swaby, Barlow and Lamboit (crew from the bomber that flew me from Cairo) out to dinner. Then the next night Mrs. Dubash had a dinner for me to meet some people – a British general (retired), major, & a captain, the collector of Sind (he's next to the governor of this state).[43] Several wives were there – full blown English women. Then Friday I had Swaby & Barlow here and last night I was invited /p.3/ out to the Dimitrom's (Greeks) met some more people and we all went to the regular Sat. night dance at the Gymkhana Club. Today at lunch we had fourteen guests – British naval officers among which a captain – British colonel who had come down from Lahore, some wives. So you see my time has been pretty well occupied. An American colonel has been with us for several days awaiting transport to Chunking.

      The set up here is almost fabulous. We can seat 18 at our table and when we have a party like today we had seven waiting on table. The house is quite large – the ceilings must be 25 feet. Servants are all over the place, some of them sleep in some rooms in back. I have a suite – a bath and a study beside the bedroom. The bath /p.4/ is tremendous and has a shower with hot & cold water. That friend of the Hunemans' sure gave me the wrong dope on the water situation. We, certainly have an adequate supply. My bearer's name is Bagararh – a fine old gent that speaks enough English and takes care of my clothes – sees that they get laundered – changes the buttons – lays them out. In fact all I have to do is get in & out of them.

---

[42] These letters, photos, and most of the other correspondence and cables from AWZ's family to him have been lost or destroyed; they were not in the collection that was recovered in 2008 from the attic of Barbara Zimmermann Johnson.

[43] Mrs. Dubash's invitation to AZ for this dinner on 20 July 1943 is in the AWZ Scrapbook, Page 12(V). She appears in several subsequent letters from AZ. Her given name, Nadia, first appears on 24 October 1943 when AZ refers to her as the "girl friend" of his Commanding Officer, LCDR F. Howard Smith. Her maiden name, presumably Russian, is unknown. She disappears from the record after February 1944, and Smith was relieved and transferred. A letter from LT Curt Winsor at ONI, 21 February 1944, says: "Cdr. S. will be recalled without an accounting at Karachi but his vouchers here have been carefully checked and he has been directed (1) to submit a certified inventory property account. He's also been directed to (2) eliminate counter intelligence activities and discharge Sheikh. (This means discharge Mme. Dubash too but they didn't want to sully the lady's name by mentioning her.) (3) To cut down the style of operation of the mess. All this by wire." Sheikh also disappears from the record at this point, but his business card, with the NLO address stricken out, is in AWZ Wartime Scrapbook, showing Sheikh's persistence. A Google search for Nadia Dubash was fruitless.

The men here seem very nice. There is another lieut by the name of Callahan, a j.g. by the name of Browning & Burns, whom I mentioned besides Lt. Com. Smith.[44] Browning has had trouble with his eye – hemorages [sic] and he probably will have to go back to the states. They are afraid he will lose the sight in one eye – pretty tough luck for a young kid. He /p.5/ was 41 at Penn & lives in Wash.

Yesterday I played tennis at the Gymkhana Club with Burns, Callahan and an officer from one of the ships here. I stacked up pretty well with the rest of them so we should have some pretty interesting games. That & bicycling will be my means of exercise.

It's after one now. I stayed up so as to make the pouch tomorrow. I think I'd better stop & get some sleep. I'm enclosing the list of "canned" cables which you can use periodically and I can do the same. You'll have to find out where to send them in Philly. They are inexpensive & should cover most of what we want to say and are just as quick if not quicker than regular ones which is the main thing.

Well good bye dearest. Here's a big hug and a kiss.

Lots & lots of love,
Al

---

20 Aug 1943 postmark Washington, DC   To BSZ from AWZ / U.S. Naval Mess / 254, Ingle Road / Karachi. Passed by Naval Censor. This envelope was originally positioned after AWZ's letter of 18 August to his son Warren. I have re-positioned it because it was written two weeks earlier.   ALS-I

U. S. NAVAL MESS
254, INGLE ROAD
KARACHI

Sunday Aug 1st

Dearest,

Received your third letter yesterday and boy, was I glad to get it. I haven't had a letter from you in over a week. This one seemed to take longer getting here. It was postmarked the 6th and I didn't get it until the 29th.

Sat. Sorry – I had to stop yesterday as a British Capt came in and I had to meet him and sit with him till we went out to dinner at Gen. Hind's (in command of the British troops here).[45]

Today your #4 & #5 came which delighted me no end. Letters mean an awful lot and you're awfully disappointed when a pouch comes and no mail for you. #5 made about the best time so far – got here in 22 days. It doesn't do any good to put air mail stamps on – that just means it will go air mail from Philly to Wash. From there on they go by our /p.2/ pouch or by Army post office (air mail) anyway.

In answer your query about "Daddy Joe and Bitz," I was just telling you I was near enough to see their houses. You'd have to be awfully good to see them from where I was.

I was distressed to hear about Ruth. Evidently that disease still seems to be a complete enigma to doctors. It must be terribly hard on Ed and Teddie not to mention Louise & Mr. and Mrs. Parsons. I wrote Ed a letter yesterday. He must be just miserable. I sent them a card from Limerick, I think. I wonder if Ruth got it in time. I'm afraid not.

You were right about Casablanca as by now you no doubt know. I couldn't say so at the time. I think from subsequent letters you will be able to gather what was what for the rest of the way. I didn't

---

[44] Com. Smith = Lieut. Commander F. Howard Smith, USN, Commanding Officer [CO], Naval Liaison Office Karachi. AWZ almost always refers to him as "Com. Smith," as if "Com" is his nickname. LCDR is the proper Navy acronym for this rank, although by tradition an LCDR may be called "Commander." The CO is, however, usually called "Captain," no matter what his rank it, or – informally – "Skipper." For AWZ to call him "Com. Smith" may be a bit derisory, for by this time he must surely know Navy traditions. On the other hand, Smith was apparently somewhat of a petty tyrant, and he may have wanted to be called "Com."

[45] Brigadier (later Maj.-Gen.) Neville Godfray Hind, CSI, MC. Correspondence with AWZ shows his wife was Marguerite "Poppie" Hind.

see Hammond – didn't get down that way at all. Can't think where he might be on a six week's trip – certainly the way things are now he should be able to communicate home somehow. /p.3/

Getting back to last night – it was a dinner for eight – Gen. & Mrs. Hind – she from Savanah Ga [sic] – both very nice – a Mrs. Smyth wife of a British major[46] – he being in the hospital at the moment (dysentery I guess) an attractive school teacher (female) from Rochester N.Y. by the name of Stroud – has been up in Lahore for two years and now is working in the consulates office here – the A.D.C. (aide de camp), & Mrs. Dubash – whom I mentioned before, Com. Smith & myself. After dinner we came back here and Mrs. Dubash played on the piano – mostly Russian songs and I took a wack at Banjo song and Sylvia also Temple Bells. Our piano came with the house – made in Stuttgart – and is not the easiest thing in the world to play on. However a good time was had by all. Outside of that, there hasn't been much in a social way since I last wrote. The Com. did have a Vienese [sic] doctor & his wife /p.4/ to dinner one night. This is quite an international rendezvous.

Today I had a ride in an amphibious truck. Quite a sensation to be riding along on land and then with no bones about it take to the water – just as easy as rolling off a log.

I'm still in the process of learning. Each of us has to know the other fellows job. We have to switch around quite a bit. Every fourth day I have the duty which only means I have to be available here especially in the evening in case any messages come in or some sailor gets in the jug.

I can't think of anything I need. There seems to be enough of most everything – soap, cigarettes, tooth paste etc. (no Hersey bars tho). I do seem to be short on white socks but I'm sure I can get them in town. We wear shorts (white) with white shirts, long white stockings during the day. I brought three outfits. They were pretty darned expensive considering - $3 for the shirt, $3 for the shorts and $1.50 for the stockings – all of pretty poor material. In the evening we wear long whites – scrupulously cleaned by our bearers.

Dearest it's getting awfully late so I must close. I love you very very much & can't wait till this is just a memory. Lots of love   Al

----

25 Aug 1943 postmark by U.S. Fleet Post Office, Washington, DC   Marked by BSZ, received 26 Aug
From AWZ / NLO to Miss Barbara W. Zimmermann

ALS-I

U. S. NAVAL MESS
254, INGLE ROAD
KARACHI

Sunday Aug 8th

Dear Babs,

Received your letter yesterday dated July 9th but not postmarked at Haverford until July 13th. Nanie's letter written the 12th and Warren's written the 15th came in the same mail. . . . [personal comments] . . . Sorry to hear Anne Lincoln "has not improved." From what you say she sounds like a selfish irresponsible young lady . . . When I see and hear about girls like that, it makes me very proud of you and Lanie. . . .

Yesterday we went to the horse races. They have them each Sat for about four weeks. . . .

There is, of course, quite a British colony here. I have met some and they seem very nice. So far, I've found it very hard to understand them but I guess I will later on. Also there are quite a number of our Army here but we are our only Navy . . .

I guess Nanie is in camp now . . . Give her my love . . .

---

[46] Maj. D. Montgomery "Monty" Smyth and his wife Joan. He is Military Secretary to the Governor of Sind. One or the other or both of the Smyths appear frequently in AZ's letters, sometimes misspelled Smythe. His first initial and the correct spelling of his name and title are on a card in AWZ Scrapbook, page 25(R).

---

25 Aug 1943 postmark by U.S. Fleet Post Office, Washington, DC   From AWZ / U.S. Naval Mess, Ibid. to Miss Helene Zimmermann   ALS-I, 1 page, double-sided
This envelope is in the place it should be for the date of postmark, but the transcription is in the correct place for the date that it was written.

Monday Aug. 8th

Dear Nanie,

So you've had another operation. I guess it wasn't very pleasant – none of them are – especially since you had to watch the other candidates go down and come up even before you got going. From what I can remember of my tonsil operation it was much worse than the one I had last spring. It's now fun having a sand paper throat for so long. Before you know it you'll be the highest in the family for most operations. We'll have you a "beautiful swan" yet. . . . [personal comments about Ann Lincoln's cigarette burning her bedspread, and he believes HZ is now at Beaver Camp, "rather than the one you went to last year." He speaks of the movies he took in Cairo on about 10 July and hopes they turned out well. Thanks for her letter of 12 July.] . . .

I must stop now and read about the "Flying Tigers," a book just brought here by one of them. He's leaving tomorrow.

        Heaps & heaps of love,
          Daddy

---

25 Aug 1943 postmark by U.S. Fleet Post Office, Washington, DC   Marked by BSZ, received 26 Aug From AWZ / NLO. Envelope is in place in the order in which it was received, but the transcription has been placed in order of the letters that were written by AWZ.
   ALS-I

**OFFICE OF THE
UNITED STATES NAVAL LIAISON OFFICER
KARACHI, INDIA**

Monday Aug 9th [1943]

Dearest,

Your #7 arrived today. I have received them all so far. I'm sorry to hear about my #4 and #5. They were mailed in Algiers. They might have been side tracked, for when I mailed them it was the time of preparation for the big push. I remember someone telling me all outgoing mail was being held up at that time for security purposes.

I don't know where you get it that I'm "enjoying it all so much." I certainly didn't think I was sounding so cheerful. I was merely telling you what I was doing or at least as much as I could tell. I'm not "enjoying" any part of it, being away from you and the children and everything. It's just something that has to be done and you might as well make the best of it. I'd give a lot to be out of it all – back to the peace and comforts of home. I miss you so – I /p.2/ hope this old war would end tomorrow. It looks like the end might be a lot closer than we thought possible with Mussy[47] out and a rumored shake up in the German high command. I just saw the "Moon to Dawn"[48] and even half civilized people can't go on treating humans the way they evidently are.

---

[47] Mussy = Benito Amilcare Andrea Mussolini (29 July 1883 – 28 April 1945) was Prime Minister of Italy, known as "Il Duce." He was deposed on 24 July 1943 and was arrested the next day by order of the King of Italy. He was later freed by a daring rescue by Nazis but was recaptured and executed by Italian partisans.
[48] "Moon to Dawn": AWZ means *Moon is Down* (1943), starring Cedric Hardwicke, about German soldiers in Norway. It is based on a novel of the same name by John Steinbeck.

I've written to all the children (yesterday), so I really haven't much news. Things have quieted down quite considerably in a social way. I'm going out on Sat. night to Dr. and Mrs. Rheinerts. He is a medical doctor from Vienna – got out just in time.

Two of the officers here are going away on trips so it will leave us pretty short handed. It will mean I will be duty officer every third day. Having our sleeping quarters connected with the office makes this a lot more simple than otherwise. We have to be here on call but can carry on normally with this reservation.

This is a very stupid letter but there really isn't much to say and it's getting late. Might as well send the enclosed pictures along. I'm getting tired of looking at them.

            Darling   all my love   Al

Send me Warren & Nanie's birthday dates. I'm not sure of them.

---

19 Aug 1943 postmarked U.S. Army Postal service, from AWZ at ONI to Miss Babs W. Zimmermann. One page, ALS-I, front and back. No header.

                                                                      Wednesday, Aug 18th

Dear Babs,

Received your #3 on Saturday. You sound as though you've been having a lot of fun visiting . . . [personal family details] . . . There 'r only three officers here at the moment. That means every third day I have to be here for 24 hours. I really don't mind as there is nothing much to do anyway. Living at your office has its drawbacks I find. I thought it was nice at first but you're far from independent. If any visitors are here you have to entertain them whether or no, and all hours of the night people are calling mostly wrong numbers but sometimes to get you out of bed for one reason or another. . . .

I'm feeling a bit low tonight so I'd better stop before I get to feel too sorry for myself – not being with all of you.

                        Heaps of love,
                        Daddy

---

19 Aug 1943 postmarked U.S. Army Postal service, from AWZ at ONI to Master Warren Zimmermann. One page, ALS-I, front and back. No header.

                                                                       Wednesday, Aug 18th

Dear Warren,

I received the letter you wrote July 22nd a few days ago. . . . [personal conversation]. . . .

There is not much chance for me to go swimming over here. There are no pools and the Arabian Sea is much too rough at this time of the year. You can go down to the /p.2/ beach though. It's about 15 miles from here. I haven't been there yet. Some of our friends have cottages, quite small ones, sort of like the ones at Monomoy Point near Chatham. You go there and spend the night and do you own cooking. I think it will be fun. At this time of the year a strong wind blows night and day from the south west. It makes the waves very high and treacherous – too dangerous for swimming. The wind is called the monsoon. . . . It brings a lot of clouds with it but it doesn't rain here. . . . In fact it rains so little here that we are surrounded by a desert. . . .

Well good bye, Warr. I hope you're all well.

                    Lots & lots of love,
                    Daddy

---

21 Aug 1943 U.S. Army Postal Service postmark. From AWZ at ONI to BSZ. ALS-I
            [no header]

Friday Aug 20th

#22
Dearest,

Your 9 and 10 came yesterday which leaves 7 or 8 unaccounted for. One came after 6 unnumbered so I can't tell which it is. Somehow or other you must have misunderstood my Cairo box – there was a necklace for each of you. The gold with amethyst and seed pearls was yours. I said 'my girls' meaning the three of you – that might have been misunderstood but I remember writing you in Cairo about your necklace. You've said nothing about it except "Babs loves her presents" which leads me to believe you thought they were hers. I'm sorry to have to take it away after she thought it was hers but it is really yours and you probably thought I'd forgotten all about you and very strange sending Babs two necklaces and you none. I still don't quite see how you figured the gold one was not yours.

I'm glad the boys liked their presents. It was awfully hard trying to find them something I thought they'd like. Warren /p.2/ as usual showed his magnanimity by letting Albie have first choice. As a matter of fact the elephant may have been made in India as I see pretty much the same things here.

Glad to hear it's been cool since June. As a matter of fact those June days did seem cool compared with some of the places en route especially Tripoli and Habanniah[49]. It has been delightful here – no need for blankets at night but cool enough to really sleep.

I certainly miss those delicious tomatoes & corn from the garden. It's nice to Timoney[50] to think of me as he picks them. I miss the other vegetables too but those especially. We get tomatoes here but very poor quality and of course no corn. The carrots are hard as rocks and the string beans taste very peculiar. However I manage to eat enough to live. My breakfast is the same as at home – the fruit juice is canned – but good – the eggs have quite a strong flavor. At lunch & dinner the soups are usually good. I eat a little fish once in a while and cold meat – today we had some /p.3/ good cold ham – but boy how I could go for a good charcoal steak cooked outside – instead of what we get from these undernourished cattle. Our cook comes from Goa – down near Bombay. He's very dark but is Portugese [sic] at least part. They settled there years ago. He cooks with charcoal which helps a bit.

Sorry to hear about Jack Kane. It sounds pretty ominous. He certainly had lost weight the last time I saw him. I guess this has been going on for some time. Chub Rogers too but I guess he wasn't having a very good time in the last year. He was a swell guy. Too bad about Bob Holden too. Seems to be a lot of bad news coming from home.

Had dinner with Mr. Ecker last night. Came down from Delhi to greet some visitors we're expecting. The American consul Mr. Macey[51] and wife were there. She is French and quite opinionated. We took turns dancing with her until about eleven. It wasn't a very exciting evening – the only one out since last writing. Two of the officers are away, Browning & Callahan /p.4/ so that leaves the Com. Burns & myself which means we have the duty every third day (and night). Browning is going back to the states because of his eye. I hope his relief comes soon. It's pretty confining the way it is. In fact the Com. has asked for two more officers to take care of the work here.

Well, dear I miss you terribly – I hope with recent developments going so well my internment here will be shortened and I'll be back with you sooner than expected. Boy! That will sure be a happy day.

Lots of love to everybody & especially you.

Al

---

[49] Habbaniyah Airbase is located in Central Iraq about 80 kilometers west of Baghdad. It is near Fallujah.
[50] Arthur Timoney was their Irish handyman.
[51] "Macey" is Clarence E. Macy, U.S. Consul in Karachi. His rank was actually Vice Consul. From William Denslow, *10,000 Famous Freemasons* (2004), on Google Books (accessed 1/25/11): b. 1886; Vice Consul in Coblenz 1921, then Dakar, Monrovia, Port Elizabeth, Tampico, and Karachi (1938-43). This is incomplete, for he continued at Karachi into 1944 and probably later. Was Consul General at Istanbul, 1947-48; retired 1948. Political Graveyard website adds that he was in U.S. Army in World War I; and was Vice Consul in Kabul in 1938, although I believe he did not reside there. Died 1984 (age 97); buried Ft. Logan National Cemetery, site T1, 1433.

---

23 Aug 1943 postmark by U.S. Army Postal Service   From AWZ / ONI to Miss Helene Zimmermann

ALS-I, 3 pp.

[no header]

Monday Aug 23rd

Dear Lanie,

Just received your two letters, that Mummy forwarded, from you. It appears that camp is a pretty busy place. [personal conversation, admonishing her for losing a tennis racquet, hoping she gets over her poison ivy and a cold, etc.] . . .

Last night we had a real live senator stay with us. Richard Russell from Georgia. He seems to be a very nice man and interesting to talk to. He arrived yesterday afternoon and leaves today.

I had lunch yesterday at the Boat Club . . . [he tells of trying to learn the "language here," gives the numbers from one to ten, and mentions some words that appear in English, such as pyjamas and bungalow, and the translation of other words such as "carriage ghāree"] . . .

        Loads of love,
           Daddy

---

25 Aug 1943 postmarked by U.S. Army Postal Service   From AWZ / ONI
    to Master Warren Zimmermann   ALS-I, 1 pg, double-sided

[no header]

Tuesday, Aug 24th

Dear Warr,

Received your letter of Aug 3rd yesterday. You certainly are full of baseball . . . [personal conversation about cricket, other sports, and a riddle that he sent, without the answer; the riddle was told to him by the American Consul]

        Loads of love,
           Daddy

---

26 Aug 1943 postmark U.S. Army Postal Service   From AWZ/ONI to BSZ

ALS-I

[no header]

Wednesday, Aug 25th

Dearest Barb,

I wrote to Warren yesterday and gave him a riddle to solve and told him I'd send him the solution in my next letter. . . . The answer is . . .

I don't think I've ever tried to describe our personnel here. The commander is a man about 50 – in fact there is to be a party Sept 4th – his 50th birthday. He is short – quick-witted and cocky – as most short people are. He knows a lot about this country, having lived here a number of years – can speak Urdu /p.2/ and loves to bask in his importance. Callahan (a lieut) is rather a well-built guy who has been here for two years – was in F.B.I. – quite clever doesn't drink or smoke – which helps the ration situation a bit – a bachelor – but quite interested in the opposite sex.

Burns (a j.g.) just got thru Georgetown Univ. before getting into the Navy and was sent here a little over a year ago. Very nice guy a bit on the proper side – the type for a diplomatic career – also a bachelor but engaged and not interested in the opposite sex here.

Browning (a j.g.) the other officer who, I've written you is going home, is quiet, a well mannered, bright chap who also just finished college (Penn) and has been here about a year. I'm sorry he is going. He's a very nice kid and has lots of ability. He also is a bachelor – but doesn't seem socially inclined.

Browning & Burns live in Wash. D.C. Callahan in New Haven and Com Smith in Akron.

The enlisted men (six of them) come /p.3/ from all parts of the states from California to Phila and one by the name of Hilley worked with a cotton firm Manney Steel in the Drexel Bldg. Our chief yeoman comes from Everett Mass and is devoted to his wife – writes pages every day that, I, at the moment, have to read as I am censor. You certainly get a liberal education when it's your turn to be censor. All in all we have a pretty good bunch. I think the opinion is unanimous that we want the war to be over & get back to previous status.

Haven't really done much unusual since my last letter. Did go to the Boat Club on Sunday at the invitation of Maj and Mrs Smyth, the military secretary to the governor. We had a bottle of beer apiece (the weekly allotment of the Boat Club) and some gin and then they came here for lunch. They were quite good friends of Maj Hunter (the Hunneman's friend) they tell me. I'm to go there a week from tonight for dinner and bridge (the first I've played) /p.4/

The Boat Club is rather attractive. Reminded me of similar clubs at home but now no boats. There is swimming and sun bathing, but the Hindus swim by the thousands a half mile away so that part of it didn't appeal to me. The yacht club, further out in the harbor, is closed probably for duration, as it is inaccessible and no boats, sailing or otherwise, are allowed in the harbor now.

Talked to a man today a Mr. Bond whose wife and three daughters are at Camp Hanounum for the summer. They've lived here for years but wife and children 3 girls were sent home parce que la guerre. The wife is head of the junior camp (Meadowbrook?)

Had a senator spend Sunday night, Senator Russell. A very nice guy – very interesting.[52]

It's been two whole months since I've left. It's that much out of the way. This war can't end too quickly for me. I miss you all so much. I'd give anything to be back with you.

Loads lots & heaps of love,
Al

---

2 Sep1943 postmarked Washington, DC   From AWZ/ONI to BSZ   Receipt not dated   ALS-I

OFFICE OF THE
UNITED STATES NAVAL LIAISON OFFICER
KARACHI, INDIA

CABLE ADDRESS:
"ALUSLO"   Sunday Aug. 15th

Dearest,

Your letter of July 24th arrived today unnumbered but I guess it is #7, #6 is dated July 18th – almost a week before. I hope I haven't missed any. On Warren's "V" mail letter of July 22nd you added a post script about addressing it differently from your #7 but as far as I could see they are both the same. Warren's first "V" mail letter came through as written as well as the one on the 22nd so he needn't worry about writing in pencil. They probably don't bother photographing the pouch mail. I also received Babs

---

[52] Senator Russell = Richard Brevard Russell, Jr. (November 2, 1897 – January 21, 1971) was a Democratic Party politician who was Governor and a long-time United States Senator from the state of Georgia. He represented Georgia in the Senate from 1933 until his death in 1971. He was a founder and leader of the conservative coalition that dominated Congress from 1937 to 1963, and at his death was the most senior member of the Senate. He was for decades a leader of Southern opposition to the civil rights movement. Russell served in the enlisted ranks of the United States Naval Reserve Forces in 1918. During World War II, he was known for his uncompromising position towards Japan and its civilian casualties. He held that Japan should not be treated with more lenience than Germany, and that the United States should not encourage Japan to sue for peace.

#3 today along with letters from Jack,[53] Sally and Jim Winsor.[54] Quite a haul for one day and I was very pleased. Babs said you've received letters /p.2/ including the 15th but excepting 4 and 5. Those are the ones I wrote in Algiers and you should have them by now unless they are lost. You asked if I was in Tripoli and Casablanca. I spent a day in Casa which I wrote you about – had my picture taken there. I spent a night in Tripoli at the "Hotel De Gink" which I know I must have written about. It was right at the airport – the city was jammed – no one was allowed in except for official business and besides I left early the next morning. That was the place where it was 120° when we arrived in the early afternoon and I slept under three blankets that night. It was about as hot as that at Habanniah (Iraq) but there it didn't cool off at night. In fact it was damn hot at night. That was where the crew got sick and we spent three days.

    I'm glad you got the lace and the necklaces but I haven't heard how /p.3/ you like yours. Babs seemed to like hers.

    Bill & Dottie are certainly to be congratulated on their achievements. You have to hand it to them. They deserve a lot of credit and that does too for the swell way they are plugging along.

    Life goes on here. Lots of visitors. Just met a girl friend of Freeman's who is passing thru. Today we had some people in for luncheon – Viennese Jew doctor & his wife (I was at their flat last night) the American consul & his 2 ton French wife, a Chinese defense man & his wife, an American by the name of Ecker here for Lend Lease and another American who is the local representative of the War Shipping Administration. It seems every time there is a party you meet new and interesting people from all over /p.4/ the world.

    Last night at the doctors there were two English women who fancied themselves as singers, a Presbyterian minister and wife and another refugee Jewess. You get to meet some of the strangest conglomeration. The commander went with me. We had a delicious dinner and afterwards did a bit of community singing – not so good. I sang a couple of songs accompanying myself on a dilapidated piano. Mrs. Evans one of the English women sang "Danny Boy" and several others rather badly. It wasn't the sort of evening you would rave about. The minister is supposed to be an expert on hand-writing. I think I'll take some specimens of yours to find out what you are like.

    Well, that's about all for now,
        Lots & heaps & loads of love
           Al.

---

2 Sep 1943   APO[55]   ONI   BSZ   Rec'd 14 Sep   ALS-I
[No letterhead]

                                                          Wednesday Sept 1st

Dearest darling,

    There is a pouch leaving in a little while and I am stealing a few minutes to write. I'm sorry it's been so long since my last letter (the 25th) but we have been very busy. A ship has come in and there have been lots to do all day long and most of the nights. It's been very interesting but a bit hard on one's constitution. Dinners ashore and dinners aboard, etc.

    We haven't had any mail for 10 days. I'm dying to hear from you. It is awfully annoying to have to wait long in between. I hope when it does come there will be lots. I sent you a cable the other day. I'll try and send one once a month. I wish you would do the same. You can use the "canned" messages at quite small cost (about 50¢ for three numbers which will let me know all of you /p.2/ are safe and well).

---

[53] John "Jack" Ott, AWZ's partner in the wool brokerage firm, Ott and Zimmermann. AWZ also referred in his letters to CDR (sel) John "Jack" Kane, but he usually added the surname Kane.
[54] Jim Winsor = James Davis Winsor III (b. 1908). He was the brother of AWZ's desk officer at ONI, LT Curtin "Curt" Winsor (b. 1905). Jim Winsor's wife was the "Sally" who was mentioned in this sentence; he was in France shortly after D-Day, June 1944. See "Winsor" file in AWZ's Wartime Papers.

They are called EFM and I believe can be sent from post offices. Use the code word AMBICA in the address in other words the address would be APO 886 AMBICA.

I hope everything is all right at home. I worry about you so. I'm not liking this one bit being away from you for so long. My heart sinks when I think how much longer it's going to be. Well, darling I've got to stop so lots & lots & lots of love.

Al

| 2 Sep 1943 | APO[56] | ONI | BSZ | Rec'd 14 Sep | ALS-I |

[No letterhead]

Wednesday Sept 1st

Dearest darling,

There is a pouch leaving in a little while and I am stealing a few minutes to write. I'm sorry it's been so long since my last letter (the 25th) but we have been very busy. A ship has come in and there have been lots to do all day long and most of the nights. It's been very interesting but a bit hard on one's constitution. Dinners ashore and dinners aboard, etc.

We haven't had any mail for 10 days. I'm dying to hear from you. It is awfully annoying to have to wait long in between. I hope when it does come there will be lots. I sent you a cable the other day. I'll try and send one once a month. I wish you would do the same. You can use the "canned" messages at quite small cost (about 50¢ for three numbers which will let me know all of you /p.2/ are safe and well). They are called EFM and I believe can be sent from post offices. Use the code word AMBICA in the address in other words the address would be APO 886 AMBICA.

I hope everything is all right at home. I worry about you so. I'm not liking this one bit being away from you for so long. My heart sinks when I think how much longer it's going to be. Well, darling I've got to stop so lots & lots & lots of love.

Al

| Postmark | Return address | Received 3 Sep 1943 | ONI | 15 Sep |

ASL-I

[No letterhead]

Thursday Sept 2nd

Dearest,

Blow me down! Are you trying to pull my leg or is Dick Ziening[?]. Where in the world did you ever get the idea I could write. Evidently your supply of Scotch isn't as low as you've led me to believe or is it that social life along the Paoli local has fallen to such a low ebb that even my letters sound interesting. I suppose Dick is going to have me move my desk across the square after the war – or was until the morning after.

I've been cheated. I've just received your #12 and #13 but no #1 and no 7 or 8 which I complained about before. Yesterday's mail was the first I'd received in 12 days. Are these letters lost or do you just skip a number once in a while?

It was shocking to hear about Jack Kane.[57] I can't believe he has gone. I'm very sorry. There was an awful lot of good in Jack that heavily out-weighed his short-comings. I guess a lot of /p.2/ tongues have been bitten, including Jack Ott's, if tongue wagers ever let consciences bother them. Poor Amelie has a bleak outlook. Ten thousand insurance won't go far. She'll have to take a new view on financial responsibility or her position will get worse and worse. I would be glad to help if in doing so she would be holding her own but there's not much satisfaction in helping people when they are going steadily down hill and becoming more involved.

---

[56] APO = U.S. Army Postal Service, A.P.O. 886.

[57] CDR (sel) Kane (*supra*) developed stomach cancer, which progressed rapidly and this letter shows that he must have died in about August 1943. He died at the U.S. Naval Hospital in Philadelphia.

Warren wrote me a lengthy letter on his activities especially baseball. He certainly is full of it. I think it's a very healthy interest and a hobby that can afford amusement and incentive for many of his younger years. What about tennis? Has he given that up? I'd be disappointed if he had. Why don't you take him on? Would do you both a lot of good. We might have a family game when I get back. I suppose the girls are about hopeless in getting themselves interested. /p.3/

As for suggestions for things to send me I really haven't any. I really don't need a thing. We were able to get a large supply of various things off the ship that has been here, to keep us quite happy for some time to come. If you can lay your hands on some tennis balls, that are worth sending, you might send them along – but don't go to any trouble about it. You can't buy them here but you still can rent fairly decent ones. Also shaving lotion would be useful.

It was a lot of fun having the ships here for several days – my first experience at being "piped up the side." The commander and I escorted the governor and our consul with wives to view the ship. Also my first opportunity to mess on a real American ship. We had a lot to do in both as a business and social way. Quite an event in the life of Karachi.

So far your giving my address to friends hasn't produced "buckets of /p.4/ mail." I've had several very nice letters from Sally,[58] one from Jim & Hukor[?] and Jack[59] once in a while sends me a dry letter mostly about business that reads as though he'd spent about ten minutes dictating to a stenographer.

Speaking of Jack, I'm very burnt up the Otts haven't tried to do something to amuse you. I would think that is the least they could do. Did he get you any liquor or can't he spare any from his Admiral's parties? Then to crow about their entertaining to you who sounds so lonely and last: I would think that Ruth's death would have some quieting influence on their gayety. I suppose Ed is left out on a limb, too.

As I write this, there are some serious malodious strains of music being wafted to my ears by a gentle breeze that Haliburton[60] would lead you to believe were coming from a cobra charmer's lute, but to me it's some wog playing on something that came in a cracker-jack box.

Heaps of love. I miss you terribly. Can't wait till this damn thing is over.

Al

----

8 Sep 1943 postmarked U.S. Army Postal Service. From AWZ/ONI to BSZ, rec'd 16 Sep 43.
    ALS-I [four thin sheets, written only on recto, ink very faded after p.1]
    [no header]

Tuesday, Sept 7th

#26
Dearest Barb,

It always seems that when I plan to write home something comes up. I was all set to spend the evening writing to you and the children but as it's now eleven thirty I'll barely have time to write to you. It takes me so long you know – now that I'm to consider myself somewhat of a correspondent. Some more friends of Freeman dropped in. We had to shake the African dust from them, scrub them up a bit and have dinner.[61] That brings us (Oh yes they had a drink or two) to practically bed time but I thought (and I really wanted to) I'd jot a few lines before I really went to sleep. I am pretty tired.

The trouble with me and my writing is – I think of so many things to say /p.2/ during the day – and then, when I sit down to write, I can't think of them. They say it's indiginous [sic] to the country –

---

[58] Sally = Mrs. James Winsor, sister-in-law of Curtin Winsor, AWZ's desk officer in ONI.
[59] John "Jack" Ott was the senior partner in the firm Ott & Zimmermann (wool brokers). AWZ was the junior partner.
[60] Haliburton [sic] = Richard Halliburton (9 January 1900 – presumed dead after 24 March 1939) was an American traveler, adventurer, and author. Best known today for having swum the length of the Panama Canal and paying the lowest toll in its history — thirty-six cents — Halliburton was headline news for most of his brief career. His final, fatal adventure was an attempt to sail a Chinese junk across the Pacific Ocean from Hong Kong to San Francisco.
[61] AWZ knows that his letters are now being circulated to his friends in Philadelphia, and it makes him careful about what he writes. "Freeman" is his friend Freeman Lincoln, a writer, son of the very successful author, Joseph C. "Daddy Joe" Lincoln. Freeman is the father of Ann and Crosby Lincoln, who appear in this correspondence.

losing your memory – but it is a confession I'm thinking of you practically all the time – "Night and Day"[62] etc. You certainly know the rest of the words and they are very appropriate. Sometimes I think I'll write anything that comes into my head without regard to penmanship, spelling, paragraphing or hanging participles and – maybe I'll be able to catch up with my thoughts. Or maybe you think I do that now. I don't blame you.

No mail for over a week – from any of you. And the children needn't think they are going to get letters from me without writing themselves. It's been over two weeks since I've heard from Babs and Nanie and I've yet to hear from Albie. You mean /p.3/ to say he can't write yet? It has seemed ages since I've seen him. Surely he can write letters by now. I love to hear from all of you and I will try to answer every letter (this to the children) but I insist on being rewarded as my letter writing time is very limited and valuable. I've read just 126 pages of a book since I've been here, trying to do my job as a liaison officer and a correspondent.

I felt like a super sleuth today. I was censoring mail and I came across five letters written by the same man to the same addressee – except the middle initial of said addressee varied. Putting the letters in calendar sequence the middle initials spelled BOMBA. Probably somebody trying to be awfully clever, trying to get something by the censor – but not by me. I am sending said letters to Wash. for further investigation.

Speaking of censoring, I came across /p.4/ a letter written by Grey Emmons' son to his mother so I added a little postscript "greetings from the censor."

Powell Browning left here last week and will be home in another month or so. He's the one that had trouble with his eyes. He probably will get to Wash. shortly after arrival in the U.S. and I asked him to look you up if he possibly could. He said he would make every effort to. He can tell you a lot more about life here than I can write. I was very sorry to see him go as he was very capable and interested in his work and a very nice guy.

Also there are some R.A.F. blokes that might be over sometime – rather improbably – but don't be surprised.

Goodbye dearest – loads of love – I can't wait till we're together again.
          With hugs & kisses,
          Al

----

10 Sep 1943 postmarked U.S. Army Postal Service from AWZ/NLO to BSZ.

This envelope contains only two mimeographed pages, printed front and back, on brown pulp paper. It is a one-day example of unclassified information, with the header:
          THE LISTENING POST
U.S. ARMED FORCES IN INDIA-BURMA-CHINA – THURSDAY 9 SEPTEMBER 1943

Sections include: Latest News Flashes, e.g.: "General Eisenhower announced from Algiers yesterday evening that Italy had surrendered unconditionally and that a military armistice had been granted Marshal Badoglio in which Great Britain, the United States and Russia Concurred." The other sections were: The Home Front, Sports in Review, Cinema, and Stranger than Fiction.

----

11 Sep 1943 postmarked U.S. Army Postal Service from AWZ/NLO to BSZ.
The next issue of THE LISTENING POST is all that is in this envelope, dated 10 Sept.
"The Eighth Army is driving north from the toe of Italy. . . Fresh landings have been effected by Allied troops north and south of Rome and other landings have been made successfully at both Leghorn and

---

[62] Cole Porter, *Night and Day* (1932).

Genoa." [Editor's note: I don't believe any landings were made north of Rome by 11 September 1943, and I suspect that the report of landings at Genoa is also incorrect.]

---

13 Sep 1943   postmarked U.S. Army Postal Service   From AWZ/ONI to BSZ

ALS-I [four sheets, written only on recto]

[no header]

Sunday Sept 12th

26 [sic: AWZ's last letter was also numbered 26]

Dearest darling,

    Your #14 arrived yesterday along with Babs #5. I was glad to hear the movies arrived. One reel (50 ft) was taken in Algiers and the other in Cairo. There will be two more taken on the way from Cairo to here and then scenes in Karachi showing our office, servants, and pet monkey. The end of the reel shows our officer staff including myself which is my first appearance and probably my last.[63]

    Right now I'm siting [sic] in a beach hut about 15 miles from Karachi on the Arabian Sea. We came out last night. It belongs to Mr. Arlo Bond, who has been in India for years with the Standard Oil. Incidently [sic] his wife and daughters are in the States at Camp Hanoum, the wife being in charge of "the Meadow." Besides Arlo there is an officer up from Columbo here for medical examination. He chipped his elbow and it had to be /p.2/ xrayed. He probably will have to go back to the States to have it operated on – lucky stiff. His name is Castle and was in the class of 36 at Penn (Psi Upsilon). It's very pleasant here – quite cool – a welcome interlude. The hut's just big enough for three – we do our own cooking, cleaning up etc – take a dip now and then. There are breakers and the ever present monsoon. The sand is not as nice as Chatham but you don't expect perfection in India. It's not at all a resort but sort of like Monomoy – about 20 huts – widely separated and a few people walking by now and then. We don't bother wearing anything when we go in. Once and a while a camel or two will go by – back and forth from a tiny fishing village about a mile away.

    I'm glad Dick Kelley is thinking of coming this way for his leave or at least Anna[64] has suggested it. It will be swell seeing him. However I think /p.3/ he'd enjoy Kasmir or Darjeeling much better than Karachi but maybe he can swing around and pay me a short visit and spend most of his time in a more pleasant place.

    Writing facilities are not ideal at this hut – no desk – I'm using a child's surf-board for a rest so I'll stop now and perhaps add some more when I get home this evening.

    <u>Later</u> I forgot to tell you #11 arrived a few days ago, so with 7 or 8 missing I have all your letters thru #14. I'm glad you are getting a freezer – that should help quite a lot. Nanie's letters you said you'd enclose failed to get in. I'm glad Babs is doing so well at tennis – we'll have some good games when I get home. I'm afraid I'm not going to want to do much travelling when I get home, at least not this kind or to these places, but we will have the rest of our lives to make up for it. I think you deserve anything you want after this is over. The parachute joke was very funny. /p.4/

    I wish you had gone away for a week. It would have done you a lot of good. You've been thru a lot in the last year – so much morbidity. I don't think the children would have minded. Tell Albie if he doesn't behave himself I'll bring home an Indian boy to take his place. It's too bad you can't use this surplus manpower. There certainly isn't any servant problem over here.

    You have my important stops Algiers and Cairo. I stayed in the former 3 days and 14 in the latter. We did stop at Tripoli over night – at the airport – a horrible place – and also at Oran, Constantine, Sousse, Benghazi but only to fuel etc. From Cairo to here we spent several nights at Habanniah Iraq where the crew got sick (I wrote you about it).

---

[63] Unfortunately, most of these movies have not been preserved. However, the reel showing scenes in Karachi and the pet monkey survived.

[64] His sister, Anna Zimmermann Kelly, and her son, Richard (then an Army captain).

I miss you very much, darling, and I hope the good news in the papers means an early return home. Nothing could suit me better.

Loads & loads of love,
Al

----

16 Sep 1943 postmarked U.S. Army Postal Service   From AWZ/ONI to BSZ

ALS-I [one page, written on both sides]

[no header]

Wednesday, Sept 15th

#28

Dearest Barb,

I've just been a recipient of one of India's gifts to humanity – a touch of sand fly fever – a light case thank goodness. I felt it coming on at the beach on Sunday. When I arrived home my temperature was 100. I went to bed, finished my last letter to you, took a couple of aspirins, had a light supper, spent a restless night to wake up in the morning with a temperature of 103. I called for a doctor, Dr. Reinitz (the Viennese that I had dinner with shortly after arriving). He gave me some pills to induce perspiration to bring down the fever, which they did by that evening. Yesterday I was between 100-101 and today 99-100 so I 'm about over it. I feel almost normal, though still in bed, propped up on pillows writing this on a board. A not too pleasant way to spend a few days but a relief to know it's not one of the other delightful diseases like /p.2/ dengue or malaria both of which start with the same symptoms as my little disease.[65]

I'm sorry it came when it did for we are putting up the U.S. Ambassador (Gauss)[66] to China with attachés for a few days and tonight there is a dinner here in their honor which includes the new general for this area Gen. Haddon.[67] I certainly am sorry to miss it for it should be quite interesting.

---

[65] The three diseases that AWZ mentions will appear again in his correspondence.

    Sand Fly = Sandfly fever, known as kala azar, or leishmaniasis. It is actually a very serious disease, not a small thing, as AWZ implies. It is a protozoan disease, transmitted by the bite of some species of sand flies (Phlebotimanae, meaning blood-drawers). The initial presentation may be a large open sore of the skin (cutaneous leishmaniasis), but it also may directly attack the internal organs. Treatment is difficult and there is about a 10% mortality rate from the visceral form of the disease.

    Dengue = a viral disease transmitted by the Aedes mosquito, which usually bites during the day. It is caused by an RNA virus and is related to yellow fever, although the mortality rate for dengue is much lower (1-5%). It is called "breakbone fever" because of the pains in muscles and bones that occur during the acute phase of the disease. Patients are miserable, with a rash, headache, and other symptoms. It is the most widespread of the insect-borne diseases in the world. There are four forms of the virus, and although partial immunity occurs to each form after infection, life-long immunity occurs only after infection with each of the forms of dengue. There is still no vaccine available.

    Malaria = a protozoan disease transmitted by the Anopheles mosquito, which usually bites at night. There are several forms of malaria, principally falciparum malaria (which can have a high mortality rate) and vivax malaria (which is less severe, but is relapsing). Immunity is not achieved except after repeated infections, so prevention from bites of the mosquito and preventive drugs are important. Treatment in the 1940s was mainly with quinine, which was (and still is) effective, if given promptly.

[66] Clarence Edward Gauss was a Foreign Service Officer in the United States Foreign Service; U.S. Vice Consul in Shanghai, 1912–15; U.S. Consul in Shanghai, 1916; Amoy, 1916–20; Tsinan, 1920–23; U.S. Consul General in Mukden, 1923–24; Tsinan, 1924–26; Shanghai, 1926–27, 1935–38; Tientsin, 1927–31; Paris, 1935; U.S. Minister to Australia, 1940–41; and he was the United States ambassador to the Republic of China before and during the Second World War (1941-44). He resigned from the post in November 1944, and was replaced by Patrick Hurley.

[67] Brig. Gen. Julian B. Haddon. From "The Army Air Forces in World War II. V (Pacific), Section II, Chapter 6 (accessed via Google, 1/25/11): "The CBI Air Service Command came into existence on 20 August 1943 as the successor to X Air Force Service Command. The establishment of a separate air force in China the preceding March had been followed by activation of a separate XIV Air Force Service Command in May under Brig. Gen. Julian B. Haddon. That organization, however, never attained anything more than a tentative status, although General

I've been able to do a bit of reading and if you want to read about these amazing people get "Hindu Manners, Customs & Ceremonies" by Dubois and Beauchamp published by Oxford University Press. I'm astounded at what goes on in the world today – how the other half lives. It's really an eye-opener.

This is a bit awkward so I'll stop. Tell Babs I'll get around to answering her letter later on.

           Loads & loads of love,
            Al

Boy – you sure miss home more than ever when you're sick.

----

17 Sep 1943 postmarked U.S. Army Postal Service. From AWZ/ONI to Miss Barbara W. Zimmermann (BWZ), Chatham Hall, Va., U.S.A.

                         ALS-I

           [no header]

                       Thursday, Sept. 16th

#5
Dear Babs,

 Your "5 or 6" arrived last Friday. I spent the week-end out on the beach on the Arabian Sea about 15 mi from here, contracted one of the delightful little tropical diseases, came home and went to bed where I've been ever since. The fever is about spent . . .

 I'm glad the movies arrived. There are two more reels on the way.[68] I'll try to give you the highlights of all I've taken so far. The first reel was taken in Algiers about a week before the Sicilian invasion. There were so many troops etc that the city ran out of water – you can imagine what that would mean – one scene is a water tank truck sent by the authorities to alleviate the situation. You get a pretty good cross section of the natives as they stand in line, waiting to fill whatever tins or containers /p.2/ they could lay their hands on. The rest of the reel is self explanatory – scenes taken around the place where I

Haddon put forth every effort to save it from failure. The difficulty lay in the fact that heavy repair and overhaul still had to be done in India and in the fact that all supplies had to come into China by way of India."

 Also, from "CBI Order of Battle" website: "20 Aug 43: HQ, China-Burma-India Training Unit (Prov) activated at Karachi, India, Brig Gen Julian B. Haddon commander. Unit consisted of 489 Airbase Sq., Karachi Overseas Training Unit, and Chinese Operational Training Unit. Primary mission was to direct the training of all Air Forces units and Air Forces casuals within the CBI Theater prior to their being formally released for operational employment."

 General Haddon is in the front row of an undated group photo of Anglo-American officers that was probably taken in Delhi or Karachi in about 1943. The photo is on recto of sheet 30 in AWZ's Wartime Scrapbook (AWZ-SB), which was assembled by BSZ in about 1945. AWZ-SB consists of 36pp of cardstock paper, originally bound with heavy green leather. The bindings have disintegrated and the pages are loose. They are probably not all in their original order. The photo in which Gen. Haddon appears is 9x6½in., black and white, with 28 men in three rows. Most are in uniform, and most are British. The names are on the back of this U.S. Army Air Corps Official Photograph. In the center of the first row – usually the ranking officer, but in this case the ranking US AAF officer – is Brigadier Gen. John A. Warden. He was Commander SOS [Services of Supply], USAF [US Forces, IBT (India-Burma Theater)], 18 Dec 1944-10 Feb 1945. SOS/USAF/IBT was headquartered in Delhi. Gen. Haddon is seated at Gen. Warden's right. Warden's legs are akimbo and he is carrying a swagger stick. Haddon is clearly deferential to Warden, but he is dapper, in his squashed aviator's hat and aviator's dark glasses. Warden's uniform is a wool blend, beautifully tailored; Haddon is in khakis, with a CBI patch on his left shoulder, and he is wearing aviator's wings. AWZ is the only U.S. Navy officer present. The senior British officer present is wearing ribbons and a beret; he is a Brigadier Langlands. Two civilians are present. One is J. Harris, who was Central Intelligence Officer, Karachi (title in letter from IB Quetta to Harris, 23 October 1943, in AWZ-SB).

[68] These reels have probably not survived. Some of his movies were spliced into two reels after the war.

stayed which was back on one of the hills away from the sea. I was dying to take some pictures of the invasion fleet but that of course was not allowed.

The next reel was taken in Cairo and Mena where the sphinx and 3 pyramids are. My guide claimed he knew how to take movies so I let him try . . .

The third reel if it comes, I took from the B 24 British bomber that flew me here from Cairo. I have my doubts, that it will be much good. After we passed over the Garden of Eden (the conflux of the Tigris & Euphrates) which looked just a little greener than the rest of the desert, we crossed some mountains at the mouth of the Persian Gulf. I started taking pictures there & took add shots on the way to here. This is some of the worst country I've ever seen. The last reel was taken here & I think will explain itself. My back is getting awfully tired so I think I'd better stop. Glad to hear you are now a baseball fan. How about football?

    Lots & lots of love,
     Daddy

[Enclosed: two photos. 3 ¼ x 2 ½ on the back of which AWZ wrote: "A sort of composite picture of some of their gods. They name thousands of them." On the other, a post card, he wrote: "One of the Indian movie Queens. They make their own movies here. Those that have seen them say they are pretty bad – long and boring."]

---

22 Sep 1943 postmarked U.S. Army Postal Service   From AWZ/ONI to BSZ   ALS-I
              [no header]

                                 Tuesday Sept 21st

#29
Dearest Barb,

I feel so hopeless concerning anniversaries, or whatever they are called, when counting by months, but the day before yesterday was the end of my second month here and two days after tomorrow will be three months since I've seen you. I feel so hopeless because it has seemed an eternity but it's still only a small portion of a tour of duty. My heart just sinks when I think about it.

I sent you a cable today – just to let you know I am still kicking and not sooner had I sent a bearer down to the telegraph office than one arrived from you saying "all well at home – all our love." It was very welcome news and set me up no end. I haven't had a word from home in 10 days and I was pretty regusted. Every once in a while the mails seem to jam up some where. I certainly hope it's not that my /p.2/ family has forgotten all about me.

I've been up and about for three or four days since my sickness but I still don't feel one hundred per cent. I suppose it takes a while before the bug is convinced he should try new worlds to conquer.

There really isn't a thing new since I last wrote. Life has been very dull. I miss all of you terribly and I can't wait until this bloody mess is over. I'm afraid my letters are getting very dry and uninteresting but there just isn't anything to write about. I'm sitting here trying to "grin and bear it" but I'm afraid the "bearing it" is loosing its "grin."

Enough for now – it's getting late and I'm feeling pretty blue so I'll try to bring myself in a detective story and try to forget I'm 10,000 miles away from my loved ones.

    Heaps & loads of love,
     Al

---

25 September 1943 postmarked U.S. Army Postal Service. From AWZ to BSZ, rec'd 7 October 1943

                                    ALS-I

               [no header]

                                Friday Sept 24th

#30
Dearest darling,
> Your #7 arrived the day before yesterday. It was mailed July 22nd so took 2 months getting here. Needless to say addressing it c/o U.S. Naval Liaison Officer is not the right way. For one thing it went thru the English censors here and furthermore it must have come by ship. Continue with Mail & Dispatch etc. They reach me in 15 to 23 days. It was very welcome, although it did come late. It's the only word I've received from you for over two weeks (outside of the cable). Also it explained about the package I sent from Cairo. It thought, as I told you in a previous letter, there must have been some misunderstanding as you hadn't said a word about your necklace. But now all is explained and I'm glad you liked it. I really thought it was something special extra and I couldn't understand why you hadn't said anything about it. I wish I could see you wearing it. I'd give anything to be /p.2/ home right now.
> I sent two more packages home today but I think you should wait till Christmas to open them. I didn't mark who was to get what but I think that will be easily understood. There are 4, 2, 2 and 1, the latter for you and the others for the children. There will be some more coming thru later. If you don't like any of the stuff you can give it away. You won't hurt my feelings. There isn't very much you can buy here in Karachi. Somewhere along the line you might give the Otts something but wait until around Christmas when I will have sent everything I can find that you might like. Then if you don't particularly care for something or there might be too much of it, you can give it to them.
> A Lt. Com. Lidirer[?] passed thru here yesterday. He had been at the Naval Hosp. Phila. and had known Jack Kane. He was very nice guy (a doctor) and /p.3/ thought very highly of Jack. A Com Tayloe[?] was with him. Amelie might have met one or both at the hospital.
> Played bridge last night – a pleasant diversion – with Mrs. Smyth, Mrs. Greatbatch and an English officer. Mrs. Smyth swung the party – her husband, military secretary of the governor of Sind being in the hospital. It was a nice profitable evening (6 rupees) and a change of food which I enjoyed. I'm getting rather tired of this continual banquet menu they serve here. Oh boy, how I could go for some real hard boiled eggs, ripe tomatoes, charcoal steak and some real milk. There is enough food here and I suppose variety but there is a boring sameness in the cooking and it's not very appetizing to see how the animals have to live that supply that part of our diet – no rich grazing pastures – just city street streets and desert. Talk about goats eating tin cans, waste paper /p.4/ and old shoes – that's commonplace – but when you see milking cows chewing on an old oily rag your fondness for dairy products suddenly wanes. I don't know what they feed their chickens but the eggs are about half the size and have a very "strong" taste. There is no fruit juice except out of a can which is alright – better than none – but rather monotonous. Then you long for good fresh vegetables right out of some clean garden instead of the stuff they have to buy at the native markets that practically turn your stomach every time you pass them.
> Well, good night, dear, sleep well and as happy as you can.
>   Loads & heaps of love,
>   Al

----

1 Oct 1943 postmarked U.S. Army Postal Service   From AWZ/ONI to BSZ

ALS-I

[no header]

Thursday Sept 30th

#31
Dearest Barb,
> Right after I mailed your last letter three came in from you – 15, 16 and an unnumbered one written Aug 2nd which I take it to mean Sept 2nd, as it's postmarked Sept 3rd, and which I take to be number 17. It seems I can't do much better than a letter every five days. It's not that time flys so quickly but letter writing time is scarce.

Since I last wrote we've had a flow of doctors, two of whom knew you at Graduate – Drs. Farrell and Jones. I had seen Jones somewhere before and it turned out to be at O.C.[69] (graduate party) and Harry Wilmer's. Dr. Farrell has been taking care of the Penn football team. One of the doctors didn't have a yellow fever shot so had to stay in quarantine for 8 days which meant staying in a doubled screened house without being able to get out in the air or sunshine, the theory being /p.2/ if he had yellow fever a mosquito couldn't reach him to carry it to someone else. India has no y.f. and they don't intent to have any. They have enough other things to keep them busy, though.

Had dinner last night at Mrs. Dubash's with Harmon, Frank and our Parsi neighbors the Dinshaws. Rather a stick evening – even got down to my fork, needle, cork and dime trick which amazed them. Instead of a dime I used a ¼ rupee piece which is made of silver. The Dinshaws (2 sisters and their brother) are fairly light skinned as most Parsis are but unfortunately rather homely. They are quite nice though not brilliant conversationalists. They sat most of the evening like three bumps on three logs.

You asked in one of your letters who the fourth was at bridge at the Smyths. It was Mrs. Greatbatch (some name what?) a red head whose husband is somewhere around /p.3/ but not in Karachi at the moment. You needn't worry – you know much red-heads appeal to me. In fact there isn't anything I've seen here that does.

I'm glad Warren has had baseball to engross him. How about getting him interested in football? I hope you haven't given up the seats. You or Lanie could take him to seen what it's like visually and then he would get a lot out of listening to games on the radio. If you did give up the season tickets you can still get seats of course. Please don't try to economize now I want all of you to live as comfortably as possible and to have as much fun as possible. I'm hardly spending a cent here. My maintenance allowance takes care of most of my living expenses and there's nothing to spend money on otherwise.

I'm awfully glad to hear Babs has /p.4/ been so helpful. I hope Nanie[70] will fill her boots when Babs goes away to school.

As for another blue uniform – it it's not too much trouble you can send it along. We wear blues from Dec to March and I have only one, but I don't want you to have to go to a lot of trouble. I'll be able to manage. If you do send it you might send two more shirts, ½ dog collars size 15½ (or 15) and some black socks. No victrola – thanks we have a small wind up one – but I haven't felt much like listening to records. Sorry to hear about Ed Street and his troubles. He's had some awfully bad breaks. Let me know when the next movies arrive. (reels 3 & 4) I haven't taken any since then. I'm waiting for some interesting subjects. If you have to pay $50 a week to get a maid – go ahead. I think you should have someone and I guess we can out-bid others if it has to come to that. I would certainly think you could give up the canteen[71] until /p.5/ you're able to conveniently go.

I've run on to a new page trying to answer the various items in your letters and find it is dinner time so I'll have to close. I'm watch officer and tonight we move our visitors on their way. I'm elected to take them out to the airport.

Good bye, dearest. Keep happy and don't hesitate to spend money if that's what it takes to do so.
    Lots & lots of love,
        Al

P.S. Haven't heard from the children in a long time. I'm sticking to my rule not to write unless they write me.

The little things in the Christmas package, open (4 little things for the children) but tell them to be careful they are filled with something very small.

---

[69] O.C. = Orpheus Club. In the previous sentence, "Graduate" refers to the Graduate Hospital of Philadelphia. BSZ served on the Board of the hospital; her father, William Toy Shoemaker, M.D. (1869-1942) had been Chief of Ophthalmology at Graduate.

[70] AWZ sometimes calls his daughter Helene either "Lanie" (as in the previous paragraph) or "Nanie."

[71] BSZ was a volunteer at a "canteen" for transient service personnel.

A single undated page was located here, concluding a letter to one of his sons, probably Warren, [s] Daddy. It is the second page of AWZ's letter to Warren of 20 October, and will be filed there.

----

5 Oct 1943 postmarked APO. From AWZ/ONI to Master Warren Zimmermann (WZ). The letter in this envelope is dated Thursday Jan 20th [1944]. The letter that came in this envelope was probably the one transcribed above, which was undated and incomplete. I will place the transcription of the 20 January 1944 letter in the proper chronological order in 1944, but the letter and enclosures will remain in the envelope postmarked 5 October 1943.

----

7 Oct 1943 postmarked APO. From AWZ/ONI to BSZ, fwd from Haverford, PA, to The Homestead, Hot Springs, Va., on 18 Oct 1943

ALS-I

[no header]

Dearest Barb:

Again it's been a long time between letters. They must be changing the system in Wash. Instead of a pouch every few days with personal mail, the last two have been two weeks from previous ones. This time I'm specially anxious, for Lanie was to have her operation and I haven't heard a thing about it. As a matter of fact I haven't heard from Lanie directly since the 23rd of Aug. I'm worried. I hope nothing went wrong. You get to think all sorts of things way out here when you don't hear from home.

We've just had a couple animal incidents that I know you'd enjoy hearing about. Last night one of our guests stepped on a scorpion with his sandaled foot. The scorpion is a delightful big little animal – a cross between a lobster and a thousand legger – about four inches /p.2/ long with a tail like a lobster and a claw on each one of its thousand legs. Instead of having the mono-color of either of its progenitors it has yellow and green stripes. As out guest squashed it with his foot it reared its ugly head and bit his toe. As I hit the light to write a letter to you and Babs I found a nice little lizard sitting on my desk. Last night as I was ready to bite into a nice almond bar I noticed it was sans nuts. On further investigation I found my little friends the ants had been there. Choosey little fellows – they don't seem to be fond of chocolate – just the nuts – smart little dears – they found their way into what I thought was a water tight closet – through a tightly fitting carton and the usual wrapping around a chocolate bar. The little pigs weren't satisfied with just the one, they /p.3/ gorged themselves on the six I had tucked away, I thought, so securely. A sad loss for me – such a delicacy in these parts. However I gave them to my bearer – ants and all – to give to his fifteen children and thereby became a true "Indian giver."

On thinking over the uniform situation – don't bother sending my other blue (or the other things I mentioned). We wear them only at night in the winter (khakis during the day). With the quick service we have here I can get along very well with just one and the less clothes I'll have to go home with the better.

Don't open any of the packages Santa is sending you, before Xmas.

Don't complain to me about those shy little creatures that crawl around Haverford.

Heaps & heaps of love,
Al

----

7 Oct 1943 postmarked APO. From AWZ/ONI to BWZ, Chatham Hall, Chatham, Va., U.S.A.

ALS-I

[no header]

Dear Babs,

Here I am breaking my rule again. I said would only answer only the letters I received from you but I know you must have been awfully busy getting ready for school etc. Mummy tells me you did a great job at home helping her with housework, the canteen, and getting the boys organized – so I'll forgive you. . . . [repeats the scorpion story]

So you're really a senior. My, I bet you feel important. Well, you should. This time next year you'll be in college so I guess you really deserve your importance. As long as you don't lay down the law to us less important people we'll let you feel as important as you want . . .

By the way the way, don't open any packages addressed to you, from now on, before Xmas. . . .

Loads & lots of love,
Daddy
XOXOXOXOXO ← The censor might take these

---

12 Oct 1943 postmarked APO. From AWZ/ONI to BSZ, fwd from Haverford, PA, to The Homestead, Hot Springs, Va., 22 Oct 1943 ALS-I

[no header]

Sunday Oct 10th

#33
Dearest Barb:

Friday I received letters from all the family, Jack Ott & George Belic. Yours was unnumbered but I guess #18, written Sept 11th. Jack informed me that George Buzby had read my letters at the Sims and had made the remark that from the tone of my letters I have never been without a scotch and soda in my hand since I left. Naturally I am pretty burned up to hear people encouraging words of commendation [sic: presumably he means condemnation], especially coming from someone whose life has been so little affected by the war. Furthermore I didn't intend to have my ramblings spread around to be sniped at. I don't mind your reading extracts that may sound amusing to sympathetic ears but I do mind their getting such wide publication. For Buzby's edification scotch is as rare /p.2/ as his contributions to the war effort and there are about 10 mil that would trade with him anytime.

I'm sorry to hear you are not using the football tickets this year. I thought it would be a new interest for Warren and a healthy one, to say nothing of taking your mind off the dreary routine that it's evidently fallen into.

Glad to hear in Lanie's letter, you finally got some maids. I hope they work out all right. You certainly deserve more freedom and less work and worry than you've had.

I'm very proud of Babs doing so well in her entrance examination at Vassar and also the way both she and Lanie have cooperated. I think the bank account and clothes allowance a good idea. She'll have to learn about such things some day and she seems perfectly capable to begin now.

Lanie wrote about the arrival of the next two films but didn't go into details. I would like to hear more about them as I still have two /p.3/ more reels and would like guidance on what particularly appeals.

A friend of Freeman's arrived today who had seen him six days ago, and gave me the latest news on Crosby. He said she is much better. It was very shocking to hear she had infantile[72] and I hope she'll have a speedy and complete recovery. I'll try to find something to send her that she'll like.

Our A.P.O. here is 886 if they need it for E.F.Ms but the code word is supposed to be sufficient. I received a cable yesterday saying you were all well. It had been sent Sept 27th which means it took almost two weeks. Maybe the APO number will help but if anything urgent comes up you'd better not use E.F.M. It was swell to hear that all of you are well. I was quite worried when I didn't hear from any of you for so long especially Lanie who was /p.4/ going to have her eye operation this fall.

I sent the boys something for Xmas. I hope they fit. If they don't you could stretch them or put some padding in them. I had no idea of what size to get. I'm not trying to make shriners of them but I

---

[72] "infantile paralysis" was the lay term for poliomyelitis, also called "polio," a viral illness that caused widespread disease and deaths at that time. FDR was affected by polio and was unable to stand or walk without assistance.

thought they'd like these ? [AWZ inserts a question mark here. The implication is each boy got a hat known as a fez] as a lot of the young boys who are Muslims wear them here.

Last night our neighbors (the Dinshaws) invited me to go to the Karachi Club for a supper dance. It was quite nice – a good chance to see the elite of the Indians – Parsi, Muslim & Hindu. There were a few other "whites" there but the rest were from slightly tinged to almost jet black. Sort of sad, in a way, to see nice featured girls as black as Charleston negroes. You just know they are conscious of it. Would you like a sari – that is what most of the girls had on last night. They are about a yard wide with beautiful borders and about 7 yds long – you keep wrapping it around. Let me know if you do.

Another friend of Freeman's[73] arrived since I /p.5/ started this letter. A commander whom you met at Jenny & Freeman's as a lieut. com. He had seen Freeman about three weeks ago so did not have as recent dope as the other guy I mentioned. It's nice to meet someone with whom you can discuss mutual friends for a change. I'm awfully tired of trying to make conversation with utter strangers. We always seem to have a house full of them.

They weren't kidding when they said October would be hot. In the last few days the wind has changed completely around and now blows from the desert bringing intense heat and fine dust. It's like a blast furnace to go out in the middle of the day. Fortunately the nights are /p.6/ somewhat cooler but in this season you have to be under a net at night as the mosquitos are active and there's no sense in letting them present you with one of the nice little diseases they carry.

The days drag along and seem to get slower & slower. I miss you *so* much. I shudder every time I think how long it's going to be. Even a year out here is going to be one hell of a long time. Boy, how I wish this damn war would be over and I could be back with you.
      Loads & heaps of love,
      Al

----

18 Oct 1943 postmarked APO. From AWZ/ONI to BSZ, fwd from Haverford, PA, to The Homestead, Hot Springs, Va., 26 Oct 1943   ALS-I
      [no header]
                       Saturday Oct 16th
#34
Dearest Barb

Another week has gone by – another week nearer to getting home. Boy, I'm sure anxious to have that time roll around.

We are in the middle of the hot spell and it is hot. I was on a ship today and had to keep moving around as the heat came right thru the soles of my shoes – like walking on hot sand – bare footed.

Socially, there are two avenues here – the heel clicking British or the naïve natives. In the last week I've had a taste of both. My friend Arlo Bond of Standard Oil invited me to have dinner at the Sind Club to listen to the R.A.F. band concert. All Karachi was there in its best – a beautiful setting outdoors under some of the few trees they have here. Cocktails in the cocktail room – fowl [sic] Indian rum – dinner upstairs in the reading room (you bring your own bearer to serve when you have /p.2/ a party) – then the concert on the terrace with the audience seated on comfortable cane chairs. Everything as the British would do it – everyone on their best behavior hardly speaking to anyone else not on their own party. The guy who might have unburdened his soul to you the night before would barely recognize you. The band played as well as you'd expect a band to play that is busy with R.A.F. The soloists, one male one female, sang as well as you'd expect, being plucked right out of the governor's party. They each sang their programed number, then after faint applause their encores with still fainter applause – she – lumbering over runs & trills and he "chuckling and laughing ho ho" as though it were the most serious thing he'd ever done in life. Arlo broke out a pint of scotch for ten of us in the middle of it which added

---

[73] "friend of Freeman" is AWZ's code word for an OSS visitor, who he cannot name, who was vetted by his friend Freeman Lincoln, who has passed through London to Karachi.

to our appreciation of true art. After the proper rendition of /p.3/ Star Spangled Banner and God Save the King, all stood up in respect to the governor's departure, to the heel-clicking precise salute of his military secretary.

On Monday my friend, Aspey Dinshaw, invited me to accompany him and his sisters to a native movie house to have dinner and see an Indian made picture. I had to get a special pass from our Army MP's as it was in the out-of-bounds area and it certainly was. A friend of his owned and operated the theatre and had an apartment right in it. We had dinner in his bedroom that overlooked the street, to the noises and smells of the tenderloin. Food came from somewhere downstairs. I'm glad I didn't know where. As it was, I just toyed with the many courses and made a meal of caschew nuts that fortunately were in abundance.

The room was full of photographs, from his grandparents on down. /p.4/ His bed – a sort of four poster – had Christmas tree bulbs strung all over it. A metal crocodile was twisted around one of the posts with a red light that blinked in its mouth. In the corner a multi colored light rotated slowly. The flowers in the center of the table, covered with tinsil, I thought were artificial. The ones on his hideous sideboard, real but I was wrong both times. The movie was Mack Sennett[74] at its worst and to make it worse they explained it to me from start to finish. I'd rather have suffered in silence. All in all it was a very interesting evening to say nothing of being educational. No wonder the MP's sent a jeep to conduct us back to civilization.

Mr. Chatrubhuj R.F., one of the guests, insisted that I have dinner the next night at the Karachi Club. I enclose some of the place cards which I thought might amuse you. [8 cards, 1½ x 3in. with typed names on recto and AWZ's sometimes pungent descriptions on verso are enclosed. AWZ's card is: "Capt. Zinimerman." He wrote on the back, "I guess you didn't know of my promotion."]

At present writing I think the Indians a lot more interesting and amusing than the stereotyped Britishers.

All for now – it's late. I miss you so very much and can't wait till you're in my arms again.[75]

Loads of love, dearest   Al

[Also enclosed, a short thank you note from Kathy Oram, in an envelope postmarked at Wynnewood, Pa., 25 Oct 1943, addressed to BSZ and forwarded to The Homestead]

---

18 Oct 1943, postmarked APO, from AZW/ONI to Master Albert W. Zimmermann Jr. (AWZJr.) Empty.

---

20 Oct 1943, postmarked ONI, from AWZ/ONI to WZ.   ALS-I
[The first page of this letter was not in the envelope, but it was inserted after the letter (above) from AWZ of 30 September. It has been moved to its correct location and is transcribed here.]
[no header]

Wednesday, Oct 20th

Dear Warren,

Thank you very much for your letter of Sept 12th. That was the letter where you told about going to baseball games, gave me the Phillies batting averages, and told about your visit to Aunt Lil's[76] and going to Willow Grove.

---

[74] Mack Sennett (1880-1960) was a director of slap-stick movies
[75] "In My Arms" was the hit song from the movie, *See Here Private Hargrove*, which appeared in March 1944. However, the sheet music for "In My Arms," and the first recording by Dick Haymes, was released in 1943. It was immediately popular and must have been on AWZ's mind when he wrote to BSZ. The phrase has often reappeared, most famously in the Supremes' hit, *Back in My Arms Again* (1965).
[76] AWZ's sister Lilian (1896-~1967), who was married to James Fligg.

Too bad the Phillies didn't do better this year but maybe they will next year.

So you went to Willow Grove. When I was your age I thought it the most wonderful place. My favorite was the "mountain scenic railway." I guess from what you say they now call it "A trip thru the Alps." Another feature, that I liked, was the coal mine. Also, believe it or not, the "movie theatre" was big attraction. Those where the days when movies were first invented and the theatre at Willow Grove was /p.2/ one of the very few in the world. Of course now they must have a lot of new things that must be very exciting. I'm glad you had the opportunity to go to the place where I spent so many happy hours years ago.

I guess by now you're back in school again. Baseball is over and you'll be playing soccer and listening to football games on the radio. I hope your soccer ball is in good shape. If it isn't you can take it to Seavens and have a new bladder put in it.

I'm very homesick for all of you. I wish I could be home helping you grow up. Some day, this nasty war will be over and it will be a very happy reunion to be with all of you again.

      Loads of love,
       Daddy

----

22 Oct 1943, postmarked APO From AWZ/ONI to Helene Zimmermann, Rose Lane North, Ibid.
                          ALS-I

      [no header]

                    Thursday, Oct. 22nd

#5

Dear Lanie,

Your nice letter of Sept 16th has arrived. Glad to hear the movie films arrived OK. Sorry some seem to be over exposed. They might be the ones I took up in the plane. I would love to hear more details about how they came out. Those little white specks you mentioned are barrage balloons over Algiers (just before the Sicilian invasion).

The Pennsy R.R. certainly has had its troubles. That must have been a horrible accident in Frankford. What's the idea of calling J Z & Sons "Uncle Jim Fligg's mill."[77] He's not the only one interested in it. Some day even you might own a couple of bricks.

I suppose you are busily at work in school right now. I thought you were going to have your eye fixed before school but I haven't heard a thing about it. You're our last hope to have our name engraved on the exclusive walls of Baldwin School. Are you going to make /p.2/ it this year? Don't work too hard and get yourself all tired out.

I've made a lot of new friends here in India most of whom I'm not very anxious to get to know any better – ants, sand flies, scorpions, lizards, rats and last night a bat kept flying around screeching his head off. They all go to make life more interesting in India.

I sent Crosby a Kashmir scarf today. I hope she likes it. It is certainly too bad she contracted such a horrible disease but thank goodness they seem to know how to treat it now.[78]

Penn must have a pretty good team this year. I hear they've licked Dartmouth, Yale and Princeton. Have you gone to any of the games?

Good bye for now – keep well and happy and help Mummy as much as you can.

      Lots of love,
       Daddy

----

25 Oct 1943 postmarked APO. From AWZ/ONI to BSZ   ALS-I

---

[77] John Zimmermann & Sons was the company that AWZ's father founded for his children to manage. The sons and sons-in-law did not get along well as business partners, and the company was eventually sold.

[78] Crosby Lincoln survived polio, but she had residual paralyses of some muscles.

[No letterhead]

Sunday, the 24th [79]

Dearest Barb,

Here at the beach again with Arlo Bond – just the two of us. Came out last night and will leave this afternoon. I hope this letter is intelligible. I am back to the scrap box desk.

This morning a few minutes ago, first as I was about to start this letter, Nadia Dubash, the commander's girl friend, brought out my mail that had arrived in Karachi – quite a haul – about time though as I haven't had any for three weeks. In the lot the postmarks ran from Sept 16th to the 30th so you can get an idea of how long it takes from the States. We used to get pouches ever few days with personal mail in them, but now they must wait for two to three weeks before they send a pouch with personal mail. Very demoralizing – I liked the other system much better as does everyone else out here.

In my haul were three letters from you dated the 18th, 26th and 30th, the clippings from Life dated the 16th (don't worry I'm safe as a church, as far as any of that stuff is /p.2/ concerned), a letter from Jim Winsor and also one from Sally (damn nice of them to take the trouble to write as they do), a letter from Babs and Amelie and one from Tim[?] saying Mr. & Mrs. Beard had arranged sending me Tim's Air Express Edition – very thoughtful of them. I won't go into the new mail in this letter but will in my next except I hope you get maids and I don't care what you have to pay them and I hope you got Tarbaby. He should be a new interest and comfort to all of you.

Your box arrived a few days ago. I was thrilled with all the thoughtful things. I hope I won't need the hot water with appurtenences but it's nice to have it here. Last night Arlo and I had one of the noodle soup packages and boy it sure tasted good. The cigarette case will be very useful – nice of you to think of it after not being able to find one in N.Y. the last few days. No need to send cigarettes as there seems to be plenty here. Shaving soap, after shaving lotion (Yardley's if obtainable) seems to be scarce so if /p.3/ you intend sending anything more you might include them – also the peanuts were very welcome. We don't need salt tablets or halizone [sic][80] as that situation is well taken care of. You might send 3 sets of underwear.

Talk to Bill Stuble about my giving each child $2,000 or its equivalent before the end of the year. Even if there's a small tax I think it should be done. I'll enclose an authorization to effect this [a letter to William H. Stuble, Esq., to this effect is enclosed in this envelope]. By building up their estates gradually with little or no tax, the inheritance tax will not take such a heavy toll.

I've sent some packages home this week that should help to keep you warm this winter. I hope they arrive safely. There is a scarf, an inexpensive one, that you can give away or keep as you prefer. You'll understand when you open the packages. I sent the mate to Crosby. If you don't want to wait till Christmas you don't have to.

Let me know the size of our beds or the size of any table you'd like a cover for. There are some beautiful things here from Kashmir – bed covers, table covers etc but /p.4/ I resisted buying – I didn't know the size and I'd be a poor guesser.

My back is breaking leaning over writing this so I'd better close. Lunch is about ready anyway.

I miss you so much. I get awfully depressed. It's been 4 months now. I wonder how much longer. All my love dearest. Goodbye for now,

Al

P.S. The Otts burn me up – thoroughly selfish. I thought of getting Louise [Mrs. John Ott] something but I'll be damned if I will.

Took some movies at the beach –showing the Arabian Sea, some Muslim natives coming from the fishing village, a Muslim fisherman drinking some of our coffee after having delivered our lunch – a two pound fish for 4 annas (8¢) with Arlo in the background – a view along the beach showing the

---

[79] Amelie Kane was the widow of CDR (sel) John Kane, mentioned above. Jim and Sally Winsor were mentioned above. He was the brother of LT Curtin Winsor, AWZ's desk officer at ONI.
[80] Halazone = chlorine-based water purification tablets that were used in World War II.

painted cottages – scenes on the way back showing how they live with their cattle – 2 awe inspiring camel drivers – a street scene in the "out of bounds" area and a Sunday afternoon "hay ride."

----

29 Oct 1943   ONI   To BSZ   Not marked   ALS-I
                              [No letterhead]
                                                                 Thursday, Oct. 28th
#36
Barb dearest,

    I wrote you that I had received 19, 20 and 21. Your telling me about the party you had certainly makes my mouth water. I could sure go for some of that grilled stuff right now. Too bad one of the reels was deleted. It was the part taken from the plane and should have been quite interesting. I sent a reel off today with a picture of me for a few feet at the end, as you requested. How about pictures of you and the kids? Those films will spoil if you wait too long and besides I'm anxious as the dickens to be kept up to date. Sorry to hear the deep freeze went on the fritz. Hope it's all right now and will give you no more trouble. Can't understand the Otts. They certainly make me boiling mad by not helping to make your life more bearable – to spend so much time with the British Navy playing golf with the Admiral on Long Island etc etc and not give you a thought burns me up. They are all wrapped up in their own selfish interests. What will Jack /p.2/ do now that he hasn't his Lincoln Liberty jamborees. I understand that's all out now. I never thought they represented the slightest sacrifice on his part anyway. I haven't seen my handwriting friend again so can't give you a report. Extend my congrats to Freeman – he certainly has worked hard and deserves the promotion. Poor Crosby must be having a grim time with these continuous hot packs. By now surely the worst is over. Glad to hear our children are doing so well in school. My chest fills with pride. Keep me informed about their work, it's thrilling. You can fix it so it won't be bragging.

    Since Sunday I've been out to dinner twice. The night before last Arlo Bond took me to the Sind Club (an English club – the best in town). Afterwards we went to the Boat Club and watched a few couples dance on the veranda to a victrola. An American nurse, from the base hospital here, got herself pretty well pickled, to the amusement /p.3/ of the onlooking Britishers.

    Last night Mr. Chawda, our wool shipper in India, in Karachi for a few days' visit, took me to the movies (Andy Hardy in Spring Fever) and then to a friend's house to dinner. It was quite an interesting evening – an insight into Indian family life. Chawda and his friend are both Hindus but the latter is a more liberal one and served whiskey and meat. For Chawda he had a special vegetable dinner. We had about four courses – much too much. I nibbled thru – probably insulting my host for not eating more. His oldest son joined us but the rest of the six children and his wife were very much in the background. They had fourteen children, only seven surviving – that's India. I'm afraid I made a couple of faux-pas. After dinner I asked to meet the rest of the family as I was interested in seeing /p.4/ what they were like – had seen one or two peering in the door. When I met his wife I turned to him and said "Does your mother speak English." Fortunately she didn't and didn't know what I'd said but it must have set him up. I tried to quickly cover it by turning to one of the boys and saying "I mean <u>your</u> mother," but I'm afraid I didn't get away with it. She looked seventy instead of his forty five – but who wouldn't after having that many children in India. The other was asking if one of the servant boys was his son.

    The discussion got to India problems and it was surprising to learn that they all were against the caste and dowry system. My host had been "approached" for the hand of his daughter and 15,000 rupees. With three other daughters to marry off you see what they are up against.

About time to eat so will close. Going on a trip tomorrow to Colombo[81] so the next letter will probably be long in coming as all mail comes thru here. I miss you and miss you and will go on that way.

   Loads & heaps of love   Al

----

6 Nov 1943, postmarked APO   From AWZ   ONI to BSZ   ALS-I/P/I
            [No letterhead]

Saturday Oct 30th

#37

Dearest Barb,

  I'm up in the air again. I left early yesterday morning on a trip that will take me down to the tip of India and over to Ceylon. There are three of us as passengers, two Lt. Com.'s in the British Navy and myself and three in the crew, all Indians. They are very nice looking guys, fairly light skinned, probably coming from the Punjab in north India and members of the R.A.F. The Britishers are good blokes. They have just recently arrived in India and are heading for the same place as I. So far it's been a very pleasant trip – a chance to get first hand information on what the land of mystery and misery looks like.

  Our first stop was Ahmedabad, the hot bed of the Congress Party. Ghandi spends most of his time there when he's not in the clink. It is quite an industrial town. Our view was only from the air as we didn't stay there long enough to ride around. On the way the country we passed over was most horrible looking – not a sign of life – the delta of the Indus River and the Rann of Cutch[82] – mostly winding streams and rivers and mud flats. I suppose in the rainy season it becomes completely inundated. Approaching the interior we got to see some green for a change – haven't seen any for months. The villages are interesting. From 10 000 ft they look completely deserted. Each one seems to have a lake or rather a mud hole where the Hindus can drink and perform their religious ceremonies. (We are coming down to land at Hyderabad and it's getting very bumpy so I'll continue later)

  Up in my blue heaven again after a 30 min stop. After Ahmedabad, yesterday, we came down to Bombay. We were supposed to stop only for two hours but a leak in the oil line was discovered so it was decided not to leave until this morning. I was rather glad as it gave me a chance to see the town, although I expect to stop off on the way back. One of our officers Al Payne came out and got me and showed me around. He was one that I knew, having been up to see us for about two weeks shortly after I had arrived in Karachi.

  Bombay is quite a beautiful city, situated /p.3/ [sheet 2] as it is on an island and surrounded by mills[?] – reminds me somewhat of Rio with its lagoons & mountains. It is much bigger than Karachi and much cleaner. There are lots of palm trees and beautiful gardens. Al Payne put me up for the night and put me on the plane this morning. We talked of seeing some of the night life but decided against it and just sat around & talked. We have flown about four hours and are now deep in the heart of India. Hyderabad, where we just stopped, is one of the richest states and is on a 1500 ft plateau. The soil looks quite rich and much of it is under cultivation. Being up in the air a bit it has a better climate. The ruler, the NIZAM of Hyderabad is one of the richest men in the world and is said to be so jealous of his high station in life that he sleeps only with his daughters. With this discordant note I'll stop.

 [AWZ shifts to pencil] Tuesday

  On my way home after a couple of days in Colombo. We landed there /p.4/ about Sunset Saturday after two more stops. A pretty long day – flying from sunrise to sunset.

  Ceylon is very beautiful. It gets lots of rain and sunshine which make the abundant vegetation fresh and clean looking. It is supposed to be the paradise of these parts. I took the last of the colored pictures (someone gave me a film of black & white – still undeveloped). The first part is taken outside Lt. Comm. Goldsborough's house (where I stayed) sharing what passes by a Colombo house and the rest

---

[81] He is going to Ceylon as a courier and will receive a briefing on what he is intended to do on the trip that is the subject of the book, *Proceed to Peshawar* (2013).
[82] Now spelled Rann of Kutch, an area of salt marshes in the Sind Province.

taken at the beach where we went Sunday afternoon (with a little snip at the end of a city street). Com. Goldsborough comes from Baltimore and seems to know the right people. He and a Lt. Babson run the office and share a house with a British colonel. My quarters were not what you'd call ideal: the bathroom was on the floor below – but was about the only place half way comfortable to stay as the hotels were jammed. I had some papaya for the first time – not so good – and a drink /p.5/ [sheet 3] of coconut milk. At the beach a boy will shinny up a tree (shown in the movies) bring down a coconut, chop the head off – all for 10¢ Ceylonese (about 3¢ American). They climb these straight branchless trees by fastening a rope between their feet (keeping them about 15 inches apart) and grab the tree trunk with the hands, then with a jack-knife motion, pressing the trunk with the rope & moving up their hands they can climb a tree very quickly. The cute little Singhalese boys in the movies washed the sand off my feet after bathing. A service provided at very small cost.

I didn't do much in Colombo otherwise spent most of the time at the office & Com. G's home. I would liked to have seen some of the night life and met some other people but my hosts seemed to be tied up with engagements and weren't particularly interested in showing me around.

I was rather glad to leave, though it is /p.6/ beautiful, the climate is hot and humid.

I used my pen to write Babs a little while ago but when I started again on this letter the ink flowed out like water. We must be a lot higher than we were. Most of the time we can't see the ground as we're above a blanket of billowy snow-white clouds. When we do see the bottom thru the holes it looks far, far away – about 15 000 ft I suppose – the altitude is making me a bit short-breathed. This is a pretty hot plane. It took us about half the time to fly back to Bangalore than it took going down.

Enough for now – I think I'll take a snooze.
[he resumes use of ink] <u>Wednesday</u>

Back in Karachi and glad of it. While the other places are more glamorous, Karachi has a much better climate, not as hot and sticky and it's a relief to get back in one's own bed.

The rest of the trip was quite uneventful. We arrived in Bombay in the afternoon. [/p.7/ I spent the night again with Al Payne – left yesterday around noon and arrived here a few hours later.

A cable from you was waiting for me, saying everyone was all right and not to worry. The address is what confused me – Hot Springs – have you gone there? Or is this cockeyed Indian telegraph system goofy? I hope you are there & getting a nice rest. I suppose I'll get the details later. The mail is about to leave so I'll finish this disjointed letter.

Lots & lots of love,
Al

----

11 Nov 1943, postmarked by APO   From AWZ/ONI to BSZ, received by her 27 Nov. "Xmas presents"
written on envelope by BSZ   ALS-I
[No letterhead]
                                                                                                    Wednesday, Nov. 10th
Dearest,

Received your #22 & #23 the latter part of last week. Awfully glad you have a maid. I hope she works out and is contented. Too bad about Aunt Catherine.[83] I hope she died peacefully and didn't have to suffer long. Uncle Ralph will be left pretty much in the cold if Dick is transferred.

Yes, I play tennis about 3 times a week. The nearest to a kindred spirit is Harman Burns who unfortunately is being transferred to go to sea. He is a very nice boy of 25 and is going to his new post by way of Wash. If he gets leave he promises to come up and see you in Haverford – he will call by telephone. If he doesn't get leave you might go to Wash and see him. He's taking another bag I got from Agra and a silver cigarette case I got in Ceylon – filled with straw belts and two small star rubies for the girls. If you make no connection I asked him to drop them off at the Lincolns and you will get them eventually. I think you'd /p.2/ enjoy talking to him. He's been at this post over a year and is the nicest guy here.

---

[83] Catherine was probably the wife of BSZ's mother's younger brother, Ralph Lambert Warren.

I am sending about everything else to you so that you can divide the loot as you see fit. You might give Sally Winsor, Amelie & Dot, or anyone that has been nice to you, something but it's up to you. Sally has been very thoughtful about sending post cards & a couple of letters. Jim has also written several times.[84]

Practically everything I'm sending is by airmail and should arrive in plenty of time. I'm afraid there isn't much for the boys but there isn't much here to get them. One thing I wanted Warren to have is a magic brass bowl that tells the story of the daughter of Vishnu (Hindu god). When the evil nurse tried to drown the baby in a lake all the water ran out. When you fill the bowl – as soon as the water touches the baby it will completely drain out the bottom. It comes from Ceylon.

The elephant seeds that have arrived have /p.3/ tiny elephants in them if I haven't mentioned it before (one for each child). The ivory god I sent to Warren is Ganesh the god of happiness. He rides around on a rat (he's sitting on it) and as rats Ganesh's representatives are in every house – happiness is in every home. The Hindus will not kill them. If they get too numerous they catch them and let them go. When Ganesh was a baby, the brilliance of his mother's glance reduced his head to ashes so Shiva, his father, replaced it with an elephant's head, being the first living thing that he found facing the north.

Albie's ivory piece is Buddha, founder of Buddhismm. A rich Indian by birth, he devoted his life to religion establishing what later proved to be the religion of the largest following on earth.

The rest of the stuff needs no explanation other than the 3 straw bags, 6 nickle match boxes, 4 brass trays, 2 cigarette cases and 3 dolls (you might give one to Anna) come from Ceylon. The rest are from India. The very soft scarfs & the shawl are from llamas /p.4/ and are called Pehmina shawls from Kashmir. The other wool things and the luncheon sets & table cover are also from Kashmir. They have larger table covers or bed spreads. If you're interested send me the size etc. Also if anything particularly appeals, let me know and I can duplicate it.

Yes I call Mrs. Smyth, Joan, Mrs. Greatbatch Betty and Mrs. Dubash, Nadia, but don't let that worry you. None of them hold a candle to my one & only.

Glad the Rauch's have been so thoughtful to you & Sydney. I didn't think so much of Dr. Jones but thought Farrel very nice. Haven't done much of anything since I returned from Colombo. Going out tonight to a musical evening at the Reinitzes. I sort of dread it as similar evenings have been pretty grim.

We've had a Mr. Preston staying with us. He is at Lourenço Marques (Africa) as consul general and was in charge of the exchange at Morina Goa.[85] Nice guy.

Day after tomorrow I leave for the North West Frontier by Indian train 48 hrs to be away about a month so you can look for some delay in my letters [and this] means I won't get any from you which will be pretty tuff. Must go now to meet a plane.

Lots & lots of love, Al

AZ wrote two letters to BSZ about the North-West Frontier Province (NWFP) trip. The first letter was written in Peshawar on 25 November. He started the second letter in Wana on 6 December; he added to it in Ft. Sandeman on 9 December; and he completed it in Karachi on 15 December.

BSZ transcribed AZ's two letters about his trip and made copies that she passed on to her friends. One carbon copy and two original typed pages of her transcription are in the folder entitled "AWZ Maps, Typed Letters, Notes"; it is described in my Word file with that title. BSZ omitted some parts of AZ's letter, and I believe my transcription (below) should be regarded as the reference copy. She was unaware of the spelling that was in use at that time; for example, she refers to Kurran rather than Kurram, to Khotel rather than Khotal, Zhab instead of Zhob, and so forth. AZ's spelling was sometimes phonetic and inaccurate, but he had maps to refer to, so he knew how the words were usually spelled. I have

---

[84] Jim and Sally Winsor were mentioned above. Dot was BSZ's older sister Dorothy (Shoemaker) Boericke. Amelie (née Sexias) was CDR (sel) John Kane's widow.

[85] A Google search confirms that in 1943 the U.S. Consul General in Lourenço Marques was indeed a Mr. Preston, but I have not been able to find his first name. The hidden web page refers to "Consul General and Mrs. Preston."

attempted to type what I think he meant to write. For example, he surely knew that Miram Shah should be spelled with one "r" but he sometimes spelled it "Mirram Shah" in error. AZ's original letters are in the box of letters, which I have tied up again with string.

----

2 December 1943, postmarked by APO   From AWZ/ONI to BSZ, marked by her: "Rec'd Dec. 9 and "NW Frontier Trip #1"

ALS-I[86]

[no header]

Thursday, the 25th [November 1943, Peshawar, North-West Frontier Province]

Darling,

Your 24th? [October] was forwarded to me here in Peshawar along with the questionnaire. I was certainly glad to hear from you as I feel pretty lost up in this neck of the world. A letter from Babs, Clarence Lewis[87] & Stan Welsh also arrived, and were very welcome. I don't know when I'll get time to answer them. Babs' letter was very cute. Tell her she doesn't have to worry about writing short letters. The main thing is to hear from her. [Marginal note] Hope you enjoyed your stay at Hot Springs. You deserved all the rest & relaxation you could get.

I started on this jaunt about 10 days ago from Karachi. Spent 48 hours in a train getting here. The first part to Lahore was in an air-conditioned car, sharing a compartment with a Col. Fagin, another American Army officer, and a British Army officer. Fagin knows Nonnie Hunter quite well and could understand why he hated Karachi. He seemed to think he was a pretty pampered individual. The trip was quite pleasant. There is no way of getting thru Indian trains so one has to /p.2/ get off to get to the dining car and then wait for another station stop to get back. We were very lucky to be in an air conditioned car as the route took us over the Sind desert which compares to our B. A. – Baralochi[88] trip. The windows are double so we kept fairly clean.

From Lahore on it was a bit different – no luxuries whatever – not even a dining car. I had to fight my way onto the train even tho I had a reservation. I found later it's just another Indian way of getting baksheesh (tips). During the night there was all sorts of commotion – people getting in and out of the compartments which was supposed to be I class but was nearer III. Having started off with a couple of Sikhs I finally wound up with a London limey, who had been in India ten years recruiting candidates for a London music school, and a British colonel and his wife as my bunk-mates. The limey's bearer was also there spending /p.3/ [sheet 2] the night sitting on the floor.

On Indian trains you carry your bed with you in what's known as a bedding roll. If you don't own one you can rent one from Cooks. I borrowed Com. Smith's. Your bearer (if you have one along – I didn't) un-rolls your roll on one of the four long seats and then is supposed to spend the night in a bearer's compartment or in III class, to appear the next morning, help you dress and roll up your bed again. The system is about the only one that would work in India. You at least know something about

---

[86] A photocopy of this letter was given to me by AWZ's daughter Barbara in 2008, and I transcribed it, with annotations, in 2008-09. The original letter is easier to read. I have now made a few minor corrections in my original transcription, and additions of some words that were off the margin of the photocopy. Some letters are difficult to decipher; AWZ's "i" is often very similar to his "e," although he usually puts a dot over his "i" and of course there is no dot over the "e." I am not sure if he knew at this time that Bromhead is spelled with an "o," nor do I know whether AWZ spelled the title of the ruler as "mehter" or "mihtir." I have also made a few light corrections in AWZ's punctuation and corrected a few unintended errors in his spelling.

[87] Clarence = 1st Lt. Clarence J. Lewis Jr., USMC. He was AWZ's next-door neighbor. He was assigned to the OSS and served in India and China. See "CI History" / "CI Marines" by CW05 C. A. Menges, "The History of Marine Corps Counter Intelligence" at http://www.mccia.org/History/ci_toc.htm (accessed 3/1/11). Section 1 shows officers by rank and alphabetically listed within ranks. Clarence Lewis appears immediately above William B. Macomber, who served in France and Burma. After WWII, Macomber served as a U.S. ambassador in the Near East and then was the Director of the Metropolitan Museum of Art in New York City.

[88] Buenos Aires and San Carlos de Bariloche, Argentina.

the cleanliness of the bed clothes and your chances of picking up new travelling companions (very small ones), are thus minimized. It was quite an experience and one must have to appreciate a foreign country. Too bad the colonel's wife wasn't a bit younger and more attractive. It would have added a bit more zest to the intimacy of our bed room. /p.4/

I was met by Major Sir Benjamin Bromhead on my arrival and after a good hot bath at my hotel was wisked to the Peshawar Club for lunch.[89] It being Sunday, lunch at the club was quite festive. The English colony was out in full force, drinks and lunch being served to the accompaniment of a good Indian orchestra playing most of the good modern tunes. I was quite surprised to see such nice looking people in this part of the world.

The city is divided into two sections – the native and the cantonement. The cantonement is surrounded by a barbed wire fence with gates on the entrance roads. This is a protection against the plundering tribes that inhabit this part of the world.

The Norwest Frontier Provinces have a peculiar set-up for government. The Administered Area is governed with Indian laws. A boundary extends thru the provinces that separates the administered from the non-administered area or tribal area and then comes the boundary between India & Afghanistan /p.5/ [sheet 3] known as the Durand Line more or less an arbitrary one set up around the turn of the century. The Tribal Area has its own local governments ruled by mehters, nawabs, Walis or mulliks[90] according to where you are in the Provinces. The tribes have never really been conquered and the present set up seems to be the best solution of a bad situation. The terrain is practically all mountainous and lends itself very well to guerilla warfare, necessitating a large force & enormous expenditures to beat unwilling tribesmen to submission. The tribes have their own laws with offenses against property taking precedence over lives.

The various sections have scouts or militia oftentimes officered by the British to defend the frontier and help preserve peace. They also have Political Agents who try to advise the local rulers but otherwise the local rule is undisturbed.

Sir Benjamin has been an officer in one of the scout outposts and is now /p.6/ in the Public Relations Dept. dealing with these tribes. He is about forty-five, rather large with typical dry English humor. He went to Sandhurst (British West Point) and has been in India for years, mostly on the Frontier – a very interesting guy. He married an English girl in Buenos Aires in 1938 who is here now about to have another child. They have two others.

Peshawar is on the edge of Tribal Territory, and is the house of H.E. the governor[91] of the Provinces. It is the Indian end of the Khyber Pass[92] – quite primitive in many ways – ox-carts, goats, water buffaloes all over the place.

On Monday we made a trip into Tribal Territory to Yusef Khel. Major Enders,[93] the military attaché at Kabul (Afghanistan) was supposed to be with us but didn't arrive till the next day. (The

---

[89] Sir Benjamin Denis Gonville Bromhead, O.B.E., I.A. (1900-1981), was the 5th Baronet Bromhead. He was about 43 years old at this time. Bromhead and his wife Nancy and their two daughters were living at the Services Hotel on Fort Road, near the Governor's House. It was the grandest hotel for government officials and their wives in Peshawar. It is not clear where AWZ stayed, but he probably did not stay there because he refers to it as the Bromhead's hotel (p.7). The Peshawar Club (also known as Peshawar Garrison Services Club) was on the Mall, near Fort Road. (See folder for "Bromhead," including later correspondence with "Benjie" and Lady Nancy Bromhead; Google Maps of Peshawar; and www.thePeerage).

[90] Now spelled *mehtar*, *nawab*, *wali*, and *mullick*.

[91] Governor of North-West Frontier Province = Sir George Cunningham (see *infra*, and Google images).

[92] Khyber Pass = One of the great passes of the world, it was first traversed by a European when Alexander the Great passed it in the 4th Century B.C. on his invasion of India. Lowell Thomas was the first American to take a motor vehicle across the Khyber Pass. It was not particularly difficult, compared with the traverse of the Lowari Pass that was the goal (successfully achieved) in a jeep by Enders, AWZ, and Bromhead.

[93] Gordon Bandy Enders was the author of *Nowhere Else in the World* (1935) and *Foreign Devil: An American Kim in Modern Asia* (1942). He was born in Iowa in 1895, but as the son of a missionary, he grew up in northern India, near the border with Tibet. He was wounded as a pilot in World War I, and later served as an advisor to the Panchen

secretary to the minister[94] at Kabul is Charley Thayer[95] from Phila, brother of George Thayer.) We had lunch with the mullik of the lower Mohmand Tribe (one of the peaceful ones). /p.7/ [sheet 4] We started off having our hands washed by pouring water over them as we sat at the table. There were no implements, everything was picked up by hand and you weren't supposed to use your left hand. We had roast chicken (pretty tuff), native bread (large doughy pancakes) and lots of hard-boiled eggs, cooked just right and very good. For dessert there was fruit (casawba melon) and cookies. An interesting time was had by all. I was more or less a curiosity being American and in the Navy.

At all the entrances to the Tribal Territory there is a gate for proper identification both ways and a place to park your shootin-irons when you are entering the land of law and order. Caravans of donkeys, mules and camels pass thru taking grains, poultry, etc., to the towns, bringing back piece goods, and other supplies. They would never think of traveling thru most of the Tribal Territory without a gun.

Lama and Chiang Kai-Shek. Soon after completing the historic journey along the Afghan-Indian border in December 1943, Enders returned to Kabul to speak with FDR's personal representative, Maj. Gen. Patrick Hurley, who was en route to Teheran. Hurley had previously been at the Teheran Conference of Stalin, FDR, and Churchill, which concluded while Enders and his companions were on their trip along the Afghan border in November-December. Enders was then assigned to intelligence duty in Delhi. He later served in Korea, received the Legion of Merit, and retired as a colonel after serving with Army Intelligence (G-2) at Fort Meade, Maryland.

Enders had met the Governor of the North-West Frontier Province in December 1942, on his way into Kabul, and I believe he probably first proposed making a trip along the border at that time. Ten months later (in October 1943), this trip was approved by the governor, to be led by Bromhead, using Enders as the driver, and with AWZ to go along for rather vague reasons. In his first meeting with the governor, Enders used a card of introduction from a "Colonel Benson," who was Lt.-Col. Rex (later Sir Rex) Benson, the British Military Attaché in Washington. Benson was a cousin by marriage and close friend of Stuart Menzies (later Maj.-Gen. Sir Stuart), MC, who was referred to as "C" when he was head of the British Secret Intelligence Service, MI6. Both Benson and Menzies had served in India, and both were very familiar with the North-West Frontier and the border. (See my transcription of unpublished letters of Gordon Enders, and on conversations and correspondence with his nephew and niece in my files.)

[94] U.S. Minister to Kabul = Cornelius Van Hemert Engert, born in Vienna, 31 December 1887, son of John Cornelius Engert (Dutch origin, Russian citizen) and Irma Babetz, a Hungarian Jewish physician. He came to America with his mother in 1904, attended high school in California, and became a naturalized citizen. University of California at Berkeley, B.Litt. (1909) and M.Litt. (1910). Law school, 1908-11. LeCote Memorial Fellow at Harvard, 1911-12. Began work for the U.S. Diplomatic Service in Turkey in 1912 and rose through the ranks to become the first U.S. Minister to Afghanistan from 1942-45. In 1922, he became the first American diplomat to visit Kabul at the time he was serving as Second Secretary to the Legation in Teheran, Iran. His 200+ page report on Afghanistan in 1923 was used as a reference by the State Department for many years. After retirement from the Foreign Service in 1945, was involved in many non-profit organizations and international service programs. Awarded C.B.E. Died in 1985. His papers were donated to Georgetown University by his son, Roderick. His biography was written in 2006 by Roderick's daughter, Jane Morrison Engert, who pointed out many inconsistencies between his official biography and the records that she found. (See Jane Morrison Engert, *Tales from the Embassy: The Extraordinary World of C. Van H. Engert.* (Westminster, Md.: Heritage Books/Eagle Editions, 2006); conversation with his son Roderick; and the Engert files in Georgetown University Special Collections.)

[95] Charles Wheeler Thayer (1910-1969) was a graduate of West Point, but he had resigned his commission and was a Foreign Service Officer in 1943. His brother George was a friend of the Zimmermanns, but they did not know Charles before the war. The Thayer brothers were first cousins of John "Jack" Thayer, Jr., who was a good friend of the Zimmermanns. Jack Thayer and his father had been on the voyage of *Titanic* when it sank. Jack survived but his father, John Thayer, Sr., perished, along with John Jacob Astor. Astor and Thayer Sr. were the two most prominent (wealthy) individuals who died when *Titanic* sank. Astor's son, Vincent Astor, was picked by his friend and neighbor, Franklin Roosevelt, to be head of Naval Intelligence in New York City, and as a Reserve Commander (later Captain), he was tasked by FDR to identify civilians in the Eastern U.S. (such as AWZ) who could be good Naval Intelligence Officers. Charles Thayer later went on to be head of the OSS mission in Yugoslavia, and then returned to the State Department as a Russian expert. His sister married Ambassador Charles "Chip" Bohlen. Charles Thayer wrote several books that provide useful background for the trip that is being undertaken in November 1943 by AWZ, Bromhead, and Enders. Charles Thayer is buried in the cemetery of the Church of the Redeemer in Bryn Mawr, Pa., not far from AWZ's grave.

We returned in time for dinner at the Bromheads' hotel and were joined by /p.8/ Enders who had arrived from Kabul.

The next day we started out on a trip thru Swat, Dir, and Chitral states to the north of here. A trip that took about a week almost to the Wakhan (the narrow strip of Afghanistan between India & Russia).

The first day we had lunch with the Wali of Swat, a venerable man who is said to have pushed his brother over a cliff. We were greeted by saluting soldiers lining the streets, a blare of trumpets at the "palace" gate and the old Wali himself at the door, along with his son the Waliard and the "prime minister." His palace is astounding – Persian rugs, modern furniture, up to date plumbing, etc., etc., in the middle of mud huts and buffaloes. A very good lunch with all the courses and trimmings of India – my first try at poulow (a milder curry). This part of India is about 95% Mohammedan and meat eating.

Along the way we passed the scene of Gunga Din who risked his life going from one of the many forts in this country /p.9/ [sheet 5] down to the Swat River to get water for the marooned garrison. Churchill as a sub-lieutenant fought in these mountains and Auchinleck in 1935 was stationed near here.[96]

That night we spent with the Political Agent, Major Pachman, at Malakand (one of Churchill's first books was about Malakand – get it & read it – it's supposed to be good). He and his wife and two daughters live at the fort in a very nice house over-looking the Swat Valley. As with the other P.A.'s, they live a rather lonely existence, living so far from civilization and seeing very few white people for months. They have a baby ibex as a pet.

The next day we drove to Dir at the foot of the Lowari Pass.[97] We had dinner with the Nawab and Waliard (heir to the "throne"), spent the night and started early next morning for the tortuous trip over the pass in Maj. Enders' jeep – the first time a car of any kind had attempted the feat. The trail is no more than a donkey caravan /p.10/ route, one of two, the other being thru Gilgit, that give the state of Chitral communication with the outside world. After squeezing between a stone wall on one side and the edge of the trail on the other, which sometimes dropped several hundred feet to the valley below, creeping over the icy and snowy places and coaxing the jeep around the many zig-zags we arrived at Drosh about 7.30 that evening. We had gone from 4,000 ft to 10,000 ft and down again to 4,000 ft. Quite an exciting trip. There were four of us, we having picked up an officer of the Chitral Scouts at Dir. This was a help as he was the only one able to speak Chitrali, which was needed at several places to instruct the natives on how to help us get around the zig-zags. Up to this point Pushtu was the language which Sir Benjamin could speak. Urdu (Hindustani) is not understood in this part of India.

We spent the night and most of the next day at the British officers mess /p.11/ [sheet 6] at Drosh and went on to Chitral late in the afternoon.

---

[96] "Gunga Din" by Rudyard Kipling was in AWZ's library in Haverford. The Zimmermanns had a complete set of Kipling's works. "Gunga Din" is in Kipling, *Verses, 1889-1896*. New York: Charles Scribner's Sons, 1898. They did not have a copy of Kipling's *Kim*, which was published in 1901, and which Gordon Enders referred to when he called himself "an American Kim" (*infra*). "Kim" also became the nickname of the notorious British turncoat (Russian spy), Harold "Kim" Philby, and also of Kermit "Kim" Roosevelt, Jr., grandson of President Theodore Roosevelt. "Kim" Roosevelt, as a CIA officer, engineered a coup in Iran in 1950 which is one of the first regime changes caused by the U.S. in the Cold War. The effects of Kim Roosevelt's coup seem to be never ending.

Auchinleck = General Sir Claude Auchinleck, Commander-in-Chief, India. AWZ may have met Auchinleck, and if so, it was probably in 1944. He certainly came very close to him at some point in time. The undated snapshot that AWZ took of "The Auck teaching a sepoy to shoot" is in the envelope of loose photos taken by AWZ in India.

Winston Churchill, *Malakand Field Force* (London: Longmans, Green and Co., 1898). Churchill provides a vivid description of the terrain, the flora and fauna, and the people who live in this area.

[97] Lowari Pass = The ancient invasion route into Chitral from the south. It is usually closed by snow in mid-November each year.

Chitral is surrounded by snow-capped mountains, the highest being Tirichmir,[98] 25,600'. We were very fortunate to see its top as it is usually in the clouds. But the 2 days we were there, the weather was perfectly clear, giving us wonderful mountain scenery to look at all the time.

As soon as we arrived, amid blaring trumpets again, bands playing and much saluting, we had tea with "His Highness,' the mehter,[99] and then attended a polo match in our honor. This is where polo originated. After the game we went to the Assistant Political Agent's house for drinks and then to the "palace" for dinner. After dinner the Chitralis showed us some native dances in the "ball-room" of the palace. All of the dancers were men, of course. In fact, we didn't see a woman, except on the road, the whole trip. The women are kept secreted away from curious eyes. /p.12/ The ones we did see would turn their heads and cover their faces. Most of the time they were working hard carrying wood if they were local inhabitants or doing most of the carrying, mostly on their heads, moving their goods & chattels from the high lands to the low lands for the winter. You'd pass any number of families with their worldly possessions (which didn't amount to much) livestock & all walking along the road.

Blow me down! The next day we went horseback riding to watch "hawking." I drew a spirited ex-polo pony. I took a dim view of his thinking he was still on the polo field and it took all my strength to hold him down to a slow trot.

Hawking consists of going to a high place where the river is narrow, getting about 200 men to beat the brush, rousing the birds to flight (chicors – like partridge), releasing the hawks (falcons) as the chicors fly by, and watching the kill. Not my idea of real sport. /p.13/ [sheet 7] It took all morning – the efforts of 250 men with the net result of 10 chicors killed. The mehter had to have his body guard along, horsemen and others to serve tea about every 10 minutes, to say nothing about the falcon trainers who devote their lives to this sport of kings.

After lunch we went for a jeep ride about 13 miles north. As soon as we got back the mehter had arranged a markor shoot. Markors are mountain deer that come to lower altitudes in the winter. We rode in his car to a summer bungalo, spotted two of them across the river. Maj. Enders killed one but didn't have time to see it as we were leaving for our journey home. It was all very interesting & exciting and something very few have had the opportunity to do. In fact, I was the first Navy man in Chitral, and there had been only one other American there before us.

The trip back was much the same /p.14/ as going up. The mehter of Chitral and the nawab of Dir gave us chagas (native coats) in honor of our visit.

We arrived back Tuesday pretty well tired out. Tomorrow Enders is taking me up the Khyber Pass and Saturday the Viceroy[100] of India is visiting Peshawar and we are going to attend the garden party to meet him. Last night we had dinner with the deputy commissioner and the Bromheads. The D.C. is

---

[98] Tirich Mir = The highest mountain in the Hindu Kush. AWZ's photograph of it was probably the first time it was filmed by an American (see his photos of the trip). I spoke by phone to the retired British Army officer, Col. Anthony "Tony" Streather, CBE, who was the official government representative when a Norwegian party made the successful first ascent of Tirich Mir in 1950. He was one of those who ascended, and he later was famous for climbs elsewhere in the Himalayas, most notably on K-2. He knew of Sir Benjamin Bromhead, but he did not know anything about Bromhead's trip in 1943.

[99] Mehtir of Chitral = In 1943, the new mehtir was Muzaffar-ul-Mulk, who succeeded to the throne in the summer of that year. He died in 1948, his son Saif-ur-Rehman was mehtir of Chitral until he was killed in an airplane crash at Lowari Pass in 1954. The title passed to Saif's son, Saif-ul-Mulk Nasir, who held the title until 1972. The title and privileges of the rulers of the former states of Pakistan, including Chitral, were abolished in April 1972. Chitral is now a district of the Malakand Division of the Northwest Frontier Province of Pakistan. The present population of Chitral town is about 20,000.

[100] Viceroy of India = Field Marshal Archibald Lord Wavell. Wavell served in the North-West Frontier Province as a young officer. He traversed the Lowari Pass himself, on horseback, at that time. He was familiar with the pass and its place in history, from the time of Alexander (whose troops crossed it) to the present. AWZ would meet Wavell again in Karachi, at a party in the Viceroy's honor in 1944, given by the Governor of Sind, Sir Hugh Dow (See invitation in AWZ-SB).

sort of a prime minister to H.E. the governor. He is a very light-skinned Mohammedan as are his wife and daughter. We had a delicious dinner and Scotch for a change.

So you see I've been pretty busy and have been seeing many interesting things and meeting interesting people. I expect to be back in Karachi about the 10th of Dec. We will proceed south from here ending up at Quetta in Baluchistan.

I wish you could be with me in all these new experiences. I miss you terribly.

Lots & lots of love,
Al

----

16 December 1943, postmarked by APO   From AWZ/ONI to BSZ, marked: "NW Frontier Trip #2."
ALS-P

[no header]
Monday, Dec 6th [Wana, South Waziristan, Tribal Territory]

Dearest Barb

This will have to be in pencil as my fountain pen has run dry and there's no sign of any ink in this place. It's been over a week since I've written you. I just haven't had time. We've been on the go ever since. I sent the last letter from Peshawar to Karachi to be mailed but this one will have to wait till I get there as mail out of here is far from reliable. We decided to rest up today as we are pretty worn out. The name of this place is Wana in South Waziristan about 10 miles from the Afghanistan border. It is occupied by a brigade of British and Indian Army.

Since I last wrote I've seen many strange places I never expected to see. Beginning from my last letter – I went up Khyber Pass to the Afghan border. The pass is one of the main routes between India & Afghanistan. It is quite a sight to see and I would hate to have missed it, being as close to it as I was. Maj Bromhead took me up and patiently explained the points of interest.*

That afternoon the gov. of Northwest Provinces had a Garden Party to meet the Viceroy. Maj B. arranged for Enders & me to be /p.2/ invited. It was a grand affair in a beautiful setting on the Government House lawn. All the mulliks (tribal leaders) from the surrounding country were there along with important Britishers of the community. The Viceroy had heard we'd been up the Lowari to Chitral in a jeep so [he] came over to where we were standing and asked all about it. Enders did the talking as he has a flare for it and modesty isn't one of his virtues. By the way he has written 3 books. You might be interested in reading at least one of them. His first name is Gordon. I'm sorry not to know the titles. (Foreign Devil is one I think)

After several days in and about Peshawar we started south to visit more of the northwest provinces, mostly in tribal territory. Our stopping places have been with the Scouts, militia or Army whoever was responsible for keeping peace, sleeping in the forts and eating in their officers mess.

Our first stop was with the Kurram Militia at Parachinar. Here we climbed the Paiwar Khotal pass to Afghanistan.[101] We then spent a few days with the Tochi Scouts at Miram Shah, the South Waziristan /p.3/ [sheet 2] Scouts at Jandola and arrived here last night and are staying with the Army.

Most of the way we have been escorted by Kassadars (local policemen). They line the road on either side (about ½ mile apart). Others ride in lorries ahead & behind us. It makes one feel quite important having so much a fuss made of you but it's the only way to travel in this country with comparative safety. Some of the local boys just don't like strangers and might decide to take a crack at them or try a little highway robbery or kidnapping.

Since we've been here, this is what has happened – a road engineer's lorry held up – the supervisor kidnapped being held for ransom. At Dasali, where we spent a night, about twenty shots were fired from a hill into the Army camp fortunately not hurting anyone – we were in the Scout post ½ mile

---

[101] Kurram = The Kurram River crosses from Afghanistan to India (now Pakistan) just to the south of here.
Parichinar = The great plain that extends down into India from the Afghan border. In AWZ's photos.
Peiwar Khotal = Fort on the Afghan Border. Twenty miles north are the Tora Bora mountains. Ibid.

away – ten Hindu girls were kidnapped from Bannu along with a Hindu shopkeeper. He is being held for 7000 Rs. ransom. (The girls aren't considered worth ransoming.) Another lorry was held up and three of the highwaymen were killed, one an Afghan. The RAF bombed some outlaws who were living in caves. /p.4/ Such is life on the Northwest Frontier. It forms one of Britain's big problems and costs millions a year to keep in as good order as it is at present.

Enough for now will continue this later.

* [AWZ's footnote] A Sikh general was killed in battle (200 years ago) – fearing [sic] this would sway the hopes of the enemy – they pickled his body and stuck it in one of the windows of the fort at Jamrud, entrance of Khyber, where they could periodically could show it to the enemy.

Thurs Dec 9th [BSZ typed "Friday, Dec. 9." This is incorrect. The travelers left Fort Sandeman for Quetta on Friday, 10 December, according to his Notes.]
We are now at Fort Sandeman in Baluchistan staying with the Zhob Militia. Tomorrow we leave for Quetta, then after a couple of days I take the train for Karachi. I'll be glad to get back. While it's been very interesting it's been rather hectic, none of these places have modern plumbing – I'm tired of bathing in a wash tub and washing in a basin. Also it's been quite cold because of the altitude (between 4 to 10,000 feet). It's not so bad here – it's only 4500. My place in Enders' car is in the back seat. It's an Army car, open of course, and the breezes & dust make life far from comfortable, especially in back. I finally picked up a cold, one of the good ones. Today we rest. I hope it will do some good. /p.5/ [sheet 3]

Most of this country is pretty God-forsaken. You marvel that anyone can scratch a living out of it. That's partly the trouble – some can't – so they take to plunder & pillaging.

Yesterday we saw a crowd right off the road near an Indian Army camp. We immediately thought there was trouble but it turned out to be a cremation. A Hindu Army recruit had been killed and his friends had built a bonfire and were disposing of his remains. Hindus are cremated – Muslims buried as is – in shallow graves – Parsis dedicate their dead to the elements and animals – a delectable dish for the vultures. I saw a Hindu funeral in Bombay. The body is carried on a stretcher on the shoulders of mourners without the benefit of a coffin. I'll end now with this morbid description of Indian burials.

Wednesday the 15th

Back "home" again and am I glad. Five letters from you and one from Warren awaited me, adding greatly to the joy of being back "be it ever so humble." Karachi isn't a bad place at all after one has seen other parts of India.

I won't go into your letters now but will finish my trip and get this off in tomorrow's /p.6/ mail. It has been impossible to send letters along the route and since my last was from Peshawar three weeks ago I'll get this off and answer your recent ones as soon as I can get caught up around here.

The rest of the trip was quite uneventful – more of the same thing. At Quetta we stayed with the Governor of Baluchistan[102] at the "residency." The "residency" consists of tin shacks and tents – very elaborate however. Quetta was the scene of one of the earth's worst quakes in 1935. 30,000 people were killed out of 120,000. (They also had a minor one in 1940.) Practically the whole town was levelled. They are building a new residence and in the meanwhile the governor & his family live in tin shacks and visitors live in tents. You'd be surprised how comfortable they've made them. We were royally entertained by His Excellency. Had a dinner party for us the night before we left. Took us on a chikhor (partridge) shoot on Sunday with picnic lunch etc. I'd have enjoyed my stay more if it weren't for this damn cold which wasn't helped any by the dusty (a la Argentina) train ride home.

Good bye dear – I miss you so much.

Heaps & heaps of love,
Al

---

[102] "Governor of Baluchistan" = Lt. Col. (later Sir) William Rupert Hay was the commissioner of Baluchistan. His title was probably Commissioner, rather than Governor.

----

20 Dec 1943, postmarked APO   From AWZ/ONI to BSZ.   ALS-I

[no header]

Sunday, Dec. 19th

Dearest darling,

I'm afraid my last letters have been awfully dull, concentrating on my trip and telling you how much I was missing you and how anxious I am to be back again with you. Believe me this job is no cinch from that angle. It gets worse as it goes on. I can't bear the thought of its being much longer. If only this damn war would stop

The complexion of this place is rapidly changing. Callahan has received orders to report back to Wash., lucky stiff, when he goes I'll be the oldest officer here, outside of Com. Smith and there are rumors that he's to be changed.

Curt Winsor[103] came thru while I was away. I'm sorry to have missed him, but I'll see him about the 7th of Jan. when he leaves India.

A new officer arrived when I was away, by the name of Rethbord. He /p.2/ went to Oberlin a few years ago. Seems quite nice. His wife is with her parents in Glenside.

By now I hope Burns has been in contact with you and given you some first hand info. about the place.

I'm awfully glad you had the two weeks at Hot Springs. Even though the weather wasn't nice, it was a complete change and should have done you a lot of good.

As to cables – continue sending them even tho they take about a week your letters take sometimes a month and it's good to get up to date news even "canned." They have discontinued ones from this end temporarily (till after New Years). I'll send you a straight cable tomorrow for Christmas.

The package that came was the one with the hot water bottle etc. (I /p.3/ must have written you about it.) It was very welcome & useful. The other hasn't arrived as yet.

Glad to hear about Warren's good report and to hear Lanie is doing so well.

Too bad about Tince[?] Coffin's mother. She looked robust enough. I wonder what the matter was. Give Tince my love.

Sorry to hear about your gas situation. Of course the "little" car is half mine – at least the company paid for it. It was bought with a business idea in mind to run around the docks. I think it's despicable for the Otts to hog it all themselves and by the way did you ever see anything of the six cases of scotch that Jack bought in N.Y. thru Strauss shortly before I left. I was supposed to get half and I told Jack to get it to you anyway he could? Let me know – I'm about ready to /p.4/ give these "true" friends a piece of my mind. As for the football tickets I quote his letter, "since Barbara was not going to subscribe to your football tickets this year I got her to subscribe so that you wouldn't be off the list and I am giving them to Charlie Nico." Is he trying to make a hero out of a culprit?

---

[103] Lieut. Curtin "Curt" Winsor was from Philadelphia. His brother, Jim, who was mentioned above, was friend of the Zimmermanns, but apparently Al and Barbara didn't know Curt (or didn't know him well) before the war. He worked at ONI as a desk officer and was responsible for Naval Liaison Officers in the Far East, including AWZ. He lived at 2776 35th Place, Washington, DC, near the Washington National Cathedral. He had recently remarried. His first wife, Elizabeth, was the daughter of William H. Donner, a very wealthy industrialist. She had previously been married to Elliot Roosevelt, son of the president, by whom she had a child, William Donner Roosevelt, who was thus FDR's grandchild and Curt Winsor's step-son. That marriage had been annulled, and Elliott Roosevelt had remarried. The half-brothers, William D. Roosevelt and Winsor's son Curtin Winsor Jr., were associated with the Donner family foundations for the rest of their lives. Curt Winsor's letters to AWZ and notes regarding him and his family are in Chapter 6 in this book. Winsor wrote on 9 March 1945 that the CNO (FADM King) ordered an investigation of the problem with "the mess" in Karachi which involved AWZ and his former CO, LCDR Smith, who was given an unsatisfactory fitness report when he left Karachi.

About Lanie – of course she shouldn't be wandering about at night. I'll try and write to her soon, to ask her to be more reasonable. I hate to think of anyone adding to your troubles.

On underwear – I thought I had some at home – but if not get what you can. Will also need brown & black socks, handkerchiefs, pyjamas, a few collars and a couple white shirts (if I haven't mentioned these things before).

Tell the children I'm awfully sorry not to have written to any of them for over a month. Just couldn't do it on the trip. Will get to it soon. Happy birthday to Warren (a little late) and to Albie.

Loads & heaps, heaps & loads of love,
Al

1943/1944

2 January 1944, postmarked APO   From AWZ/APO 886 to BSZ.   ALS-I → P
[no header]

Tuesday Dec 28th [1943]

Try the new address on envelope
#41
Dearest Barb

Time drags very heavy but seems to fly in between letters to you. It's been [shifts to pencil] over a week now since I've written to you. But I have written to all the children in the meantime. My pen always seems to run dry writing you. I'm tucked in bed for the night and the ink is down stairs where there are some guests. It might be a bit embarrassing if I put on a bath robe to go down and get it and get caught.

Since I last wrote we've had our Christmas. A pretty grim affair – everybody wishing they were home, but we did the best we could. We had a small tree with a few lights – furnished by the Red Cross (there are about a half dozen girls here doing "good"). I wrote Babs about the party we had so won't go into it again.

I finally got rid of the worst of my cold but I still cough – mostly at night which rather disturbs my slumbers. The activities over Christmas didn't do it any good, but it was a relief to get out and see some "old" friends and meet some new ones.

Christmas Eve was spent with the Harris's[104] /p.2/ a very nice couple – English – who have been in India for years – he mostly in government work. After dinner we went to the Boat Club to a dance – music by the RAF band. The next day – Christmas – the Intelligence School had a cocktail party – then we had our dinner and that night I went to the Sind Club as a guest of Arlo Bond, for dinner and another dance. Then the next day Mr. Markley manager of Standard-Vacuum Oil Co's branch here had a cocktail party. So you see things have been pretty gay and tiring so tonight I excused myself and came up to write to you and relax and not sit around trying to make conversation. Our guests weren't very interesting – Mrs. Dubash and a British colonel.

Day before yesterday a Charles Thayer came in to see me. He's from Villanova – is a cousin of Jack Thayer and a brother of George etc. etc. He's been in Kabul as secretary to the legation and living with Gordon Enders (who made the trip with me). He is now on his way to /p.3/ London to be on some

---

[104] John R. Harris, Esq., was Central Intelligence Officer, Karachi. He was the recipient of the message of 28 October 1943 from Intelligence Bureau Quetta, requesting that a "U.S. Naval Officer from Karachi" be assigned to the trip that Bromhead and Enders would take on the North-West Frontier, starting on about 10 November (carbon copy in AWZ-SB). Harris is in the photo with General Haddon (fn, AWZ letter of 15 Sep 43). He is lean, fit, wearing a white civilian shirt and shorts, holding a pipe. His face is partly shaded by his pith helmet, but he appears to have a moustache. He is probably the man in a white shirt and blazer, with slacks, sitting beside AWZ, who is in service dress khaki, in the group photo of intelligence officers in India (unnamed, undated) in AWZ-SB. Of course, AWZ doesn't mention Harris' role in intelligence in this letter.

mission there. He seems quite nice and has had some very interesting experiences. He's coming for dinner tomorrow night and leaving the next day. One of his problems is what to do with four large dogs (2 shepards & 2 Afghans) and a Chinese servant. I think we can get the dogs to his family in America on a ship out of here but have not yet worked out anything for the Chinese whom he hopes to take to London with him. He is flying but the servant has to go by ship if at all. Too bad I can't ship him over to you as I understand he is perfect.[105]

I'm sending a box of stuff home – things I picked up on my trip to the Norwest Frontier. There are two "pastines" for the boys – made in Afghanistan – sheep skins. I don't know that they'll ever wear them but they are the real McCoy and could be made to fit easy enough. If there is too much wool it can be trimmed down with scissors. Then there are two Pathan caps (kullahs) and "lungis" to wind around [marginal: The tail of the lungi keeps the heat of the summer sun off one's neck and wipes one's nose in the winter.] /p.4/ them. I will try to have one done up so you can see how it goes but I'm afraid it will be crushed getting there. You'll get the idea though. With a dirty shirt and night gown they'll look just like young Afghans or Pathans (who live on the Indian side of the border). There are two jackets for the girls. These are made by the Afridi tribe of the Pathans. They inhabit the part around the Khyber Pass. For you there is a piece of Kashmiri cloth I thought would make a nice skirt or something. There are also four Chitrali caps one for each of the children.[106] These come from the northernmost part of India near the Pamirs, the "roof of the world." The other object is a "chagra" given me by the Nawab of Dir on our visit. I have another – given me by the Mehter of Chitral that I'm keeping here for a bath robe. You can do with this what you want – give it away if you can't use it. One would be enough for me.

Good-bye & Happy New Year. I miss you <u>very</u> much and hope & hope I'll be home soon.

Al

[marginal note: You'd better disinfect everything that comes from here]

1944

8 Jan 1944, postmarked APO, from AWZ APO 886 to BWZ.
Marked by BWZ: "Rec'd Feb 2" ALS-I

[no header]

Friday Jan 7th

Dearest Barb,

The pictures of the children arrived the day before yesterday, the day I arrived back from Colombo. They are wonderful – awfully well done and very welcome but where's the picture of you? I wish you & whoever took these pictures take as good a picture of you, or were these taken by you with the contax and enlarged? If so you did a swell job.

Your letter of Dec. 5th written in the Union Station at Wash. also came on that day, evidently brought by a friend of Freemans whom I just missed seeing by a few hours. For goodness sake don't try

---

[105] Thayer mentions the dogs and his servant in his books, and he also describes a bear-hunting trip in eastern Afghanistan shortly before he left for London. This would have been about the same time that AWZ and his companions were on the other side of the border, in the same general area. As an expert on Russia, he may have used "the bear" as a euphemism, for he surely would have been on the lookout for Russians as well as wild animals. His books include *Bears in the Caviar*; *Hands Across the Caviar*; *Diplomat*; and *Guerrilla*. The dogs were lost when the ship on which they were traveling was sunk; I do not recall what happened to the servant.

[106] The Chitrali cap or hat is a distinctive piece of headwear, which is described and illustrated in many citations on the internet: "The Pakol (also spelled Pakul or Khapol, from the Khowar language of Chitral) is a soft, round-topped men's hat, typically of wool and found in any of a variety of earthy colors: brown, black, gray, or ivory. Before it is fitted, it resembles a bag with a round, flat bottom. The wearer rolls up the sides nearly to the top, forming a thick band, which then rests on the head like a beret or cap." It is said to be similar to a Macedonian hat, which suggests that it is a descendant of headgear worn by the troops of Alexander the Great when they passed through this region.

to economize on railroad transportation by riding in coaches. By all means take Pullman wherever you go. With Warren Bros. paying off their bonds at 100 there should be enough profit to keep you all in Pullmans the rest of your lives /p.2/ let alone the enhanced value of the preferred and common stocks. So don't worry about finances – at least not yet. You didn't say how Crosby is. I hope she has completely recovered but I suppose that's too much to ask after such a horrible disease. You mention the happy days of our apartment in Wash. I certainly look back at them with great longing, thinking how lucky we were to have them and hoping the time will come very soon when they can be repeated.

I received your cable a few days after Christmas and was so glad to hear you are all well. Please let me know about Lanie and her operation. She should be having it about now or is it all over. I'm awfully anxious to hear how it came out and if everything is all right.

Since my last letter New Years has come and gone. The eve was celebrated by the usual formula. The Army had a party early in the evening for about 50 people including some of the nurses from the hospital here. Later a crowd of us went to the Gymkhana club where they were having a dance. Everyone was pretty high and a good time was had by all.

The next morning, early, I stepped on a plane for Colombo. When we landed at Bombay Curt was at the airport along with one of the men from our office. He had just come up from where I was going. I had only 20 minutes with him as our plane stopped only for freight and gas. He seems to be enjoying his trip and looks a lot better than the last time I saw him. He is arriving tomorrow from Delhi and expects to spend several days here before leaving for home. My how I envy him.

I spent the night in Bangalore* at the airport a rather pleasant evening in the city with an RAF fellow passenger on the [marginal note, probably referenced by the asterisk: saw a yogi with his head buried in the ground like an ostrich. He'd been there for at least 10 min – amazing! what] /p.2/ plane. We landed at Colombo the next day around noon and after 24 hours there (not having a very exciting time) I started back again spending the night in Bangalore and again going into town (about 10 miles from the airport) but this time with a British major by the name of Wilder who was coming up here to visit his sister. [Editor: AWZ was probably sent to Columbo to be debriefed on his trip to the NWFP]

Our plane went U.S. (unserviceable – not a slam at the United States) on arrival at Bombay, in fact the pilot had to land without instruments so we got on another plane only to find after we were just about to take off that we couldn't land here because of a sand storm so we spent the night in Bombay with Al Payne, arriving here the next morning.

Last night I had dinner with Wilder and his sister & brother in law, the Coughlans, who are very nice. He is with the Bombay Company and might some day be of some use in getting us a better wool connection in India.

It's now 12.45 PM so I'd better get some sleep. So good by dearest. I miss you dreadfully –
All my love,
Al

----

14 Jan 1944, postmarked APO   From AWZ/APO 886 to BSZ   ALS-I
[no header]

Thursday Jan 13th 44

Dearest Barb,

Your #33 Nov. 30th arrived today with the notification from Reader's Digest, a letter from Anna[107] and one from Jack Ott. I received a letter from Dick a couple of days ago. He is in Delhi.

In your letter you speak of trying to get peace and quite [sic] to write. That's the way it's been with me. Curt arrived on Saturday. The two R.A.F. blokes that flew me here from Cairo arrived Sunday, four officers are here awaiting transportation, a navy captain arrived tonight and so it goes – all

---

[107] Anna = His oldest sister, Anna (née Zimmermann) Kelley, wife of Richard Carlyle Kelley, mother of the "Dick" who is in Delhi.

contributing to make life here a bit hectic and time consuming. You just don't have time to write – I'm way behind in my correspondence – but even though it's 1230 I wanted to get this off to you because it's been a week since I last wrote. I was all set to go to bed early, have dinner in my room and write /p.2/ letters when word came that a captain was at the airport. As the commander was out I had to change my plans, get dressed and play host.

I'm still of the same opinion about Curt. I think he is quite a difficult person. He seems quite happy in his work and looks a lot better than he did. He was here till Tuesday night when I put him on a plane for home. My heart was filled with jealousy & envy when he stepped on the plane.

The RAF boys (Swabe and Barlow) left last night. They are very nice and helped our supply situation a bit by bringing some "stuff" from England. I'm getting to be so air minded that when they asked me to come along with them for the test flight yesterday afternoon I accepted. They promised to let me fly the plane, a B24, but /p.3/ the first engine they started blew a hose connection, which would take several hours to fix, so probably my only opportunity to become a pilot vanished. It would have been quite educational as they promised to show me the city from all angles. They might show up in Phila. one of these days and look you up – one never knows – not really very likely but they do seem to get around.

Jack's letter enclosed a copy of Andy Anderson's letter to Hecker. He seemed very appreciative of the time he had in Phila. Evidently he is laid up and is going back to England to recuperate from burns he received on his ship. Someday I guess we'll hear the story of how he got them. I suppose it wasn't at a just a tea party.

I'm glad to hear Warren is getting /p.4/ as much fun out of football as he did of baseball. His little active mind needs something to keep it busy and a boy should be interested in athletics. Too bad there was such a slim attendance at Thanksgiving. I hope next year it will be entirely different.

Nothing has arrived as yet in the way of packages but I'm no exception. Nobody in our office has received any.

My taxable pay from the Navy is $2524.70 for 1943 which you will need for making out income tax.

I'm enclosing some pictures taken a couple of weeks ago to show what we all look like. Our present staff is minus Callahan (who said he might come and see you or at least call).

It's now 1.30 so I'll have to get some shut-eye. Good-bye, dearest. I miss you very much and how I'd like to trade places with that pad.

Heaps & heaps of love,
Al

From AWZ/ONI to Master Warren Zimmermann (WZ). This letter is in an envelope postmarked APO on 5 Oct 1943. The letter that originally came in this envelope was probably the undated and incomplete letter that is transcribed above in the sequence for October 1943. I have placed the transcription of the 20 January 1944 letter in the proper chronological order in 1944, but the letter and enclosures will remain in the envelope postmarked 5 October 1943.

ALS-I

[no header]

Thursday Jan 20th [1944]

Dear Warren,
Your letter of Nov. 21st, along with one from Lanie and Mummie, was carried all over India, even down to Ceylon before it came to me. Curt Winsor, who brought them over, couldn't find them when he arrived here, so I had to wait until he finished his trip before I got them.

It's too bad that the first time you saw Penn play she was defeated. It must have been an exciting game though. Did you see Penn play Cornell on Thanksgiving day? How was the game and who won? I haven't heard yet.

I'm sorry I was a little late in sending you birthday greetings but when it came around I was where I couldn't send any letters or cablegrams.[108] I hope you'll forgive me. The baseball game you received sounds very interesting. It must be a good game. Maybe it will arouse Albie's interest in baseball so he'll be a good player like you. But I guess you're /p.2/ pretty well wrapped up in football to think much about baseball but the time will soon come.

Your victory V was very good with the hugs and kisses in it. Was that idea yours? I think you were very smart to think it up.[109]

How did you like your presents at Christmas? I certainly missed being with you, especially to see your expression when you saw what Santa Claus had brought you.

The pictures Mummie took on the terrace are very good. I have them on my bureau where I can see you every day. Where are the big sun flowers back in the garden? If you only had a parrot you could feed it the seeds. Timoney[110] must be working very hard to grow such nice big sunflowers.

Well, good bye, Warr, I hope I'll be hearing from you soon again.

Lots & lots of love,
Daddy

[Enclosed: 2 newspaper clippings, from an unknown Philadelphia paper or papers. "Cox Discloses Stockholders / George D. Widener, Fitz Eugene Dixon, Jr., Phils' Investors" (undated) and "Wool Dealer Holds Interest in Ball Club" (29 August [1943?]). The former article lists stockholders who have purchased the Philadelphia Phillies. One of them is "Ott & Zimmermann, wool firm, Drexel Building." The latter article provides more details about the purchase of the Phillies, mentioning Ott and Zimmermann: "The wool firm is headed by two partners, John Ott, who has been identified with the carpet wool trade for many years and Albert Zimmermann, now on leave serving as lieutenant in the Navy and is now abroad."]

---

21 Jan 1944, postmarked APO 886   From AWZ/APO to HZ   ALS-I
                                    [no header]
                                                                Wednesday Jan 19th

Dear Lanie,

Your letter of Nov 21st finally arrived, brought by Curt Winsor. He took it all over India with him, not being able to find it in his possessions where he first arrived. It was a very nice letter and I enjoyed reading it very much. I'm glad to get all the information about malaria. You certainly were very thorough in your description of its germ's life history. I strongly suspect that you copied it out of a book – a good way to learn your lessons and write to your father at the same time. However I don't much care about its life cycle as long as part of it does not take place in me. Fortunately there's not very much malaria around here.

I'm very pleased to hear you've joined the French Club. You'll never regret being proficient in a foreign language, especially French.

With all the actors and actresses /p.2/ we are developing in our family we should be able to run a stock company of our own one of these days.

Were your marks good enough to make the honor roll? They sounded awfully good to me.

I'm anxiously waiting to hear how you all enjoyed Christmas and about your operation. I hope you liked the things from India especially the evening bag now that you've become a member of the "Friday Evening." Did you put up as much fuss about the stitches being taken out as the last time or didn't you have the operation at all?

---

[108] Warren's birthday was 16 November; at that time AWZ was on his trip in the Northwest Frontier Province.
[109] The "victory V" was a sort of letter.
[110] Arthur Timoney was their handyman. A native of Ireland, he was devoted to the Zimmermanns.

I certainly liked the pictures Mummy sent. My mouth watered for that glass of milk you were pouring. I haven't had a good glass of milk since I left home. You don't dare drink milk out here. How about you taking Mummy's picture & sending it to me?

      Lots & lots of love,   Daddy

---

24 Jan 1944, postmarked APO 886, from AWZ/APO to BWZ   ALS-I
      [no header]

Sunday Jan 23rd

#45

Dearest Barb,

  I was amazed when I looked in my little book and found it's been over a week since I've written you. I'm very sorry. This week has been a very busy one – lots of transients that had to be catered to. Then our C.O. left in the middle of it leaving me in charge. Also, I wrote to all the children as well as to some other people, so I haven't just been frittering my time away. I'm falling way behind in more "other" correspondence – more people seem to be writing. Recently I've received letters from Anna enclosing pictures taken at Marcian's wedding of you, Warren & Albie going through the line and of Warren handing out confetti – both very good. Hecker, Walter Johnson, Ellie Severens and George McNeely have also written –all very nice but someday requiring answers. /p.2/ There's really not much new with me. Last night one of the British Naval officers & wife had a party – dinner at their apartment and then to the Gymkhana to dance – a very nice party but I'm jealous of the British – most of them have their wives right here. This assignment wouldn't be so bad if you were only here, but it's no fun at all the way it is.

  One of our visitors was a guy with your maiden name who said he used to live in Germantown – son of Wm. Brock S. – a rather disagreeable type whose departure wasn't regretted by any of us. You probably know who he is – he must be related in some way.[111] Besides his general indifferent attitude at being a guest of our mess he swore at me when I called him a second time to catch his plane in /p.3/ the middle of the night.

  No sign of any packages as yet. I guess they will be showing up one of these days. I'm enclosing a list of the things I've sent you in the last few months so you can let me know what arrived. I wouldn't be surprised if some of the items got lost as you haven't mentioned getting them

  I'm awfully glad Powell and Harmon took the trouble to come and see you. It's too bad they came so close together. Powell left here months ago by ship. Didn't Harmon bring you a package? You said nothing about it?

  Your last letter, the one written Dec 23rd arrived, the night before the night before Christmas, arrived several days ago. It's too bad the boys had grippe. I hope they got over it in time. Believe me, I /p.4/ certainly missed all of you at Christmas and would have given anything to be there. I'm dying to get your next letter to hear all about it. It was nice of Sally to drop in with the others. She and Jim [Winsor] have been very thoughtful about writing. I've heard from George Belic – he's in Cairo. One of the men passing thru brought a card from him. I'm not a naval attaché – they don't have them in India.[112]

  Glad to hear Clarence[113] is getting his commission. No doubt I'll see him if he's heading where the children thought he might be.

  There's really nothing more to write about except that I love you dearly and miss you greatly. I'm very proud of my wife and family and don't hesitate to let people here know that I am. I've shown the newly arrived pictures with great pride. I wish I had a recent one of you.

---

[111] BSZ has no known relationship to William Brock Shoemaker, who married Ella Morris de Peyster in New York City (*New York Times*, 15 December 1905). There are other Shoemaker families in Pennsylvania.

[112] India, being a part of the British Empire, did not have a Naval Attaché, because Naval Attachés were only stationed in capital cities' embassies. The Naval Attaché for Britain and the British Empire was in London.

[113] Clarence Lewis, AWZ's next-door-neighbor, who was in OSS.

Lots & lots & lots of love,   Al

[enclosed, 1 pg in pencil, listing 63 items or collections of items, mentioned above]

---

31 Jan 1944, postmarked APO 886, from AWZ/APO to BWZ, marked rec'd 8 Feb   ALS-I
[no header]

Saturday Jan 29th

#46

Dearest darling,

Here it is our anniversary and we so far apart. I hope Caldwell[114] followed through. I wrote some time ago to supply the usual. That and the cable, I'm afraid, are my only contributions for this momentous event. Eighteen years and glorious one too. What more could a man wish for – a beautiful and adoring wife and four of the best children I've ever seen anywhere and the wherewithall for all of us to live comfortably. It's a bitter blow to have this separation – and it has seemed an eternity – but under the circumstances, I guess it was the thing to do. We'll have the rest of our lives to make up for it and that we will do. Here's a big hug and a kiss sweetheart for the best wife ever.

A whole stack of mail came several days ago postmarked from the 3rd to 8th of November – almost three months ago. I wonder where in the world it could have been. There were two letters from Bill Stuble, a letter from Babs, one from Timoney /p.2/ Christmas cards from Connie Young & Sally Winsor, the letter from Bud Patterson enclosing Andy's tragic story and a letter from Jack [Ott].

It was too bad this bunch took so long in coming as Bill wanted a reply from me before the end of the year about the gifts to the children. I wrote him yesterday saying whatever he did was all right. My main purpose is to not increase my estate further but to increase the children's estates if it can be done without too much tax. He also mentioned that you were not getting any information about the business from Jack. I am pretty bitter about this. We had a distinct understanding, Bill, Jack & I, that you were to receive reports and be personally informed on what was going on, especially when he was contemplating a major move like buying a lot of wool. In fact we all thought it good business not to increase "spot" or "to arrive" wool above the figure it then was ($60,000). I doubt from what Jack /p.3/ that that is our present position. It is not a hard and fast figure to kept but at least you should have been consulted if it was to be exceeded.* I've written Bill about this and will speak about it in my next letter to Jack.
[* marginal note: I appreciate everything Jack has done and is doing in a business way. We have had a very successful year and he deserves a lot of credit. But under present conditions it's foolish to stick your neck out too far, to get hurt if things go badly]

Since I last wrote your letter of Dec 30th arrived telling about Christmas etc. You gave a very full account and from what you say I'm sure the children had a happy one but I'm afraid you didn't. I have another present that I got in Ceylon for you that was either for Christmas, Anniversary, or birthday but having inadvertently left it in Bombay, I missed sending it by Curt [Winsor]. It's something I didn't want to trust to the mails. You'll get it someday.

Jack's presents to Louise [Ott] don't sound in good taste but after all their lives have been very little affected by the war anyway. It's about time they began thinking about you. I don't suppose their neglect has made you feel any kinder toward them. I'm anxious to hear about the /p.4/ scotch and company car situation that I wrote about before.

Please don't let Babs worry about being popular at parties. That really is quite inconsequential. She's a swell girl and that is important. She's getting sweeter every day.

It was nice of Bill Craft's wife to write as she did and I hope we'll get to see something of them when this mess is over.

---

[114] Caldwell = Philadelphia jewelers.

There isn't much to say on this end. Life goes on about as usual. Com Smith has been away for the last 10 days leaving me in charge. Capt. Thornton[115] spent a night with us. He's the one that was so fond of Jack Kane. Another captain passed thru and next week we are to have a commodore[116] and a general. All very well but they require attention and take away evenings etc. – That's why my correspondence has fallen off. It's certainly not that I don't want to write more often – it's the closest thing /p.5/ that I can do to be near you – that and reading your letters. I'm feeling especially behind in "other" correspondence but that will have to wait. You come first and if letters take away some of the sting of this separation you'll get them.

That's about all for now, dearest. You are my dearest and I love you very much.

---

[115] CAPT Thomas A. Thornton appears several times in the correspondence of AWZ, first in Cairo, 7 July 1943 (*supra*). He was from Philadelphia, and was a friend of the Zimmermanns' friend, Jack Kane.

[116] This must be Commodore (later VADM) Milton E. Miles, the only U.S. Naval person in that part of the world at that time who was referred to as "commodore." Miles was the senior U.S. Naval officer on land east of Cairo. His letter re AWZ and LCDR Smith (on 22 August 1944) is enclosed with a letter of 9 March 1945 to AWZ from LT Curt Winsor, in "Winsor" folder.

From his biography in the website for his papers in the Stanford University Special Collections: "Milton E. Miles (1900-1961), who was nicknamed "Mary" for the film star Mary Miles Minter, was a U.S. Naval Academy graduate (Class of 1922) who initially followed the traditional path of Navy officers. He became a bold captain of Navy ships in the Asiatic region in the 1930s but was re-assigned to the U.S. homeland after the Navy was expelled from Chinese waters by the Japanese. As a Commander stationed at Navy Headquarters in Washington in December 1941, he was asked to use his expertise in Chinese affairs and was sent by Admiral King to China as a Naval Intelligence officer with the euphemistic title of U.S. Naval Observer, Chungking. He was so successful that from 1942-43 he was also appointed as Chief of the O.S.S. (Office of Strategic Services) for the Far East, reporting to both William Donovan and the Director of Naval Intelligence. He was promoted to Captain and later to flag rank while in China, and from 1943-45 he served as second-in-command of a joint intelligence operation called the Sino-American Special Technical Cooperative Organization (SATO), also known as the Sino-American Cooperation Organization (SACO), which was headed by Chinese Lt. Gen. Tai Li. In 1944-45 he was also Commander, U.S. Naval Group, China. He was promoted to the rank of Vice Admiral on his retirement in 1958." From http://content.cdlib.org/view?docId=tf4j49n6xd&doc.view=entire_text&brand=oac (accessed 6/14/09).

Miles' role in the OSS and his tension with William Donovan, head of OSS, is described in Richard Harris Smith, *OSS: The Secret History of America's First Central Intelligence Agency* (Guilford, Conn.: The Lyons Press, an imprint of The Globe Pequot Press, [1972] 2005), 231-237; and Maochun Yu, *OSS in China: Prelude to Cold War* (New Haven, Conn.: Yale University Press, 1997), 49-51, 55-58, 82-88, 117-130. Smith and Yu agree that Donovan decided that Miles could not serve "two masters" (the Navy and the OSS), so on 9 November 1943, Miles was relieved of responsibility for OSS in China, and on 5 December 1943, he was officially removed and replaced by an Army colonel as OSS Chief in China. At the time he passed through Karachi he was either en route to China or was proceeding from China to points west (which would have been Cairo); AWZ does not say which it was, although Miles' papers may give the details. We do not know the name or position of the general who accompanied Miles.

It is clear from Miles' letter of 22 August 1944 in the "Winsor" file (AWZ's Wartime Papers, and in this book, Chapter 6) that Miles formed an opinion of AWZ, and it was probably at the time of this brief overnight visit on 31 Jan-1 Feb 1944. Miles wrote, "My personal opinion of Z is that he is a very nice fellow personally but he does not have much on the ball either from an administrative or naval standpoint. However I don't know much about him and withhold judgement on him because I might do him an injustice." Miles is, however, forthright in his criticism of AWZ's desk officer at ONI, LT Curt Winsor. He is furious with what he calls the "clandestine correspondence . . . subversive to discipline" that Curt Winsor sent via courier to AWZ. Winsor believed that Miles tried to get him (Winsor) fired because Miles believed Winsor was violating proper Naval standards in his work as a desk officer at ONI (copy of Winsor to Col. W. L. Bales, 9 Mar 1945, sent by Winsor to AWZ). Winsor wrote a carefully worded explanation of his relationship to AWZ to his supervisor at ONI, and he was allowed to continue working at ONI for the rest of the war. In the end he promoted to LCDR, as was AWZ.

Loads & loads & loads of it,
Al

P.S. I almost forgot. The package from Flecks[?] arrived and was very welcome. I've shared the fruit cake with the boys and they all claim it's the best they've tasted. The butter scotch is very good. I nibble on them now & again. Last night I had a few figs and they also are very good. In fact the whole box is very tasty and it's a pleasure to eat some real quality food. Next time you're sending something you include my old favorites – pretzels. That you very much for the package. It's just as welcome even though a month late.

Again loads & loads of love,
Al

----

3 Feb 1944, postmarked APO 886   From AWZ/APO to HZ   ALS-I
                                                  [no header]

Wednesday Feb. 2nd

Dear Lanie,

        Your letter of Jan 2nd arrived today and a nice long one it was. I was awfully glad to hear about your Christmas. You didn't say when you were going to have your eye operation but I judge you had it during the Christmas vacation. I hope everything went well and you didn't feel too miserably.

        I'm glad you liked the evening bag. I thought they looked very attractive and unusual. Your new slant on the elephant joke is very amusing. It must be quite baffling to the "smarties" that think they know it. My Christmas package arrived several days ago. The things in it are very good. I wish I had been home to enjoy Christmas with all of you. It's the first I've missed and I don't like the idea at all. /p.2

        They want me to sing for the British Y.M.C.A. here so tomorrow I'm to meet an accompanist. I don't know how we're ever going to get together as I have no music and there are no decent music stores in town.

        It rained today for about five minutes – the second time it's rained since July. We were about to play tennis when it started. It didn't spoil our game as the sun soon came out and dried the courts.

        Night before last we had a general and a commodore stay here. They came very late and left very early so we didn't have much chance to talk with them.

        I'm sorry this isn't much of a letter but there isn't much to say.

Lots of love,
Daddy

----

6 Feb 1944, postmarked APO 886   From AWZ/APO to BSZ   Rec'd 19 Feb   ALS-I
                                                    [no header]

Wednesday Feb. 5th

47 [prescript: I hope you gave Timoney the usual $100 bond at Xmas. I think he deserves it.]
Dearest,

        Two of the Christmas packages and the pictures came at last. Thank you very much. You were very thoughtful. You are a very good knitter of sweaters and it fits very well. I loved the poem & apropos card. The pictures are wonderful. Who said you didn't take a good picture? Now I can have a real look at you every time I want to, which is all the time. I like the one in the darker dress particularly. The clock is very useful and neat. An alarm clock is a rarity in this part of the world and sometimes necessary. The little leather folder is very nice and I'm glad to have all of you in a handy little thing to carry around. The other items, books, soap powder, shaving lotion, will come in very handy especially the latter as it's very hard to get out here. /p.2/

        Your letter postmarked Jan 17th arrived yesterday as did a letter from Babs of the same date with the new address. Lanie's letter of Jan 6th arrived two days ago with the old address so I guess the new one is a bit better. Eighteen days isn't so bad in fact it's damned good compared with the batch that took almost three months.

Your party sounded very nice. How come the Elys? and Mac Greene? Boy how I wish I could have been there. I'm awfully sorry you are so lonely. It's very sweet of Babs to be so considerate of you in my absence. She's doing right well by me too and writes very nice letters.

I'm glad you're joining the Acorn Club or at least are up for membership. I don't think there's any doubt about your getting in. It will be a nice place for you and the girls to use /p.3/ when you're in town. I'll let you even take me someday and how I wish that were soon.

The Kelley's sent a very nice box of candy and Sally & Jim [Winsor] a box from Fleckes[?]. All very thoughtful.

Last night the vice consuls had a cocktail-snack party and who should be there but Buzz Sloan, Frances's brother.[117] It was very nice to see someone from home town. The party lasted on into the night and a good time was had by all. The consulate seems to have a good supply of liquor. Some newly arrived Red Cross gals were there. I think there are about ten around town by now – some look pretty cute.

Last Saturday night I had a party here & then to dance at the Gymkhana to pay back some debts. I had the Harris's who had me /p.4/ for Christmas eve, the York-Tor's who have had me to dinner several times and the Stairs. My "girl" was Mrs. Greatbatch whom I have mentioned before and who arranges our flights (I don't think I'll be taking any more now that I'm executive officer[118]) via R.A.F. to Colombo. Don't worry I'll still take you and how.

Tomorrow, Sunday, I'm going out to Sandspit, a beach, with some Britishers. It's on an island and you go by launch. I haven't been there before. They say it is very nice – a change anyway. I feel awfully mean enjoying or I mean having mild weather while you freeze in Haverford. I hope you are able to keep the house a bit warmer than last year.

Well, that's about all I know – except I miss you more and more.
      With loads & loads of love,
         Al
Weren't you going to have Lanie's eye operation about now? You said nothing about it in your letter.

----

14 Feb 1944, postmarked APO 886	From AWZ/APO to BSZ	Rec'd 24 Feb	ALS-I
	[no header]		

Wednesday Feb.13th

48

Dearest Barb –

Here it is your birthday and me way out here where I can't give you a big hug and a kiss. It's awfull. I'm certainly going to have to make up for a lot of lost time one of these days and how I hope that day will soon come. I don't like this separation one bit.

No letter from you all week. I haven't heard from you since Feb. 4th, nine days ago. I know it's not your fault. You've been very faithful but it is discouraging not getting mail that is due you.

We had a very pleasant day on the beach last Sunday. Our party consisted of Lt. Com. & Mrs. Stores (he's stationed here) Mrs. Rivett-Carnack (her husband's in the Middle East) and a Lt. Sturt (British Army). We left the west wharf about 10 in the [marginal note: Enclosing some pictures I took around our cozy nest. Pretty professional looking what?] /p.2/ morning in a launch filled with Britishers and their families – a lot of cute kids – there was one Greek couple that had two beautiful daughters (about 5 and 7) and arrived in about a half hour on a spit of land similar to Outer Beach at Chatham. Porters lugged our lunch and swimming things to a hut – made mostly from packing boxes – nothing very elaborate. We sat around for a while – then went swimming in the Arabian Sea with surf boards – had some drinks and then lunch prepared by the girls – slept – had another swim – then tea – I supplied the

---

[117] Frances Rutan and Buzz Sloan, in fn to AZ's letter of 31 December 1944.
[118] He first mentions in this letter that he is now Executive Officer of the Naval Liaison Unit, Karachi.

fruit cake from Sally's Christmas box – then home by seven. It's the first time I've been to this beach. It's very nice – called Sandspit. The girls were quite attractive – about the best there is to offer here – about 27ish /p.4/ but I'll still take you. I missed you so – I really did.

Tuesday night one of my RAF friend's Squadron Leader Broudhead had a birthday party at the Boat Club. I felt honored being the only American on it. We had drinks at the British Officers Club – then more drinks at the Boat Club – supper (fried eggs baked beans and birthday cake) then danced to a victrola with loud speaker. Not a very exciting evening but none of them really are for me.

Last night a British naval officer invited Bunker and myself to the dinner and the dance at the Gymkhana Club which was the usual Saturday night affair. So there you have my activities up to date.

This morning a CPO (Chief Petty Officer) arrived who had gone home for /p.4/ an operation. He said he had some news from Curt but would see me later in the day to tell me about it. I'm rather anxious to know what it is.

My bearer just now brought me mail – a letter from you, one from Jack [Ott] and one from Martha Z. so I'll stop to see what's what. Your letter the one before this, although written 9 days later, beat this one by 9 days so I guess the APO does much better than the ONI.

I've read the letters – Jack's was a lot more newsy than usual – mostly business. Martha's with a short note from Dick told about the New Year's party and, yours, the best of all – a nice long one filling in the gap that have been missing.

It was nice of the Ziesings to invite you on their party and to drop me a /p.5/ line. It was very considerate of all of them not to do the usual at the midnight hour.

I'm glad to get the news about Lanie's eye. I hope it means an operation will not be necessary and it is responding to treatment and exercise. I hadn't heard and was getting anxious. I thought silence meant things weren't just right.

I don't know as I'm crazy of the news of Tat but I guess that's his business and he should know what he's doing.

I'm glad Jack [Ott] divided the liquor – now if he'd only do something about the gas I'd feel much better about it. Can't you get a taxi to take you around? Or are they completely out?

That's about all – must go down to lunch.
     Heaps & heaps & heaps of love,
       Al
The situation here is still up in the air.

----

21 Feb 1944, postmarked APO 886 From AWZ/APO to HZ   ALS-I
           [no header]

                        Wednesday Feb. 19th
Dear Lanie,
  Many thanks for your very nice Christmas presents. I was getting a bit low on handkerchiefs so yours came just in time – in fact "just what I need." The two books "American Gun Mystery" and "The Glass Key" look very intriguing. I will get at them later on when I have some reading time. It's about all I can do now to keep up my correspondence. I enjoyed the pictures immensely. There's quite a picture gallery now on my dresser – very convenient to see at least a picture of all of you whenever I want to which is always. Any part you had in supplying the miscellaneous items /p.2/ that didn't have a card – thank you.

I'm sorry to hear you had so much sickness in the house. The flu epidemic must have been pretty bad. I guess now that you've had some real cold weather the germs have vanished.

It's my duty tonight and a dispatch has just come in so I'll have to decipher it. There really isn't very much to say anyway.

Here are some pictures I bought down town this afternoon showing some of the odd sights you might see over here.
       Lots & lots of love,
         Daddy

So your operation has been postponed or maybe you won't have one after all.

----

21 Feb 1944, postmarked APO 886   From AWZ/APO to BSZ   ALS-I
[no header]
Wednesday Feb. 19th [perhaps misread by me, or misdated by AZ; 19 Feb is Saturday]
#49
Dearest Darling

    The "Timoney" box finally arrived and also the pictures (and underwear from DePinna). Both very welcome and thanks a million you and the children did a swell job for me. The underwear fit perfectly and came just in time, the various presents were grand and the pictures superb. I don't know why you think they turned out so badly or were such a disappointment. I loved the apropos card. I think they are great and am very pleased. Now I have a swell collection that I can look at and swell with pride. I love all of you so much. Your presents will keep me occupied for months to come – gosh I hope it isn't months.

    Some film (Kodachrome movie 5 rolls of 50 ft) was brought by a chief yeoman given to him by Curt. I guess it's the /p.2/ film you mentioned in your letter of Dec. 14th (that just arrived today). I didn't know you were sending it so wrote to Curt [Winsor] asking how much I owed him. Anyway it's the wrong load for my camera. It uses magazine load. I can dispose of this easily enough but unfortunately after your trouble I can't use it. I have plenty of black and white which I'll take and send on if I can find some suitable subjects. I wish you'd give me better hints of what particularly appealed in those I've already taken. You get used to seeing strange sights out here that may be interesting but I don't want to waste film. It's hard not seeing the results.

    Another batch of old mail came in today. Beside the letter from you I mentioned /p.3/ above another came enclosing the letter from Sunny (Severn's nurse) which I'll try to deliver, also a very cute, amusing Xmas card from you, a letter from Jack Beard, a card from the Orams and the Orpheus concert program with remembrances that Louise [Ott] collected but magnanimously gave you credit and a letter from Jack about business. All were dated around Dec. 15th over 2 months getting here.[119]

    I took some movies on my trip thru the Norwest Frontier with a borrowed camera and old film that had to be re-rolled from 100 ft to 50 ft rolls and then back again. Something went wrong – none of them turned out – very disappointing.[120] I did take some still that I'll send along some day. The pictures enclosed were taken on the trip near Quetta in /p.4/ Baluchistan. I took them more or less for business and am sending the negative to Jack in case he wants to make enlargements for our wool friends. The guy in the photographic shop wanted me to put the good one in an amateur contest.

    I'm trying to write a story about the trip illustrated by the pictures I took, but it's taking a long time to do. I seem to have so little time for it and I labor so over writing.

    If you want my opinion I think Lanie should go to C.H.[121] I think she's like it and would do her a lot of good.

---

[119] Orams = Jim and Kay Oram. He was a lawyer, later with the Pennsylvania R.R., and she was a teacher at the Haverford School. They stayed in the Zimmermanns' house and watched over the four children when AWZ and BSZ were out of the country on vacation prior to WWII. Notes from their daughter Cathy are in "Correspondence with Family and Civilian Friends," one of which is referred to above on 16 Oct 1943. Jack Beard = another friend. Orpheus Club = AWZ's singing club, which sent a card of remembrance, signed by club members, in his Wartime Papers.

[120] A few feet of the 16 mm movies taken on the Trip were, in fact, useful, beginning with the Lowari Pass, and ending in Quetta. There were no still photos taken on the return trip from Chitral to Quetta, and none in Peshawar. The photos taken in Baluchistan of the "fat tailed sheep" and others of sheep are in the trip sequence and are indeed very good.

[121] C.H. = Chatham Hall, Chatham, Va. Babs went there, graduating in 1944, and entered Smith College in the Fall, as a member of the Class of 1948. Lanie went to C.H. in the Fall of 1944 for her last two years in high school, graduating in 1946, and then she, too, went to Smith, where she graduated in 1950.

This week has been rather routine. Sunday we did have our next door neighbors, the Dinshaws for supper. /p.5/ Aspey and his two sisters Meta and Paren – pale hands but not very lovable. They are interesting though – of the éclat of the younger Indian society here. Friday, the War Shipping Administration man, Mr. Getchell, had a drinks and supper party. Yesterday some more of Freeman's friends arrived – two Thais quite interesting. I got up in the middle of the night to see them off as it was my duty.

Has Albie been receiving my various postals of Indian life and scenes around here? Nobody has said anything about them and I'm wondering if they got thru.[122]

I guess its sounds as though we are very gay – entertaining and being entertained.[123] It is diverting at times but I'll take my normal /p.6/ life anytime. I miss all of you so much and am anxious as the dickens to get back. Maybe that time is nearer than we think.

Heaps & heaps of love,
Al

Keep an eye on the New Yorker. We entertained a cartoonist by the name of Steinberg[124] who might do something of life around here. Did you see his cartoons in the Jan 15th issue on China?

---

28 Feb 1944, postmarked APO 886　　From AWZ/APO to BSZ　　Rec'd 7 Mar.　　ALS-I
[no header]

Sunday, Feb. 27th

#50 – that's a lot of letters
Dearest Barb,

I haven't received any letters from you written after Jan 17th. Two have come after that one but they were written before. I did get letters from Warren (written Jan 3rd but not posted until Feb. 1st) Albie (postmarked Feb. 4th) and Babs (postmarked Feb. 3rd). I can't imagine where your letters are. I know you must have written since Jan. 17th or at least I hope so. You've been very faithful. Jack wrote me on the 31st of January sending the original to the old address and a copy to the APO. The original arrived several days ago but the copy has not put in an appearance. So, I'm up a tree to tell you how to address my letters. /p.2/

---

[122] Enclosed with this letter: A 3¼ by 2¼ in. black and white photo purchased by AWZ of a Hindu "Burning Ghat," showing a naked female body on a pile of logs, being readied for cremation. On the back, AWZ wrote, "Throw this away if you think it's too morbid. But this is India. If they didn't dispose of the dead this way, think of the graves (16,000,000 a year). "

[123] AZ is often defensive about his social life, but it was even more active than he described in his letters. He did not mention the invitation to dinner on Friday, 18 February 1944, from April (Mrs. Geoffrey) Swayne-Thomas that is in the AWZ Wartime Scrapbook in its original envelope, loose between pages 20-21. "My dear Al, / This weather needs something to liven it up, don't you think? So I'm having a small party on Friday 18th & do hope you will be able to come – bring your voice too, as there will be a guitar player here. Come any time after 8 p.m. & stay for a fork supper – Geoffrey has been about to come in & write to you personally all week, but each day he has been too busy - So I hope this notice is not too short for you – Au revoir – Very sincerely yours / April (Swayne-Thomas)." Geoffrey and April do not appear in AZ's letters to his wife.

[124] Steinberg = Saul Steinberg (1914-1999). AWZ may have known that Steinberg was working for military intelligence, but he often offered clues to BSZ in her letters for further study. The cartoons of 15 January 1944 that AWZ referred would be interesting to see. A double-page of Steinberg's "India" cartoons from *The New Yorker* (24 February 1944), 6-7, are in AWZ Wartime Scrapbook, page 18(V). Other Steinberg cartoons from China are reproduced in Joel Smith, *Steinberg at the New Yorker* (New York City, N.Y.: Henry N. Abrams, Inc., 2005), 52-53 (from the *New Yorker* [5 Feb 1944]). There is also a full-page cartoon that may not have been published, showing a portrait of a U.S. Naval officer in China, undated, on Ibid., 197. The senior Naval officer present in China, who was probably the subject of this cartoon, would have been Commodore Milton Miles.

There was a friend of Paul Bakers who was flying back to the states. I gave him a present I got for you on my last trip to Colombo. He is going to give it to Paul's wife in Miami who will mail it up to you. If all goes well, you should be receiving it about the time you receive this. I hope you like it. I would be interesting in learning how much its appraised for in the states.

Tonight I am to sing for the Y.M.C.A. This will be my first public appearance in India, my Indian debut. The audience with be B.O.R. (British other ranks – meaning no officers) and a few local women who feed and entertain the boys as you do at home. I think there are to be three others soloists /p.3/ [sheet 2] all of us singing about three numbers a piece. My selection was limited to the songs the lance corporal accompanist knew by heart of by ear as there is no music to be had – but we did find a score of Iolanthe so I chose the signature song for one of the numbers. The other two will be Wagon Wheels and Easter Parade with Someday I'll Find You as an encore when as and if.

This week has been a fairly active one, in fact, too active to suit me. I don't want to appear to be a social lion but people do invite me to do things and it's very hard to refuse and one of the jobs of a liaison office is to become better acquainted with the localities and certainly not /p.4/ hurt their feelings. Monday, Tommy Weston had two of us over to his flat to play bridge. Tuesday Com. Stares had a drinks-supper party. Wednesday we had some people here to dinner. Thursday I went to the Coughlans for dinner (he's second in command in the Bombay Co. – with whom we might some day hook up in the wool business). Friday was my duty and last night our next door neighbors – the Dinshaws – invited us to a dinner dance at the Karachi Club – too much to suit me but it helps to pass the time away. We manage to get some tennis in almost every day. Twice this week we played with the Dinshaw girls who are supposed to be the best in Karachi. /p.5/ [sheet 3]

Baker and I are about even and they are so it makes a quite exciting mixed doubles match. Most of the sets are won by small margins – Paren and I managing to win the best of three. This all doesn't sound much like a war's going on does it?

Two of our officers are being transferred to Bombay – Baker & Refbord. I hate to see them go especially Baker as he besides being a good egg makes a very good tennis companion. I don't know who'll be coming here to take their places – if anybody.

I don't know whether I've told you but we're studying Urdu (Hindustani).[125] We have a "professor," Vaswani is his name, came twice a week to try to drum knowledge into our heads. Besides learning the language, or /p.6/ trying to, we have general discussions on India, its people and problems etc.

Here it is eight months since the fatal day in New York – eight <u>long</u> months.[126] How I wish this whole business would be over and I could come back to you. How I long for your loving embraces and just to be near you. The news about Finland[127] is encouraging. I hope it sets a spark for all the satellite countries to get out from under.

Good-bye, dearest. I love you very much and hope I'll soon be able to physically demonstrate it.
Heaps & heaps of love,
Al

---

4 Mar 1944, postmarked APO 886   From AWZ/APO to George McNeely, Esq., Bryn Mawr, Penna.,
U.S.A.   ALS-I

[no header]

Friday March 3rd

---

[125] The folder of "Instructional Materials, Notes, Tests" in AWZ Wartime Papers includes a one-page "Elementary Urdu Test – August 1944" which AWZ apparently took in August or September 1944. He does not mention taking this test in his letters in August and September 1944, but the handwritten notes and typed lessons strongly suggest that he took the test and passed it.

[126] Eight months = AWZ departed New York on 24 June 1943.

[127] Finland = The situation in Finland was far from solved in March 1944, for either Finland or the Allies. Early in 1944, Finland had first to rid itself of the Germans, and it then faced an invasion from Russia.

Dear George:

Thanks a lot for your letter of Dec. 8th. It took quite a while getting here and I'm afraid I'm taking quite a while answering it.

I'm enclosing a couple of views of life here to bring you back to the days when you were in Aden. I imagine they do about the same things there as it also is a Muslim country. One, as you no doubt can discern, is the familiar operation of barbering and the other shows the simplicity of taking care of your innards – a tray of bottles and vials, probably pinched from some drug store, with a very capable practitioner standing by to administer your ailments – in other words the family doctor. Why bother with the frills and expense of a Dr. Boles? /p.2/ I'm afraid I can't give you much of a clew to what's happening in the goat skin business. All I know is that there are a lot of goats around here and what they live on I can't imagine. The cows eat all the dirty shirts and tin cans.

You are right about this being a better place to live in than Aden. We get quite a few boys passing thru that have just been there and by comparison, from what they tell me, this is paradise. As a matter of fact, I have no complaints at all about the climate. The food could be a bit better but after all that situation isn't any too good at home.

My main complaint, of course, is being away from home and family and that is something pretty hard to take.

I haven't been able to find out anything about Paulo Gullino but may be able to some day.

Best of luck – give my love to Allie – Sincerely Al

[enclosed, 2 black and white prints, 3 ¼ by 2¼ in., both perhaps taken by AWZ, unlabeled. One shows a man in a turban in front of a little table with an array of bottles; the other is a man shaving another man, while both are sitting on a city street.]

---

6 Mar 1944, postmarked APO 886   From AWZ/APO to BSZ.   ALS-I
[no header]

Saturday Mar 4th

#51

Dearest Barb:

Two letters arrived from you this week, one undated and the other one dated Feb. 10th. There must be some more somewhere as the previous one was dated Jan 15th. These evidently came via Freeman but not very fast. I can't understand your not getting a letter from me between Dec 27th and Jan 13th as I sent one on the 5th of Jan telling about my trip to Colombo. Of course I didn't forget you. I'm thinking of you practically the whole time and writing at least once a week.

Your trip to Wash. sounds very gay. I hope you had a good time. I'm glad you met Mrs. Bond and saw Harmon again. What about Harmon's girl? He thought he might get married when he got /p.2/ home. I haven't heard if he did or not. I think the Hal Putmans are very nice. I'm quite sure you met him on one of the squash parties. Too bad Freeman isn't coming this way. I thought he might but I'm sure where he is going is a lot better place for him than if he were headed out here. Let me know when Peachy is due to arrive, name of ship if possible. I'll meet her myself or have someone from one of our other offices meet her. I'll try to arrange Dick's meeting her in Delhi – funny he should bump into Enders.[128]

---

[128] Dick, Peachy, Enders = This statement confirms that Maj. Gordon Enders is in Delhi at this time, March 1944. "Dick" is his nephew, CPT Richard Kelley. Virginia "Peachy" Durand appears in AWZ's photo album in Kandy, Ceylon. She was in the OSS. From Elizabeth P. McIntosh, *Sisterhood of Spies: The Women of the OSS* (Annapolis, Md.: Naval Institute Press, 1998), 209: "The first contingent of OSS women to reach their headquarters in Ceylon sailed on 9 March 1944 from California, aboard the one-time luxury liner *Mariposa*, which in the days before Pearl Harbor delivered tourists to Hawaii. They crossed the Pacific with a destroyer escort, landing in Bombay on 8 April. They departed for Ceylon ten days later and arrived in Colombo on 26 April. OSS HQ in India was in New Delhi until it was relocated to Kandy in October 1943, when Mountbatten moved SEAC HQ from New Delhi to Kandy, Ceylon. Kandy was almost indescribably beautiful, a delightful place to work, with a jolly nightlife. The OSS contingent in Ceylon eventually grew to 595. Some OSS personnel remained in New Delhi. There were nine

Do you think I should write up the trip thru the NWFP? I don't think the Navy wants it & I don't think it will be good enough to be published and I hate to put in the time. If you <u>really</u> think there's a hope I'll go ahead /p.3/ [sheet 2] with it.

By now you've seen Callahan. What did you think of him? I got along fairly well with him but he wasn't very well liked by a lot of the boys that have been stationed here.

Sally wrote me that Jim [Winsor] was whisked away. Too bad. I'm afraid she'll be a lost soul without him. By the way did you ever give her a present from here. She very nicely sent me a box from Flukes at Christmas. Not to just pay her back – I did have the idea of giving her something before I got her nice remembrance.

So I'm a great uncle. It certainly took long enough with all the nieces that have married. I guess they are all delighted it's a boy.[129] It's a calamity in this country when a girl is born.

I'm sorry Albie didn't receive the ivory Buddha I sent him. I'll get him another if I can. One of the elephants was supposed to be for Warren. I don't know how they both got addressed to Albie. Nothing has been said about the magic brass bowl I sent to Warren. You fill it to a certain level and the water runs out the bottom. I explained it all in a letter I once wrote. Can't you tell from the numbers of my letters whether any are missing or not? You seemed to have stopped numbering your letters so I can't tell if any are missing from you.

Too bad the boys were sick again. You seem to have had a lot of sickness this winter. /p.5/ [sheet 3]

I'm distressed to hear about Crosby. Poor girl – what a helluva handicap to go thru life with. I hope they find something or someway of making it well again.

I think it is a wonderful idea to take a cottage at Easthampton. I guess you've looked into the transportation problem etc. Will they let you take a car there?

Babs' letter that you sent APO arrived yesterday the 3rd. Yours arrived the 1st. She seems all sold on Smith and I'm glad you agree. All along I thought Smith was the college anyway, especially since Babs says Vassar is accelerated. I don't think she should be in any hurry to get through and that applies to Lanie too.

I loved your Valentine's Day card. You were cute to send one.

There's not much new with me. I seem to have to do an awful lot of writing to various people. I never seem to have time to read those interesting looking books I got for Christmas.

Last Sunday I sang for the Y.M.C.A. (British). There were three other soloists and the troops seemed to like it. I think they appreciated having an American officer help entertain them.

This week has been fairly quiet giving me a chance to catch up with my correspondence.

Tomorrow I'm going to the beach again with the Stares. A week from Monday the Viceroy[130] is going to pay us a visit.

Lots & lots & lots of love,
Al

We've changed to summer uniforms. Don't bother sending any of the stuff I've asked for. I could use some of my white shirts with collars attached and a few white socks.

---

women in the initial group that went to India. They were: Virginia Durand, Julia McWilliams, Rosamond Frame, Eleanor Thiry, Virginia Pryor, Louise Banville, Cora DuBois, Jeanne Taylor, and Mary Nelson Lee. All went to Kandy, except Frame, who was sent to New Delhi. Thiry kept a diary of the trip." Julia McWilliams later married Paul Child and became famous as Julia Child, the writer and TV personality, about French cooking. Durand was later one of ten women in the OSS who were flown "over the hump" to serve in China (p.226).

[129] The baby was Donald James Fleet, son of his niece Edith "Lil" Hoxie, daughter of his sister Clara, who married Albert Hoxie. Edith Hoxie married Benjamin D. Fleet.

[130] The Viceroy, Lord Wavell, will be the guest of honor at a party given by the Governor of Sind Province. Invitation and instructions are in AWZ's Wartime Scrapbook. This will be AWZ's only other meeting with Wavell, after he first met the Viceroy in Peshawar a little over 3 months earlier, in November 1943.

"Passed by Naval Censor," not postmarked. From AWZ/APO to BSZ. Rec'd 20 Mar.
[This letter was hand-carried to the U.S. and was apparently delivered directly to BSZ. It passed the censor, but it did not go through the mail. It was filed by BWZ after AWZ's letter of 13 March (*infra*), but I have transcribed it in the order in which it was written.]

ALS-I

[no header]

Thursday Mar. 9th [1944]

[52, though not numbered by AWZ]

Dearest darling:

This should get to you quicker than any letter so far. Gould Thomas a chap stationed out here is going home for a quick trip – leaving this afternoon – and will take this with him.

I just received your letter (today) written Jan 31st. It had been mis-directed. Three days ago I got your letter of Feb. 18th so I've had a pretty good week. The latter was the anniversary letter and I agree that the eighteen years have been very happy ones – I hope too it's the last anniversary spent apart. It was nice of the children to be so thoughtful. I'm sorry if I forgot to thank you for the Christmas cable. I wish you would send EFM monthly. Even though they are canned, it's nice to get the latest news.

[marginal note: I've been able to exchange a roll of color film for one that will fit my camera, so will be sending it along when exposed.] /p.2/

Too bad you're having servant trouble. If it's not [one] thing it's another. I shouldn't think you'd want the baby there, but you could help out financially if you thought she's deserving and appreciative.[131]

I'm glad to hear the children still remember me and talk about me. I love them all very much and I hope it's returned.

I'm distressed to hear about the ulcer. I'd hate to go on that diet again for four months. I hope it wasn't too bad for you. It must have been awfull not being able to resort to a bit of alcohol in your loneliness. No wonder you got so down in the dumps. Why didn't you tell me about it?

Remember Freeman's friend whom Freeman brought up to our apartment in Washington by the name of Feninaro[?] He just called me. He's out /p.3/ at the airport waiting to go home. Congratulate Clarence for me.[132]

There's nothing much new since I last wrote. Went to the beach on Sunday with the Stares (Lt. Com) Col Williams (a British doctor) and his wife (who is about to have a baby) and another couple. It was very nice. The water was just the right temperature and so was the air – The company was quite entertaining. I learned a new game called Cameroon – played with 10 dice – very good. I took 3 Rupees from Pat Stares. It's fun being with the English and getting their view point.

Sunday night I had dinner at the Cleis. He is Collector of the Province which job is a stepping stone to the governorship. They had three /p.4/ other men from various parts of the Province. I can feel pretty much at home now in such a group having travelled about this country a bit.

Got a letter from Jack [Ott] this week. He says he will send you a copy of the monthly statement and inform you of any major developments in the business. I doubt that he will, as he hasn't in at least one instance but I guess it's about the best we can do under the circumstances. I hope and guess he will be conservative from here in.

The plane is to leave shortly so I'll close with much love, dearest,

Ever aye,

Al

Very thoughtful of Morgan Suter[?]

---

[131] Their maid, Millie Craft, had a daughter, Marie, who had a baby at this time. Mother and daughter were difficult and it strained the patience of the family to have the baby there, too.

[132] Clarence Lewis, his neighbor, who was in the OSS. This may have been in response to news that he received this appointment.

14 Mar 1944, postmarked APO 886   From AWZ/APO to BSZ, which she marked on the envelope:
"Movies"   ALS-I → P
                                 [no header]

                                                                  Monday Mar 13th
#53
Dearest Barb:

The last three days, I've been on a photographic rampage.[133] I got some black and white given to me and I exchanged one of your color rolls for a magazine to fit my camera so there are six [change to pencil here] fifty (I've run out of ink) foot rolls of b & w and one of color on the way.

Here's what they're about: the races (we have them about every two weeks on Saturday at this time of the year) – a good opportunity to give you a cross section of the people here – Tommy Weston, an RAF bloke who owns several horses and the finish of the first race with his horse winning – then various scenes around the track. Another reel I took yesterday on a movies walk around town – our bungalow from a distance showing how nothing grows in an empty lot (like antafagasta) – some shops /p.2/ with one of our sweepers in the doorway (they are about the lowest in the caste system) – a look into a court in back of the main street – fencing – the nice looking boy in the suit (who became my guide from then on) comes from Burma, said he and his family walked all the way – he's a Christian & speaks fairly good English – scenes on Elphinstone St. – the main shopping street – several of the many beggars – the way natives have their hair cut and get shaved (no soap) cows & other animals walking all over main streets.

Next reel shows Clifton a nice residential section on the sea – view of the city from there (it's about 3 miles from town) the sea with several of the ever present hawks – near Clifton a beach where natives & others, not so particular bathe in the sea – a Hindu temple that seems to be a sanctuary for pigeons (5000 of them). You go down /p.3/ some steps to sort of a dungeon the ceiling of which has innumerable holes that the pigeons inhabit. You can imagine what the floor is like. You are lucky if you yourself don't [get] anointed. There are images of various gods around with a special room in the back with a well with the sacred water of the Ganges in it. Outside various worshippers, hangers on and cleaner uppers of the floor who carefully collect the droppings, probably sold at a big price as fertilizer. The priest was asleep and wouldn't be disturbed to have his picture taken so I went to the beach, took some more pictures then came back and bribed him with a rupee to get up and be photographed. He never has cut his hair and I venture to say has never combed it, therefore not /p.4/ disturbing the various animal, vegetable or mineral life contained therein.

The whole thing was unbelievable. I hope you'll be able to get some idea of it from the movies. I forgot to tell you that a hole thru the side of the rock, in the innermost part of the temple, heads to the unknown – nobody has ever returned who has entered it – said to contain cobras etc.

The next roll shows a building operation – how women work along with the men (women & children are street cleaners here – they have plenty to do) – the children of our servants who live in our compound & finishing with the garden party to meet the Viceroy.

The color roll is mostly of me – because you complained of my not being in any of the pictures and the rest of the garden party at government house. /p.5/ That happened this afternoon and was quite an affair. I'm sorry I didn't have more film but it will give you some idea – I tried to take some of the people I've been talking about but can't very well identify them – Mr. Macy (the consul) by himself – the

---

[133] Many of the scenes that AWZ describes in this letter appear in his collection of photos. Some are in the Photo Album, and some in the envelope of loose photos. It is not immediately clear if he was taking both movies and still photos on this "rampage" through town, or whether the still photos were taken at another time. There are, unfortunately, no still photos of the garden party (and of the Viceroy), so he may taken only movies there. And the movies of that event have not been preserved.

Harriss's (he in uniform)[134] – the Coughlans (civilian) also General Hind – the Viceroy & the governor & wife of Sind.

That's the story – I'm afraid they are not awfully good. The trouble with natives – they stop moving and want to pose for a still picture. Then they crowd around the camera. My Burmese boy was big help in getting them back. I'll have to wait now till more magazine film comes in so I can trade the film you sent.

Good bye dear it's getting late and there isn't much more to say except I love you and miss you and think of you all the time.

         Heaps & heaps of love,
         Al

----

22 Mar 1944, postmarked APO 886 From AWZ/APO to BSZ Rec'd Mar. 31 ALS-I
         [no header]

                        Tuesday Mar. 21st

#54
Dearest darling:

You're wonderful. I knew you could do it. Now you are a righter in your own write. I think that's a record – to have your first article accepted by a good magazine.[135] Really I think it's swell and my heart swells with pride. Also I glow with the kind thoughts & sentiments about me. I wish you would send me a copy of the whole issue. There isn't a prayer of getting one out here and I would like one. It will go by sea (if you can't get a friend of Freeman to bring it out) but will get here eventually.

You mention Jack Strubing[?] getting out of the Marines in your letter of Feb 28th. What was the matter? You said he looked well and healthy and was honorably discharged. I'd like to know the whys & wherefores.

Evidently you & Bill Stuble aren't worrying much about income tax, going out to the musicales etc. right before the dead line. I hope you had everything all /p.2/ "taped" (a British expression) so you didn't have an awful rush at the last moment. It must have been an awful mess this year. I asked the Navy to send both of you a copy of the regulations as they apply to us. We get certain exemptions. I understand you can deduct doctors & hospital bills now.

You needn't worry about letting off steam to me. I know you must be awfully lonely and depressed at times. I think you're making a lot greater sacrifice than I and deserve a lot of credit and admiration. Along that line let me quote from Jack's last letter "Barb is plenty lonely as you can imagine. Also independent. There are lots and lots of things I could do to help, but she will have none of it. I admire her attitude, but it leaves me pretty helpless when it comes to supplying a few things to which she is entitled. She is that way with everyone & and as I /p.3/ said before I admire her spunk." Why don't you let him do something? I don't know what he could do but he seems to think he could help and seems sincere.

Too bad Timoney is sick. That must complicate your life still more.

Phil Lee's nephew couldn't be in Karachi with that address – at least not to my knowledge. I'll try to find out where he is. I hope he'll look me up when he comes this way.

I don't know why you enclosed a blank stamped envelope in your letter. I guess it must have become mixed up with the enclosures. Awfully nice of Hilda Teetor to be so thoughtful and cordial. It would be fun to see them again and go out on a toot. I wrote them a postal from here.

The Fourth St Club bulletin arrived – last of the three [illegible]. Freeman's friend came Mar 1st APO Mar 4th and this /p.4/ Mar 18th so the best is APO if you can't catch somebody coming right out. I

---

[134] John Harris is the Chief of Central Intelligence in Karachi. He is in civilian clothes in other photos, and this is the first time AWZ said that he is also a military person.
[135] Barbara S. Zimmermann, "This War—And Brave Little Women," *Vogue* (1 March 1944), 137, 140. It is about her life as a wife and mother who is separated from her husband, who is overseas.

"took a dim view," British expression, of the bulletin. I didn't think it very funny. First news I had of Bruce McDonald's death. What happened to him?

Your letter of Feb. 24th also arrived since I last wrote. You sounded pretty blue. I hope you don't feel that way often. I know it must be pretty hard at times. I miss you terribly too and can't wait until this bloody thing is over. There's still no one but you. It's fun to meet new people and it helps pass the weary time but it all seems so empty without you here.

I've sent on the last two films – finishing them up at the garden party. It was a grand affair elaborately and meticulously done in the traditional British style. You'll get some idea from the pictures.

We've had a tragedy happened near here that has kept us pretty busy /p.5/ the last few days. I'll tell you about it some time.[136]

Today, I played golf for the first time. Quite an experience playing on a course without a blade of grass – where they call the "greens" "browns" really. It was fun though to swing a club and it does resemble the other game. The fairways and rough are all the same but you might find more cow tracks in the latter. There is no sand – just dirt. It's a new technique around the ~~greens~~ browns. You putt from 100 yards. You'd never get a pitch to hold. Flash Broadhead (an RAF bloke) a Mrs. Pratt (whom I had just met) and a Mrs. Geldart (American from New Rochelle who married a Britisher and is living here) played a two ball foursome. The Americans won one up.

Good night dearest. It's late.
Loads & loads of love,
Al

----

4 Apr 1944, postmarked Washington DC   From AWZ/APO to BSZ   Rec'd Apr. 5   ALS-I
[no header]

Monday March 27th

55
Dearest Barb

No mail from you all week. I guess you've forgotten all about me. But I'll go on writing just in case.

Yesterday we said good-bye to Bill Van Dusen who has been out here with Standard Oil. I'm sorry to see him go but glad for his sake. He's going home for a well deserved leave. The reason I'm particularly mentioning it is because he might look you up when he gets there. His wife left when the going was good – or at least when all wives were forced to leave – and is now at Pettie Institute at Hightstown, N.J.[137] I suggested when he gets home they drop down to see you. I like him a lot and from what I hear his wife is very attractive. I think you'll enjoy meeting them as they have been here six years and can give you an idea of what life is like in these parts.

Also my friends Swabey and Barlow – the two blokes that flew me here from Cairo – may be coming your way one of these days. They have just come over and are going back to come over again but promise they will drop down to see you before they do.

If it's not too much trouble you might ask the Van Dusens and the RAF's to spend the night. I know you'll enjoy them. They are good types and will bring most recent news from here. I've ask them to call you on the phone first so that you may appraise the situation before having them descend upon you.

---

[136] I suspect that he thought a detailed account at this point in time might not pass the censor.

[137] AWZ means The Peddie School, Hightstown, N.J. The school was founded in 1864 as a seminary for women, but in 1908 it was changed to an institute for boys only. Its name was changed to The Peddie School in 1923. Walter Annenberg, who graduated in 1927, eventually gave the school more than $100 million and it is now an immensely wealthy college prep school. Sixteen of the 502 graduates of the school who served in WW II were killed. The school became co-ed in 1970. Mrs. Van Dusen's name does not appear in the History tab of the school's website, but presumably she was a teacher there in WW II.

I don't know whether Curt [Winsor] has /p.3/ told you much about developments here but it seems we have been living in sort of a fool's paradise, the fool being the one that set it up and has paid us full per diem during our stay. Evidently we've been living on much too high a scale and the day of reckoning has come. I don't know what the future will be but at the minute I'm inclined (and so is Howard Voorhees[138]) to move to the Sind Club. I think it would be a less hectic life with a lot less entertaining etc. However that's not definite. I hope some clarification will be shortly forth-coming.[139]

We are to have some important visitors in the next day or so which means a busy time for all.

I'm enclosing a clipping from our local newspaper which I /p.3/ thought would interest you and give you some idea of the mentality of some of these people.

I've tried – in the post cards I've sent home – mostly to Albie – to give you pictures of local scenes and activities. Such are the two I enclose herewith. The Tower of Silence is where the Parsis "bury" their dead. It is where they leave them to be disposed of by nature – mostly vultures. I haven't seen it except from a distance as it is out of bounds – very few unbelievers have.[140]

The other shows the milling crowd on one of the many holidays they have here.

That's about all dearest. I hope you are well – all over your tummy trouble.

With loads, heaps and lots of love,

    Al

---

[138] LT Howard Voorhees, USNR, appears here for the first time in AZ's correspondence. He is last mentioned on 6 March 1945, having just received orders to return to the U.S. before AZ got his own orders. Voorhees did not leave Karachi before AZ left; he signed AZ's authorization for per diem on 9 April 1945. Voorhees and AZ got along very well. His origin and postwar activities have not been successfully traced. However, he was accepted by the local British community and by American VIPs who passed through Karachi as AZ's partner in social activities. I have not been able to find more about him. LT Howard Voorhees apparently had no middle initial, which may be helpful in narrowing the search.

[139] A problem with finances at the Naval Liaison Office Karachi appears for the first time in AWZ's correspondence at this point. A major investigation and audit of the Navy Mess took place in 1944, perhaps on the authority of the Chief of Naval Operations himself, ADM King. The Commanding Officer, LCDR F. Howard Smith, was eventually removed (the Navy calls it "relieved") and received an Unsatisfactory Fitness Report. AWZ succeeded him as CO, but there was apparently always a taint on AWZ, too, as a result of the financial issues, even though it appears that AWZ was officially cleared of wrong-doing. Commodore Milton Miles (*supra*), who was the senior U.S. Navy officer in Southeast Asia, believed Smith was unfairly blamed for the problem in Karachi. Miles was a protégé of Admiral King, who is said to have detested Reservists, on principle. It is not clear who wrote Smith's fitness report (i.e., who his immediate superior was), but it presumably was a Commander or Captain in Delhi. Miles implies that Curt Winsor had a hand in Smith's downfall, and that perhaps AWZ was not simply an innocent bystander. AWZ's correspondence shows that he did not respect Smith, but there is, however, no suggestion in his correspondence that he undermined his CO. Additional details are in other files and folders in the AWZ Wartime Papers.

[140] The clippings and photographs that AWZ mentions in this and the preceding paragraph are not in the envelope with this letter. Some of the photographs can probably be identified in the Scrapbook and the envelope of loose photos and cards in the AWZ Wartime Papers. Many newspaper clippings are in the Scrapbook, and some of them may be those that AWZ refers to here.

3 Apr 1944, postmarked APO 886   From AWZ/APO to BSZ   Rec'd Apr. 11   ALS-I
[no header]

Friday Mar. 31st

#56 [this letter was placed before #55, probably because it was postmarked a day earlier]
Dearest:

Your letter of Mar 9th came yesterday. I'm very proud of the reports of our two sons that were enclosed. Please don't worry about money problems. Don't stint yourself. We'll have enough to get along on. From the reports from the business – last year was a good one with a good start for this year. I'm sorry about the problems of income tax – they must be terrific this year – by now you must be able to breathe easier. Too bad about Uncle Owen.[141] Aunt Greta must feel pretty lost and lonely. Give her my love.

I've just dispatched a box of stuff. Dispose of it as you see fit. Lord Buddha should go to Albie as it /p.2/ replaces the one that didn't arrive for him. Rawan, the many headed creature, was at one time (3000 years ago) king of Ceylon. He stole Ram's wife (Ram was king of India and has since become more or less of a god). Ram went down to Ceylon with an army of monkeys – killed Rawan rescued his wife. That is why monkeys are sacred. The other image is Kali the goddess of destruction. She has her foot on Shiva the creator. The elephants in case you wonder about the slit are for place cards – an idea more or less original with me. I had them specially made. The Chitrali cap goes with the coat. The 28 year calendar might give Albie something to ponder over when /p.3/ he gets tired of adding up speedometer numbers.

We are in the midst of taking care of and entertaining & large group of officers. The CO left for Bombay quite unnecessarily, I thought, and dumped the thing in my lap.[142] Sunday night, I'm arranging a dance at the Gymkhana – quite a problem to get the gals – 60 of them – but I think we can do it.

Sorry this has to be so short but I know you'd rather have a short letter than none at all.

With all my love,

Al

[A photo taken at the garden party on 13 March, which is referred to in AWZ's letter of that date, is enclosed with this letter. The photo was published in an unknown newspaper, doubtless in Karachi. The headline of the story that happened to be below the photo is: "Will America Break Off Relations with Finland?" This shows the situation in Finland, mentioned by AWZ on 27 February, was still of concern. Wavell was blinded in his left eye in WW I, so he is rotated around to the right.]

The picture shows His Excellency the Viceroy shaking hands with Sardar Bahadur Mir Allahdad Khan Talpur at the garden party given in his honour by His Excellency the Governor of Sind on Monday.

---

[141] Owen Shoemaker (2 Jul 1860-6 Mar 1944), brother of BSZ's father, William Toy Shoemaker, M.D. He was a bookbinder, whose wife was Margaret "Greta" (nee Jack). They had five children.

[142] This brief statement is the only contemporaneous account of the departure of LCDR Smith and the assumption by AZ of responsibility as acting (or actual) CO at NLO Karachi. The expected "important visitors" that AZ mentioned in letter 55 are the "large group of officers" who were in Karachi when he wrote this letter. I suspect that the group was headed by Commodore Milton Miles. If so, they had come to evaluate the ruckus about Smith that had led to his being transferred by ONI and relieved with an Unsatisfactory Fitness Report. Miles sided with Smith and later accepted Smith to be his liaison officer in Calcutta (see Winsor file folder).

[End of Packet #1 of AWZ's letters from India. Start of Packet #2]

----

8 Apr 1944, postmarked APO 886   From AWZ/APO to BSZ   ALS-I
[no header]

Thursday April 5th

57
Dearest Barb,

No word from you since my last but I guess your trip to Chatham [Chatham Hall School, Virginia] was responsible. I hope you had a good time and it wasn't too strenuous. Travelling these days must be quite a problem. It sounded as though the trip down was pretty well taped but coming back, catching as catch can, must have been a bit difficult. No matter how bad it was it was no worse than travelling on Indian trains. I bet Warren loved every minute of it. I'm dreadfully sorry not to have seen Babs in her debut. I know she can act but she's been hiding her voice in a rain barrel as far as I'm concerned.

There was an envelope of miscellany that arrived a few days ago. It took quite a while getting here even /p.2/ though addressed A.P.O. I was very glad to get another good report of Warren's. Would some one kindly tell me when he gets all A's why he only gets "Honors" instead of "Highest Honors" and what does 1 and 2 mean? How can you get a mark of 2 supposedly meaning courtesy for "Art."

I was interested, but not very, in getting the copies of the "Half Moon Commentary." The letter from Mabs and the program from William and the letter from Ann who has a two year old brother Ralph were very touching but who the hell is Ann. Also the letter from Marie Craft was very nice. She certainly seems like a nice girl indefatiguable [sic] when it comes to letter writing. Very cute though.[143]

Our visitors have come and gone. /p.3/ Quite a treat for sleepy old _____. They gave us a chance to get some good meat, nestles chocolat bars, canned Planters peanuts etc. I had a ride in one – on the bridge – quite thrilling. The best of all was having a real American meal in the captain's cabin. Boy it certainly was good. As I was the senior officer present I felt quite important taking care of their requirements and desires. We arranged a dance on Sunday night which was declared a great success. I thought we'd be short of girls but they turned out in great style and everyone seemed to have a good time.[144]

---

[143] Mabs = Mabel Shoemaker (b. 1935), younger daughter of BSZ's brother Jesse Warren Shoemaker and his wife Catherine (née Wheelock). William = Mabs' brother (b. 1928). Ann = Ann Wood Shoemaker (b. 1937), daughter of BSZ's brother Robert Comly Shoemaker and his wife Ann "Nancy" (née Okie). Ralph = Ann's younger brother, Ralph Warren Shoemaker (b. 1941). Marie Craft, daughter of Millie the maid, is mentioned above.

[144] AWZ does not say who his "visitors" were, but they were undoubtedly from a U.S. Navy ship or ships. It was probably a small group of ships, based on the following information: AWZ had to find 60 "girls" to entertain the crew, and he was the senior officer present in the wardroom when he was the guest of the captain (i.e., the CO of the ship), which means the skipper was probably a Lieutenant, junior grade, or at most a Lieutenant who was junior in date of rank to AWZ. He also writes of a "thrilling" ride in "one of them," which surely means that there was more than one, and that they were fast-runners. AWZ also, for the first time in any letter, deliberately avoided naming the place where he was located – indicating the location only by an underline (____), which was intended to satisfy the censor. Putting all of this together, I believe he played host to a flotilla of four or five PT boats. About 800 PT's were put into service in WW II, and they ranged widely over the Pacific and Atlantic Oceans. I have not found any reference to the transit of a group of this size through Karachi in March 1944, but it is conceivable that they were en route to England to participate in "Overlord," the invasion of France that occurred on 6 June 1944. PT boats had crews of 12-17, so a group of 60 might be the men from the crews of 4 or 5 boats who could be allowed to go ashore for a night of "liberty."

LCDR John D. Bulkeley, who had won the Medal of Honor as CO of PT-41 in the Philippines, was in command of PT Squadron 102 in Australia in November 1943, when he was sent to England to organize and command PT boats for the OSS in preparation for the invasion of Europe. Bulkeley took his boat across the Atlantic, but other PT boats may later have gone west from Australia, across the Indian Ocean, the Suez Canal, and the Mediterranean Sea. They would thus avoid the perilous crossing of the Atlantic Ocean, which was infested with

After the excitement was over another job fell into my lap – this time I was the only one here – Howard [Smith] had gone to Delhi. This one was not /p.4/ so pleasant, similar to one we had almost a month ago.

This brings me about up to date. I hope I'll be getting a letter from you soon. It helps a lot to hear from you and how much you miss me. I think of you all the time and long for the day (or night) when I'll be back in your arms. There's no one like you for me.

   With load and heaps of love,
    Al

----

13 Apr 1944, postmarked APO 886 From AWZ/APO to WZ.   ALS-I→P
          [no header]

                       Wednesday Apr 12th

Dear Warren

Who do you suppose is at the foot of my bed – right now – while I'm writing this letter to you? It's Chippy our monkey! He evidently broke his chain – that is supposed to keep him in his tree – and came up to pay[145] a visit to me. He was very shy at first – every move I made – he would start to run away. I got into bed and started to write this letter to you, gradually he came closer and closer and finally got into my bed. I've had to stop writing – he's been getting more friendly crawling all over me, getting into my hair, always looking for what monkeys are always [shifts to pencil] looking for – but this time not finding any – /p.2/ or at least I hope not (you see my pen has run dry). After all his investigations he has now found a comfortable place at the foot of my bed and gone to sleep. He's a cute little rascal – never looks you straight in the eye like a dog – he doesn't like to be patted. He's very independent and wants to be let alone to carry out his own ideas that take him almost everywhere – under the bed – on chairs – up to the window sills and even up on my chifarobe [chifferobe]. Now he is sound asleep making himself quite at home probably a great treat for him lying on a soft bed instead of a hard limb of a tree.

I've just called Mawhin to put him back where he belongs. I don't fancy him spending the night here. /p.3/ I'm afraid he's not very well trained to live in houses.

Thank you very much for your detailed description of your trip to Chatham. You must have had a grand trip with lots of fun. You didn't have much to say about the play Babs was in except that it was very good. Weren't you proud of your big sister having such an important part? I wish I could have been there to see her.

Well, it's getting a bit late and I must write to Mummy so I'll say good-by with
   Lots and lots of love,
    Daddy

Also thanks of your letter of Feb 27th. How come it didn't get mailed till Mar 15th? So you like the Dodgers better than the Phils –Tsch Tsch!

[Editor's note: In April and May 1944, AWZ completed reports on the port and landing beaches of Karachi, which he sent to Washington. They are addressed from USNLO Karachi to CNO (DNI) – to the Chief of Naval Operations (Director of Naval Intelligence). Additional reports were sent on 1 September and 31 October 1944, and in February 1945. The report on 31 October ("Facilities of Port of Karachi, India") was considered to be so sensitive that, although it has been declassified, it was marked with the notation that a copy can be made only "with permission from the Secretary of the Navy." I found another copy of the report of the Karachi Port in the Army intelligence files, but I decided not to copy it because of the Navy's concern. The titles of these reports are listed in the Bibliography, and they can be found in NARA II, in Record Group 38]

German submarines. See Bruce M. Bachman, *An Honorable Profession: The Life and Times of One of America's Most Able Seamen. Rear Adm. John Duncan Bulkeley, USN* (New York, N.Y.: Vantage Press, 1985), 48-9.

[145] In the AWZ-Photo Album, Sheet 1, #13: Monkey sitting on a branch, with chain.

BSZ removed the photo from the envelope and put it in the Photograph Album (Sheet 1, #13). AZ wrote on the back: "Dear Warren / This is Chippy the monk. He got away one time and under one of the cars, breaking his jaw. We thought we'd have to put him out of suffering but now he's completely recovered. / Love / Daddy"

---

13 Apr 1944, postmarked APO 886   From AWZ/APO to BSZ.   ALS-P
[no header]

Wednesday Apr 12th

58

Dearest darling,

    Warren's letter with your appended note came yesterday. He certainly furnished a detailed description of your trip to Chatham. He must have had an exciting time, though the train part of it must have been quite wearing on you. I'm very proud of Babs having such an important part. Does that get her in the dramatic club or is that something else again?

    Too bad about Daddy Joe,[146] but I guess he's lucky to have lasted this long with – what was the matter with him? Also it's a good way to go – no harrowing long illness in bed. I'll try and write Flo a letter.

    The Major Gen. Ruter[?] you mentioned /p.2/ didn't happen to look me up. Generals don't seem to do that. We don't see much of the Army here and especially those just passing thru.

    How is it that Harry Johnson was drafted? Are they taking married men with children and with ulcers? Seems to me they are digging pretty deep.

    Haven't done much of anything since my last letter. Did go last Friday to a party given by a couple of British Navy blokes. It was a drinks-snacks affair – not very exciting. Outside of that life has been very dull and time drags slowly on.

    My letters must get duller & duller but there's nothing much I can do about it – there just isn't anything to say except that I miss you more and more and I feel lower & lower being away from you. Why does this damn war have to go on? They are just prolonging it to spite happy people like us.

    Loads & heaps of love,
        Al

---

[146] "Daddy Joe" = Joseph Crosby Lincoln, d. 10 March 1944 (cf. note on AWZ letter, 24 June 1943).

21 Apr 1944, postmarked APO 886   From AWZ/APO to BSZ.   ALS-I
[no header]

Thursday April 20th

59

Dearest Barb,

    Two letters dated Mar 19th & March 26th and the pictures of the children (2 Kodachromes) arrived yesterday and today. Quite nice after not having heard from you for quite a time. You started numbering your letters again and then promptly stopped with your letter of the 26th so I don't know if there is one in between or not. I love the pictures of the children. I'll put them in a nice bright window. I'm glad Sandy Sims is home – must give Joe & Nancy quite a thrill – he's quite a hero. I finally got a letter from Tat yesterday from San Fran. It sounds as though they were enjoying life /p.2/ and were quite impressed with your article in Vogue. Your fame is nationwide. I hope you've sent me the finished product so I can proudly show it to my erstwhile friends.[147]

    Joe Eastwick wrote me a very nice note as also Hecker and Ellie Severn. Dick Kelley sent me the results of J Z Sons for 43. They at least made the dividend which I think is very good.[148]

    I'm thrilled with what you say about Babs. I guess she really is talented. Warren must have been a delightful little traveler on the trip. A well mannered polite boy is so much nicer than otherwise.

    Do with the stones what you'd like. I think the Petri idea is a very good one. I'm rather disappointed in the valuation you got on the /p.3/ sapphire. I thought it would be at least 3 times what you said – that's what I paid for it – but that shows how much you can rely on these people out here. They asked twice what I paid for it and gave me quite a time whittling them down. Anyway it's from Ceylon. They can't take that away.

    I'm awfully glad you're planning to go to Easthampton. It sounds like a grand idea and don't be afraid to spend money to make yourself comfortable.

    I hope you told Peachy and Ann Carter Green to look me up or drop me a line. If they arrive in some other port, I wouldn't know when they would get here as our courier service is not being run from here anymore and we are not in as /p.4/ close touch with the other offices[149] as before. ~~It is now ten days after Lanie's birthday.~~ Peachy should be here by now from what you said but I haven't heard anything. I'd love to see her.[150]

---

[147] Sandy, Joe, and Nancy Sims = Joseph Patterson Sims, a Second Bass in the Orpheus Club and a noted artist and editor; Sandy is probably John Clark Sims Jr., an Orpheus Club member on active duty, who must be related to Joseph; and Nancy is probably the wife of Joe or Sandy.
Tat = H. Tatnall Brown Jr., an Orpheus Club member on active duty.

[148] Dick Kelley = Richard Carlyle Kelley (1882-1976), husband of AWZ's older sister Anna.

    J Z Sons = John Zimmermann's Sons, the carpet and fabric manufacturing firm that was established by AWZ's father, John Zimmermann Sr.

[149] By "offices," AWZ refers to the intelligence coordinators in Delhi (where his superior officer in the Navy is based) and Ceylon (Joint Intelligence Center South East Asia Command [JICSEAC]).

[150] Both "Peachy" and Anne Carter Green are civilians, destined to work at SEAC HQ in Kandy, Ceylon. Ann Carter Green appears later, as a Red Cross Worker (see 30 June 1944). Peachy appears in the AWZ Photo Album (AWZ-PA, group 7, #14), relaxing with two Naval officers on a verandah in Kandy. Many other young women worked in Kandy, and it was both an exotic and romantic place. Julia Carolyn McWilliams (1912-2004), Smith College '34, worked for the OSS at Kandy. She had worked directly for William Donovan, head of the OSS, in Washington. "In 1944 she was posted to Kandy, Ceylon (now Sri Lanka), where her responsibilities included 'registering, cataloging and channeling a great volume of highly classified communications' for the OSS's clandestine stations in Asia." While in Ceylon, she met Paul Child; they later both worked for the OSS in China and they were married in Pennsylvania in 1946. AWZ probably met her in Kandy, but he could not have known of her future fame as the author of cookbooks on French cooking and television star. AWZ did not mention her in his correspondence.

There's not much new in my life. We still seem to get a flow of visitors and with only three of us we don't get much chance to do anything but stick close to here. I went out to Sandspit on Sunday which was very nice but outside of that nothing.

Two Pan Am pilots, Jack Kuntz and a boy by the name of Mathews, were here for dinner tonight – one a close friend of Paulus Browning.[151] They are coming back and would be very glad to bring some knick-knacks that you might want to send like golf balls – tennis balls* etc. You could send them to Lt. Paulus Browning.

            care John J. Kuntz
            3045 Indiana St
            Miami 33 Florida

* The new kind made of "Buna" synthetic rubber – not the reworked ones. /p.5/
Powell is stationed in Miami and would see that Kuntz or some other pilot would bring out the package.

I was glad to hear Brownie was still in the Navy. I thought they might board him out on account of his eye. His sister, I learn, is practically an invalid with some strange paralytic disease and is staying with Mrs. Kuntz.

Good bye, sweetheart. I hope your life isn't too miserable. All I can do about it is to send you my love with a big hug and a kiss and hope that they will soon be delivered personally.
    Always devotedly,
      Al
Why did you switch to M & D? I think it's much better to stick to APO.

[Enclosed without comment, but apparently to be humorous: an undated newspaper clipping, probably from a Karachi paper, showing a photo of King Victor Emmanuel of Italy, with an apology from the editor for having run a picture of King Gustav of Sweden instead of the Italian king the previous day.]

---

24 Apr 1944, postmarked APO 886 From AWZ/APO to BSZ.   ALS-I
          [no header]

                             Tuesday Apr 23rd
#60
Dearest Barb,
  The third letter you wrote in a week came yesterday. It took just a month getting here. It was thrilling to get such a flow of mail from you especially when there was a dirth before it.

I will continue to write up my NWFP trip but I'm afraid it's going to take a long time for me to finish it. There just isn't the spare time and letter writing consumes most of what there is.[152]

It's been only three days since I wrote you so there's not much news. Friday I played bridge at the Blackwells (an older couple). My partner /p.2/ was a Mrs. Pettigrew whose husband is a colonel in the British Army. In a bridge evening here you arrive about eight – sit down immediately to play bridge – eat at nine-thirtyish and continue after dinner. It was fun and I managed to come out ahead.

---

[151] After the war, AWZ got to know the founder of Pan Am, Juan Trippe, in Easthampton, where Trippe and the Zimmermanns had summer homes. Juan Trippe's son and Warren Zimmermann were friends at Yale. With mutual friends such as appear in this letter, their conversation probably included recollections of AWZ's wartime experiences in India, his many trips on military aircraft, and of his memorable trip on the North-West Frontier.
[152] It is clear that AWZ was trying hard to finish a publishable story about this trip before he returned to America. I am indebted to him for his careful notes, several drafts of a trip narrative, the legends that he inscribed on the backs of the photos, and for the permission he obtained from the ONI and the British in Peshawar (Bromhead) to publish the photos and a story about them. (Permissions from Curt Winsor at ONI and from Bromhead are in their respective file folders in the AWZ Papers, and in Chapter 6 of this book.)

Yesterday I went to the races. After losing on every one of the first five I picked the sixth and got even. Had two 10 rupee tickets on a five to one shot in the last race that came in second – but there was an objection and it was sustained – so I won and that was fun.

You say you'd like more ivory. What about the salt & peppers I sent? Should I send more or would you like more gods, elephants, etc? What should I get Babs? You've seen practically all the good stuff from here.

Heaps & heaps of love,
Al

[Enclosed, a clipping from *The [Karachi] Sind Observer* (12 April 1944), entitled "Heavy Congestion in Karachi": The housing problem in Karachi is becoming acuter day by day. . ."]

----

7 May 1944, postmarked APO 886   From AWZ/APO to BSZ.   ALS-I
[no header]

Thursday May 4th

#61
Dearest darling Barb,
This is true devotion. It's late and I'm very tired but I must write you as it's been a week since I last wrote and I've never let more than a week go by without writing. A couple of transient officers are staying with us. Howard (Voorhees) and Phil (Halla) invited Betty Greatbatch and Patty somebody or other for dinner and now they've all gone to the Gymkhana to the regular Thursday night dance.

Two letters have come since I last wrote – Apr 5th & 12th – and both were very nice. Please don't worry about paying $1000 for a cottage at Easthampton. I don't think it's much for the whole summer and anyway the money is there to spend and I can think of no better way to spend it. I hope you all have a wonderful summer and I wish like Hell I could be with you. The whole /p.2/ thing sounds as though it would work out very well with Kay & Jim [Oram] and Marie [Craft, the maid] living in the house while you're away.

By the way while I think of it Babs seems to be completely sold on Smith. I think it would be a great mistake to try to influence her otherwise. She's old enough to make a choice of this kind and from the little I know Smith rates as high if not higher than Wellesley any day.

Now to answer some of your questions. The present C.O. [LCDR Smith] set up this place but evidently didn't do a very good job of it, or rather he may have done too good a job. Anyway, it looks as though we will stay here with some slight changes in the lease, number of servants etc.

I'm glad to hear Warren is filling out a bit. He's the only one in the /p.3/ [sheet 2] family that can stand a little flesh on his ribs. He has the build of an Indian boy. They also don't stand around long enough to put on fat. I hope Albie catches some of Warren's incentive to sport and the great outdoors. Does he still think the sun is too hot for a small boy? He'd better not come to India. He'd be indoors all day.

I'll try to get Babs a graduation present but it's awfully hard. You've got the best that's sure, but I'll keep trying and might run across something I think she'll like.

Yes, I still see Arlo Bond. He went with us last Saturday to Bahadur on Minora Island to a regatta of the Indian Naval Training Schools here. There are three of them for Indian boys and this was their annual rowing competition. In between there was a diving exhibition by a guy who was a comic diver in the Aquacade, now an instructor /p.4/ in one of the schools.

Shortly after you get this you'll be getting a box of miscellany, among which are some slippers for you and the girls and sandals for the boys. Of course the sizes are mere guesses but if you want more send along your foot print. They are quite cheap – all the same price $2.25 a pair.

Did you ever give Sally Winsor anything out of what I sent home? She sent me a very nice Fluke's package at Christmas and if you haven't already given her something I'll try to find a suitable recompense. It was darn nice of her to think of me.

Getting to your second letter, what made you decide to have Albie's tonsils out? Did they flare up all of a sudden? Too bad he had to go thru the experience but we all have to have disagreeable /p.5/ times in this life. How about Lanie's eye?

I hope you were able to go to Palm Beach with Natalie. That should have been fun. I shouldn't worry a bit about being a Mrs. Lewis.

So you're inviting me for dinner when you're going to break out the ivory elephants. I'll come but you'll have to arrange my transportation. I'm afraid that's the difficult part about it. You ask if I do think of you a lot. How silly. Of course I do. I'm thinking of you practically the whole time and longing for the day we'll be back together again. It's almost a whole year now since I've seen you and I've never spent a longer one. They'll have to arrange to have this war over damn soon.

If you could arrange to have /p.6/ some pyjamas sent out (2 pr) it would be very nice. The ones here are pretty bad. Don't bother getting new ones. I must have some at home which would do. Perhaps Powell Browning could have a Pan Am pilot bring them out.

Good by dearest. I miss you to beat Hell.

Loads & heaps of love as always,
Al

---

10 May 1944, postmarked APO 886　　From AWZ/APO to BSZ　　Rec'd May 19　　ALS-I
[no header]

Tuesday May 9th

#62
Dearest darling,

Two letters have come from you since I last wrote – not because of the way you sent them but because of their speed in getting here. The one you sent from Palm Beach took 21 days (Apr 17) and the one you sent from home (Apr 26) 12 days almost a record.

I'm awfully glad you went to Palm Beach and I hope you had a marvellous time. You certainly deserve it. Living in a pucca [pukka] (Indian word) bungalow (also Indian) in a place like P.B. must have been a great change from that **** place that we stayed in on our honeymoon. It must be much nicer being in the "swim" than feeling like a couple of rank outsiders the /p.2/ way we did.

Please don't worry about financial matters. Do what you want to do – pay the bills as they come in and don't worry if you have to sell something or if it comes from capital account. Make sure it comes from my account as my theory all the way through is to build up your and the children's estates. If worse comes to worst we can use their income to keep them going but, in the meantime, use my income and principal. 'Nuf said – now don't worry a bit – I never expected to end this war with more money than I started and there's enough to keep me going and doing the same as I've been doing the rest of my life.

Sorry you didn't get any fun playing roulette at Bradleys. Did you play our old numbers? They had a benefit here for the Greeks and I had about the same luck playing roulette as you did. Funny – it must have been about the same time. I lost about $25 but it all went to the dear old Greeks.

Too bad you missed Swabby [Swabe]. I think he would have come to see you and I'm quite sure you would have enjoyed him. His base now is somewhere in Canada but he seems to find time, between trips, to go down to New York. He probably phoned from there. He most probably will be back and I hope he will call.

Why don't you give up these extra /p.4/ duties you've taken on – like the Monday canteen etc? All they do is give you a lot of worry and trouble. Why not let these officious women run it? You have enough to do bringing up a family of four children in these difficult times – you with low blood pressure and an ulcer. I don't want you to get run down and it looks as though you were doing too darn much.

I was thrilled with Babs report, Albie taking to the piano, Lanie still plugging away at it and Warren making the "Varsity" on the nine year olds. I'm very, very proud of my family. I want to keep hearing about what they are all doing. If Warren needs a first-baseman's glove, get him one for my birthday. You can't play a /p.5/ good game at first base without a pucca mit.

You don't have to worry about my wishing I were home. There's nothing I'm wishing more. I'm not moving to the Sind Club. Evidently we are to carry on here but they seem to be taking a hellava long time telling us how we are to do it.

I wrote you on Thursday – this is Tuesday – nothing much has happened in between. I played golf late Sunday afternoon with Doris Pettigrew who promptly beat the pants off me.

I'll try and drop Babs a line or two. She wrote me a very sweet letter. So good by dearest, I love you and can't wait to hold you in my arms again.

        Heaps & heaps of love,
         Al

---

18 May 1944, postmarked APO 886 From AWZ/APO to BSZ   ALS-I
         [no header]
                              Thursday May 18th

[unnumbered 63]
Dearest darling,

Your letter of May 1st arrived yesterday. I'm thrilled to hear Albie is interested in the piano and seems to be so talented. Jack Ott even wrote that he is good so I guess he must be. By the way, what in the world is Jack going to California for? Sounds like an awfully funny place to go in these days of difficult travel unless he has something very urgent to take him. I can't conceive of what that urgency could be.

The trip to Jenkintown must have been a lot of fun. My mouth certainly waters when I think of those nice juicy steaks cooked outside – to say nothing /p.2/ about the other things – pukka tomatoes, corn, pretzels, noodle soup etc. etc. that you must be having.

I'm sorry I'm not there to enjoy the daffodils & mertensia that you planted for my homecoming but it's not my fault I'm not there. I'd be home in a minute if they'll let me. Boy! Would I?

With only three officers here, we don't get much of a chance to step out much but I am able to get tennis and some golf in in the late afternoon.

Government House is having an "At Home" on Monday. It's rumored that Noel Coward is going to be there.[153]

On Wednesday I am having the military secretary (Monty Smythe) and the A.D.C. (Roger Collett) to the governor with wives here for dinner – then to our Officers Club to dance. They have a fairly good orchestra and dancing is on a roof, under the stars. Maybe Noel Coward will join the party.

We're getting into some pretty hot weather. The end of May and June are supposed to be the hottest times. So far it's just been disagreeable in the middle of the day – the nights fairly cool.

What about Nanie? Is she going to Chatham next year or what? You haven't said whether she's been entered or not. /p.4/

I hope I'm able to get something for Babs before graduation. So far I've had no luck but the craft shop say they are getting some jewelry in from Nepal this week that should fit the bill. Nepal, as you will see on the map, is the other side of India and is the "real" Shangri-la and produces or has produced some very nice workmanship in jewelry. It will be old not like the cheap looking new stuff you find around here.

Good-bye, dearest, don't think I'm liking this separation any more than you. I <u>want you</u>.
    Heaps of love as always,
        Al

Got a very nice letter from Jack Beard today.
[A newspaper clipping is enclosed without comment, dateline Nagpur, March 25, headlined: "Monkey Killed and Taken Out in Procession." It tells of a violent "red-faced monkey who after being killed for its menacing activities" was taken in a procession accompanied by incandescent lights and a band.]

---

[153] Noel Coward did not come to this event.

23 May 1944, postmarked APO 886   From AWZ/APO to BSZ   Rec'd June 7   ALS-I
[no header]

Sunday May 21st

[unnumbered 64]
Dearest Barb

Just a note to tell you I'll soon be the oldest living U.S. Naval Officer here. Our C.O. has his orders to go to Calcutta leaving me in charge. I hope I'll be able to run the place to the Navy's satisfaction. They have made a very satisfactory arrangement, financially, in having us (Voorhees & myself plus a Coast Guard officer [Halla]) stay here. We really don't need more officers than we have but it looked as though it was going to be pretty expensive to run this place with so few.

I got Babs' present off /p.2/ yesterday. I hope she gets it in time and likes it.

Jack [Ott] evidently is going to California on business but personally I shouldn't think it worth while.

We are having a rather large farewell party for our departing C.O. this evening so I think I'd better get going on preparations etc.

I still miss you to beat Hell, and can't wait till they get this thing over with. Maybe by the time you get this they will have taken more constructive steps to reach that goal.

Heaps & heaps of love,
Al

---

23 May 1944, postmarked APO 886   From AWZ/APO to BWZ   ALS-I
[no header]

Sunday May 21st

[unnumbered]
Dear Babs

Your letter of April 30th came the day before yesterday. In it you said graduation was to be May 30th. I didn't realize it would come so soon. Graduations usually come in June. Anyway I had some jewelry picked out and they were making a handbag to match. In as much as the time was so short I stopped them making the bag and am sending along the frame. You can have a bag made at home. It will probably [be] a lot better made anyway.

The jewelry comes from Nepal. As you're such an expert on India you'll know it's an independent state in northern /p.2/ India in the heart of the Himalaya mountains, probably the exact spot of Shangri-la. You don't have to put on all the pieces at once and look like a Christmas tree but there are combinations to vary your costume. I hope you'll like them. I'm sending them to Chatham, hoping they'll get there before graduation. If not I hope the school will forward them to you.

A funny thing happened on Saturday when I went to get them from the shop. They were spread out on a table. Lady Dow, the governor of Sind Province's wife, came in the shop and greatly admired them. She didn't /p.3/ realize I had bought them and thought she would buy them right from under my nose. She was very disappointed when she said "I'll take these" to find they were already sold to me.

I was amazed at what you said about Warren when you took him to see "Pride of the Yankees." He's so enthusiastic and gets so excited about things he's interested in.

I'm awfully glad you are ending your carreer [sic] at Chatham with an "A" rating. That is very commendable.

Good bye Babs and many congratulations on your graduation.

Lots of love,
Daddy

---

30 May 1944, postmarked APO 886   From AWZ/APO to BSZ.   ALS-I
[no header]

Monday May 29th

[unnumbered 65]
Dearest Barb

I have just seen the March Vogue with your article in it. I think its swell and I'm very proud of you. Ruth Geldard, an American who has married a Britisher in the British Overseas Airways, subscribes and let me borrow her copy. I've proudly showed it to the boys here and will show it to my friends. They will think I'm hot stuff having a wife clever enough to have an article published in a world famous magazine.

No letters since I last wrote but an EFM came day before yesterday to which I immediately replied. I can't understand /p.2/ why there should be such a gap in my mail. I've written at least once a week to you and write to one of the children about once a week.

I'm as well as can be expected. All right physically but awfully fed up on being away from you for so long. I was glad to hear everybody was "well at home." The EFM took long enough getting here. You sent it, evidently, on the 5th of May. I'm sorry you were worried and hope the delayed letters have long since arrived.

Our office is a mere skeleton of its former self. Just two of us and three enlisted men /p.3/ compared with almost three times that at one stage of the game. Besides, the native help has been reduced considerably. We don't get as many transients as we did which suits us as they were pretty much of a bother and nuisance to meet and put on planes besides our furnishing an information bureau service for them during their stay.

Noel Coward didn't show up so I guess I won't meet him. I think he has gone to Ceylon.[154]

Yesterday I went to Sandspit with Doris Pettigrew and her children. She is very nice, quite attractive, plays golf, tennis and bridge. She's been /p.4/ busy having a baby (about five months ago) and has two other children, boys five & four that are very cute. Her husband is a colonel in the Baluch Regiment of the Indian Army and is stationed at Dera Khan.[155]

Saturday, we had a party here with some of the people we met at the Government House party. The Smythes were here, the ADC to the governor, the Bartons (newly arrived – he's head of the Marine Airport) an American Major Karl Word[?] and "stranded" wives to make 14 in all. We had dinner, then went over to the American Officers /p.5/ Club to dance, then back here again for a night cap. I guess it was a success as everybody seemed to have a good time.[156]

Last week the liquor I ordered from America arrived so we are well stocked and can do what entertaining we've a mind to without having to depend on our ration.

All this must sound pretty gay to you and I'm sorry to appear to be having such a "good" time. While it is diverting and I can't deny some of it is fun, I miss you dreadfully /p.6/ and only wish you were here to enjoy the parts that are fun.

Babs graduates tomorrow. I'm awfully sorry to miss it. They just can't believe I have a daughter old enough to be getting thru school. I'm very proud of my children and talk about them a lot to the various people I've met.

I haven't heard from you for about 10 days. I hope something will be forthcoming in tomorrow's mail.

    Lots & lots of love, dearest,
     Al

---

[154] Noel Coward would have been going to SEAC HQ, the office headed by Admiral Lord Louis Mountbatten at Kandy.

[155] Dera Khan = Probably Dera Ismail Khan (aka Dera I.K. or D.I.K.), now in Punjab, Pakistan. It is about 45 miles east of the border of Baluchistan, and it is the last major city in Punjab on the route between Karachi and Quetta, capital of Baluchistan. On the "Sketch Map of the North West Frontier of India" in AZ's papers, Dera Ismail Khan is seen on the west (right) bank of the Indus River on the border between Baluchistan and Punjab, and a smaller town, Darya Khan, was in Punjab, on the opposite side of the river. Alternatively, although less likely, this could be Dera Ghazi Khan, which is further downstream on the right bank of the Indus River.

[156] One of the photos in AWZ's collection is of a party that could be this one, labeled on the back with all of those in attendance.

---

5 June 1944, postmarked APO 886　　　From AWZ/APO to WZ.　　　　　　　　ALS-I
　　　　　　　　　　　　　　　[no header]

Sunday June 4th

[unnumbered]
Dear Warren

　　I received your letter written May 1st yesterday. It took over a month getting here. You addressed it to Mail & Dispatch Sect. etc. It would be better if you used the Army Post Office address here APO 886. Mail seems to reach me a lot sooner that way.

　　I'm sorry my guess on your foot size was wrong and the sandals didn't fit you. When I get your foot size I'll have a pair made and send them along. Unfortunately I can only send one 10 oz package a month. It's a new ruling starting the first of June.

　　I'm glad you talked to Flight /p.2/ Lieutenant Swabby. Too bad Mummy wasn't home. I believe he would have come over to Haverford if she had been. You and he would have gotten along very well together. He pilots B-24s and has been all over and has many interesting tales to tell. Maybe next time you'll see him.

　　Since when have you been collecting autographed pictures of movie stars? How is it you have all girls and no men? Are the girls more to your taste?

　　You said you [made] the first team. Do you mean of your class? I'm glad you like first base. I used to play it up at Camp Tecumseh. I hope you like your glove and it's a good one. It makes quite a difference.

　　　　　　　　　　　　Lots of love,
　　　　　　　　　　　　Daddy

[Enclosed: Postcard showing "Emperor's Humayon's Tomb – Delhi / Built by his wife in 1555 A.D." AWZ wrote on the back: "Dear Warren / This is what some of the Indian buildings look like. I haven't seen it yet but maybe I will some day."]

---

5 June 1944, postmarked APO 886.　　　From AWZ/APO to BSZ.　　　　　　　　ALS-I
　　　　　　　　　　　　　　　[no header]

Sunday June 4th

#67 [Is one missing?]
Dearest Barb,

　　Your #3 came yesterday took only 12 days and beat #2 which has not yet arrived. #1 arrived on Thursday taking 25 days. So you see how screwy the mail situation is.

　　I'm sorry you think I'm not writing enough. I seem to be spending an awfull lot of time writing and now you want me to spend more time. I could write more often but I'm afraid my letters would be duller than they are now. Anyway I'll try it and see what you think.

　　The bill for washing the walls sounds exhorbitant [sic]. $28 a day per man for that kind of work sounds like robbery.

　　Awfully sorry to hear about /p.2/ Martha Ziesing. How long has she had arthritis? Does it mean she's going to be laid up for a while? It must be an awful blow. She gets such fun out of life and loves to be active.

　　I wish you'd go into more detail about the pictures I've sent. I don't know how many you've received and what you particularly enjoy seeing – what phase of this crazy life etc. I still have four rolls of color [Kodachrome] that I haven't been able to exchange for magazine load as yet, but I've been promised some by the shop here.

　　In yesterday's letter you sounded awfully depressed – more than usual. Last night /p.3/ I was on a party – given by Dr. & Mrs. Reinitz – and I wanted to come home and write to you but I'm afraid it would have hurt their feelings. I've turned down a couple of their parties and I think they are a bit sensitive, but I did want to reply to your letter as soon as I could. You were on my mind the whole evening.

I'm disturbed to hear Ted is down in the dumps again. What seems to be his trouble? Is it his physical condition that is worrying him? I hope the outlook is better than outlined by Dr. D'Elseux. Do they need any financial help? I wrote to Ted about a week or so ago /p.4/ in reply to a long letter I had received from him. He didn't say anything had been troubling him but maybe this is a recent development. I don't quite remember Dr. D'Elseaux & his wife at Warren's. There were quite a few people there that night.[157]

When did you take on the BM hospital cafeteria? Don't you think you have enough to do already? I certainly think so. Don't get yourself all worn out.

Beginning June 1st, we are limited to one package home a month not to exceed 10 oz., so don't expect much from now on. I'm glad I've sent what I have – from now on it looks as though the pickins will be /p.5/ pretty small. However send along the shoe sizes and I'll see what I can do.

I'm very tickled with Albie's accomplishments. I hope he keeps it up.

I did like your poem and sorry I didn't mention it in a letter. You seem to be seeing quite a lot of Natalie these days – lunches, theatre etc. I guess she must be pretty lonely too and I'm glad you get together with her and do things.

I'm afraid Ed Lord's going to be a bit disallusioned with his outfit. The stories we get here are not too encouraging but maybe he's not headed this way. You've got to hand it to him. He certainly is sincere and full /p.6/ of spunk.

Friday I played bridge at John Blackwell's – His wife is away in Kashmir. The other two were Doris Pettigrew and Vere Birdwood.[158] It was good fun and I enjoyed it. John is #1 at Burmah Shell and

---

[157] Ted = Theodore Shoemaker (1889-1962), older brother of BSZ and thus AWZ's brother-in-law. He had serious clinical depression, and also suffered from tuberculosis. His wife, Mary Louise Beiderbecke, was the sister of the great jazz musician, Bix Beiderbecke, who became an alcoholic and died of pneumonia in 1931, when he was very young. Warren = J. Warren Shoemaker, brother of Ted and Barbara (BSZ). In the next paragraph, BM refers to Bryn Mawr Hospital.

[158] Vere Birdwood = Lady Elizabeth Vere Drummond Ogilvie Birdwood, later Baroness Birdwood. AWZ does not mention this, and he may not have known it: her mother-in-law was the aunt of AWZ's companion on the NEFP trip, Sir Benjamin Gonville Bromhead; and her husband, Christopher Birdwood, was in the Waziristan Campaign with Bromhead in 1919-1920.

She was the daughter of a knight and wife of Lt.-Col. Christopher Bromhead Birdwood, who was then the heir presumptive to the Baronetcy of Birdwood. He was the son of Field Marshal William Birdwood, 1st Baron Birdwood, and Janetta Bromhead, daughter of the 4th Baronet Bromhead. Janetta Bromhead was the older sister of Benjamin Parnell Bromhead, father of Benjamin Gonville Bromhead, later 5th Baronet, who was a companion of AWZ and Gordon Enders on the NWFP trip in November-December 1943. Vere Ogilvie Birdwood became Baroness Birdwood in 1951 upon the death of her father-in-law. She and Lt.-Col. Birdwood had two children (Sonia, b. 1933, and Mark, b. 1938). She was divorced two years later and Baron Birdwood remarried. She was invested M.V.O. in 1958 and C.V.O. in 1972. She died 1 May 1997. On 6-23-11, on the display of Lt.-Col. Birdwood's medals, offered for sale, it was reported that: "Vere Birdwood, whom he divorced in 1954, had a few words to say on 'marrying in to Probyn's Horse' in after years: 'The life itself was excessively boring, trivial, claustrophobic, confined and totally male orientated. The army wife was not expected to do anything or be anything except a decorative chattel or appendage of her husband. Nothing else was required of her whatsoever.' She further recalled the 'very strong unwritten law that regimental officers could have little affairs with wives of other regiments, but to do so with a wife in your own regiment was much frowned up. So strongly was this law obeyed that in a Frontier station, when the husband was away campaigning, it was generally considered wise for the wife left behind to have a young officer to sleep overnight in the bungalow as a guard. As far as I know this privilege, if you can call it that, was never abused'."

Field Marshal William Riddell Birdwood, 1st Baron Birdwood, GCB, GCSI, GCMG, GCVO, GBE, CIE, DSO (13 September 1865 – 17 May 1951) was a First World War British general who is best known as the commander of the Australian and New Zealand Army Corps (ANZAC) during the Gallipoli Campaign in 1915. . . . He commanded the Northern Army in India until 1925, when he was promoted to field marshal and made Commander-in-Chief of the British Indian Army, which he remained until 1930. . . . In 1938 he was raised to the peerage in recognition of his wartime service as Baron Birdwood, of Anzac and of Totnes in the County of Devon . .

has a very nice house and serves a delicious dinner. The people that have been around here a while certainly know how to get the best food available and have the best cooks.

Tonight there is a cocktail party for the retiring naval officer in charge – Capt. Bayfield – so Howard and I are going. And so it goes.

Good bye, dearest, don't worry and don't get too depressed.

Heaps of hugs and kisses,
Al

[marginal note on p.5: When do you go to Easthampton & when do you return?]

----

8 June 1944, postmarked APO 886   From AWZ/APO to BSZ   Rec'd June 15   ALS-I
                                      [no header]

Wednesday June 7th [159]

#68
Dearest Barb,

Your #2 arrived today taking practically a month. I don't know where you get this stuff about writing twice a week. Your last three letters are dated the 4th (May) the 12th and the 20th and you complain about my writing when I write at least once and sometimes twice a week.

Too bad you had such a fall at the Otts. You must have been awfully uncomfortable for a while but by now you must be all better.

I'm awfully sorry to hear about Martha. That must be a grim outlook to be in bed for so long /p.2/ and then never be right again. I'll try to write her a letter one of these days.

What is the matter with Chatham Hall as far as Lanie is concerned? Why doesn't she want to go? I would think she would be thrilled at the opportunity. Don't tell me she still has some crazy idea about going to Chicago W. Of course, I don't think she'd be making any mistake in finishing up at Baldwins.[160]

Jack Kuntz arrived yesterday bringing the package with him. I tried the tennis balls yesterday and they are swell – just right for these courts. The lotion, handkerchiefs and film are very much appreciated. Thanks awfully /p.3/ for your trouble. I couldn't be more comfortable now as far as material things go. Jack also brought some canned peanuts which were very welcome. He didn't mind bringing the package a bit and I gather I got it in less than a month from the time you sent it. I'm sorry you haven't sent your foot prints as he would gladly take back a package for you. I think I'll buy some sandals tomorrow anyway and send them back. If they don't fit any of you, no doubt they'll fit somebody and I hate to miss the opportunity due to our recent limitation.

. Birdwood died at Hampton Court Palace on 17 May 1951 and was buried at Twickenham Cemetery with full military honours."

From ThePeerage.com. (accessed 2/10/11): Children of Field Marshal William Riddell Birdwood, 1st Baron Birdwood and Janetta Hope Gonville Bromhead: Hon. Constance Jean Gonville Birdwood+ d. 19 Oct 1975; Hon. Judith Horatia Birdwood; and Lt.-Col. Christopher Bromhead Birdwood, 2nd Baron Birdwood+ b. 22 May 1899, d. 5 Jan 1962. He fought in the Waziristan Campaign between 1919 and 1920 [his cousin Sir Benjamin Bromhead was there, too]. Children of Lt.-Col. Christopher Bromhead Birdwood, 2nd Baron Birdwood and Elizabeth Vere Drummond Ogilvie: Hon. Sonia Gina Ogilvie Birdwood+ b. 25 Nov 1933, d. 2 Sep 2006; and Mark William Ogilvie Birdwood, 3rd Baron Birdwood+ b. 22 Nov 1938. He and Elizabeth Vere Drummond Ogilvie were divorced in 1953.

[159] This is the first letter written by AWZ after 6 June, when Operation Overlord began with the invasion of Normandy – so-called "D-Day." He mentions "the recent news" in his last sentence, which he hopes "means the war will end soon," suggesting that by the time she receives this letter, more will be known.

[160] Baldwins = The Baldwin School, Bryn Mawr, Pa. Babs and Lanie both went there as day students until they finished 10th grade. They then spent the next two years at Chatham Hall School, Chatham, Va. Babs graduated from Chatham Hall in 1944 and then matriculated at Smith College. Lanie graduated in 1946 and she, too, then went to Smith.

You haven't said much about playing golf so I don't suppose you've used the golf balls I had /p.4/ left in the wine closet. I think there were about a dozen and a half. Save some for yourself for this summer and if there are any above that would you send them via the same method. You can tell Jack (who always is so anxious to make sacrifices for business connection) that he would make a great hit with Bill Cullen (Bombay Co.) if he sent him some golf balls as he is very keen.

I think Jack Kunz would be very glad to do it anyway but doubly so, as we have rescued him from the Army's Killarney Hotel and put him and his flight engineer up at our mess and he is indeed grateful. They have no water at all over /p.5/ there at the present moment for anything but drinking.

Sunday night we went to the Naval party at the British Officers Club to say farewell to Capt. Bayfield (NOIC – Naval Officer in Charge). Afterwards we had about ten officers back here for a ham sandwich (we get hams from our ships that come in here). A good time was had by all and we were invited to lunch aboard one of the British ships which we did yesterday. It was very enjoyable. Most of the people around here, as the captain of this ship, have been knocking about this part of the world for years and have many interesting stories to tell. /p.6/ He has been in Burma (Rangoon) as a harbor pilot for a number of years.

Monday I played bridge with the A.D.C. to the Fortress Commander, Eric Laminou, Doris Pettigrew and Petsa Denderius[?] whose husband is connected with Ralli Brothers, a trading firm. Last night I played bridge again with Janah Yorke-Torr, Col and Mrs. Cobbet and tonight I rest, thank goodness. I find it very difficult to turn down invitation to these various things as they usually put it in your name a night when you <u>can</u> play after you turn down one or two suggested nights.

Well it's getting late. I've tried to go more in detail about the people / writes in margin / I see as you requested but I shouldn't think you'd be very interested but here they are. Good night, dearest. I hope & pray the recent news means the war will end soon and we'll be back together again. Heaps & heaps of love, Al

----

15 June 1944, postmarked APO 886   From AWZ/APO to BSZ   Rec'd June 21   ALS-I
                                                     [no header]

Wednesday June 7th

#69
Dearest Barb

Your #4 arrived several days ago. I'm distressed to hear you had such a bad time on your trip to Chatham. It must have been harrowing to feel so punk when there was so much to do and so many people to meet. I hope by now everything is all right and you're back to normal again.

Too bad Ted is in such a state. A sanitarium, I think, would be the best place for him. He'll have a disagreeable time for a while but should respond to the modern treatment of such cases. Bill Z. seems quite all right now after his spell. There's /p.2/ no reason why Ted can't pull himself around in the same way. The main thing is cooperation with the doctors and his own will to get well, which I'm sure he has.[161]

I got a letter from Babs yesterday telling about the activities at graduation. She seemed to be quite thrilled and justly so. I'm very proud of her. I'm glad my present arrived in time. It evidently just about made it. What did you think of it?

So Albie is really knuckling down to his piano work and playing real pieces. What about Warren and Lanie and their piano lessons? Have they stopped? Also I'd like to hear more of /p.3/ Warren's baseball carreer [sic]. How many games did they play? How many did they win etc?

This week has been just like any other – doing about the same things. Spent Sunday at Sandspit with Doris P. and her children. It's very hot and sticky know and a great relief to get to the sea shore.

---

[161] Ted = Theodore Shoemaker, BWZ's brother, who had bipolar illness at this time. He later developed tuberculosis. In spite of these serious illnesses, he developed a successful road-paving business in Massachusetts. Bill Z = AWZ's brother William (1894-1978), who had a bipolar problem, too. In spite of this, William Zimmermann had a successful career in the family carpet manufacturing business, John Zimmermann's Sons.

The water, however, was very soupy. I've never felt such warm sea water but there was a nice breeze most of the day. The sea was quite rough and dangerous for the fools that go out over their depth. From now on through the monsoon it will be that way. /p.4/ I played the usual amount of tennis and played golf once.

Last week I played bridge at the Pikes (Geoffrey). He asked me if I happened to know Harvey Pike. Isn't that the name of Jack & Louise's friend they met thru Abbott Dickson? Well anyway he's a cousin of this one and this one would be very interested in finding any news about his cousin Harvey. Geoffrey is #2 at Burma Shell and very nice.

Last night I played bridge at the Dendrinas (Greeks). He's here with Ralli Bros – importers & exporters.[162] I find my bridge stacks up pretty well. I'm a bit /p.5/ ahead of the game at the moment. The fourth was Eric Laman, A.D.C. to General Hind.

Gene Markey, a Capt in the Navy passed thru here several days ago. He's been down in Ceylon and is on his way home for a visit. I wish I could do that. Wouldn't it be swell? He's noted for being the ex husband of Heddy Lamarr and Constance Bennett.[163]

What do you think of the war news? I hope it means this damn war will be over soon. I've had enough and I know you have. I'm dying to get home – a year is an awfully long time to be away from you.

With all my love, dearest
Al

----

16 June 1944, postmarked APO 886   From AWZ/APO to BSZ

[This envelope contains only a portion of one page of a newspaper, undoubtedly the Karachi newspaper from which other clippings have been made. It is an advertisement for "Navarozi Charms" which "last for one year" and are "supplied in silver locket." It is alleged to produce all sorts of marvelous results, such as in matrimony, in court, and in personal protection for fighting forces. AWZ wrote on the bottom margin: "Just let me know what's troubling you."]

----

22 June 1944, postmarked APO 886   From AWZ/APO to BSZ   Rec'd June 28   ALS-I
                                     [no header]

---

[162] Ralli Brothers: The five Ralli brothers, Zannis a.k.a. John (1785–1859), Augustus (1792–1878), Pandia a.k.a. Zeus (1793–1865), Toumazis (1799–1858), and Eustratios (1800–84) founded Ralli Brothers, one of the most successful expatriate Greek merchant businesses of the Victorian era. . . . From 1851 Ralli Brothers started operations in India with offices in Calcutta and Bombay that specialized in jute, shellac, teelseed, turmeric, ginger, rice, saltpetre, and borax, with 4,000 clerks and 15,000 warehousemen and dockers." Rallis is now a TATA industry, and that website says Rallis opened an office in Karachi after the American Civil War, but that operations in India were shut down from 1931-1948: "In late 1931, Ralli Brothers closed down in India after 80 years of existence and the business passed to Argenti and Co., acting as agents selling on commission for all of Ralli Brothers Limited."

[163] Hedy Lamarr was the stage name of Hedwig Eva Maria Kiesler (1913-2000). She was born to Jewish parents in Austria but fled from her first husband to become a Hollywood film star. She married Gene Markey in 1939 and had one child by him; they were divorced in 1941. In that year, she and composer George Antheil invented an electronic technique that they patented in 1942 which was used to scramble communications and guide torpedoes; it has since been used in Wi-Fi communications. In 1997, the Electronic Frontier Foundation gave Lamarr an award for this contribution.

Natalie Angier, "From the Lab to the Red Carpet," *New York Times* (1 March 2011), D4: "Hedy Lamarr, the actress habitually regarded as 'that most beautiful woman in Hollywood,' was a rocket scientist on the side, inventing and patenting a torpedo guidance technique she called 'frequency hopping,' which thwarted efforts to jam the signals that kept the missiles on track. [She] complained bitterly that people would look at her face and assume there was nothing behind it." An image, apparently of Lamarr, is shown.

AWZ mistook Constance Bennett for her sister, Joan Bennett (1910-1990), who married in 1932, as his first wife, Gene Markey; they had a daughter, Melina Markey, b. 1934. Joan's sister, Constance Bennett is buried in Arlington National Cemetery beside her fifth husband, Brig. Gen. John Coulter.

Monday June 19th

[unnumbered; would be 70]
Dearest Barb

Writing every few days can't produce much news but if that's the way you want it here goes. I wrote to Babs day before yesterday, and I expect she lets you read her letters, so there can't be much in this one.

I received a letter from Jack yesterday. He said he had dinner at our house with the Van Dusens. I'm waiting to hear from you on the evening. Bill is very popular here – is the life of every party he's on. His wife Lillian left of course before I got here but everyone seems to think she is a pretty swell person. I hope you had a nice evening with them. By now you should /p.2/ know about as much as there is to know about this place. One thing, you can't believe all Van says as he loves to kid and exaggerate.

Haven't heard from you since my last letter so there's none of your questions to answer.

Howard [Voorhees] and I went to the horse races on Saturday. We left early as they were delayed for about an hour when a protest decision went against a horse heavily backed by the bazar wallas and they nearly started a riot. I find you can lose money at this track as easily as you can at home, but it is something to do and you do see some amusing specimens of humanity.

Jack said you had movies /p.3/ that night and described some of the scenes, but not one word have I heard from you of what you liked or what you didn't. I haven't seen a damn one myself and it would certainly help if I could get some idea of how they are turning out and what appeals.

Yesterday being my duty I stuck to the ship and caught up on my correspondence. Today the new Liaison Officer for Colombo arrived to assume his new duties. There have been four COs there since I've been here. Evidently it's not a very healthy spot. We've just had dinner, chatted for a while and I've taken to bed and am /p.4/ writing on a board – propped up. It's pretty darn hot. I've got the ceiling fan going but even then perspiration is rolling off. Of for a good rain storm to cool things off and settle the dust. Then I guess we'd have some mosquitoes so maybe that's not such a good idea.

That's about all for now. Let me know what your Easthampton address is and when you want me to send letters there.

Good night, dearest. I miss you very much.

With lots and lots and lots of love,
    Al

----

25 June 1944, postmarked APO 886 David Lane, East Hampton, N.Y.	From AWZ/APO to BSZ [In 6¢ airmail envelope] [no header]	Forwarded from Haverford to ALS-I

Saturday June 24

71
Dearest darling,

Well here it is one whole year away from home. A year that seems an eternity. Why does this war have to go on? There's only one way it can end but still German and her stupid satellite nations still go on – killing a lot of people, distorting everyone's lives. Japan certainly couldn't last long after Germany throws in the sponge. It all seems such a waste and so unnecessary. Besides which I want to get home to you and my family.

Your letter of June 3 arrived yesterday. Jack's letters seem to get through faster. His letter of June 5th arrived about a week ago. I'm glad you're feeling better. You must have had a grim time on your /p.2/ trip to Chatham.

You must have had a house full with the Van Dusens and Babs roommate. I hope you liked having them and am anxious to hear your reactions.

I wonder if Ed Lord is heading this way. There are a lot of boys in his outfit with his training who have [been] thru here. If he does, I don't envy him his experience from what I hear. You've got to hand it to him. He's certainly has spunk.

With Lanie & Babs both away, life for you will be lonlier [sic] still, but maybe by that time the war will be over and I'll be back. Who can tell? It's awfully lucky we have two more to keep you occupied. I'll do my bit when I get back.

I last wrote you on Tuesday so there isn't very much to say. /p.3/ Thursday I played bridge at the Cobbets (Col & Mrs.) with Mrs. Yorke-Torr as my partner – a rather stupid evening but one of those evenings that are planned weeks ahead and you can't think of any excuse.

Last night I played golf with Doris Pettigrew (the Pettigrews are very good friends of the Van Dusens). The only ball I own looks as though it had the mumps. It must be defective. The cover doesn't split but it bulges out in a very funny way – not so funny when you try to putt it. I had dinner at Doris's. After dinner I took her to see the Indian premier of Mme. Curie. I didn't think it nearly so good as /p.4/ Greer Garson's previous pictures.[164]

It's afternoon and a bloke is coming up from Bombay to go over some of our accounts so I'd better stop and go out to the airport to meet him.

Still missing you like Hell and longing to be back in your arms.

With loads & loads of love,

Al

----

30 June 1944, postmarked APO 886   From AWZ to BSZ, Easthampton, Long Island, New York, U.S.A.
Rec'd July 7   ALS-I

[no header]

Thursday June 30th

[unnumbered 72]

Dearest Barb,

I was just about to write a scortching [sic] letter in answer to yours of June 8th but another arrived dated June 12th with a better tone so I thought better of it. I suppose you think I'm having a social whirl out here loving every minute of it. I'm sorry I asked the Van Dusens to come if that is what they've been feeding you or maybe I tell you too much in my letters about social life here which is about all I can talk about anyway. It's part of this kind of a job – to meet people and get to know them better.

It's not much comfort for me to hear "my youth has gone for good and all – and it certainly shows – There's not a giggle left in me" etc /p.2/

Why in the world don't you give up these straining extra things you do – the gift shop – the canteen? You have enough to do without them and it certainly is not worth it to anybody to have yourself all worn out.

Are you glad Walter J. is moving to Phila. Not very I expect. I'm not very thrilled with the idea either but I guess it will work out all right. Our paths have sort of parted in the last few years.

I'm sure I mentioned how much I liked your poem. I thought it was very clever. Why don't you answer some of my questions? Does Albie get my postals? Did you ever give Sally W. [Winsor] anything from me? - describe the movies /p.3/ more in detail etc.

No, I didn't have a party on my birthday. Nobody knew about it until it was all over. That's the way I wanted it.

I'm frightfully sorry to hear about Ted [Shoemaker]. I'm sure he can be straightened out given time. If you think a letter from me would help I would be glad to write one. I think I'd know what to say. There's no use of you getting all worked up about it – there's nothing you can do except help financially which of course do if they need it.

I'm very proud of Lanie making cum laude. Does that mean her name will be engraven on the walls of Baldwin school for posterity! many congratulations /p.4/ I really think she's made a wise choice to go to Chatham. She should have a very pleasant time, under nice surroundings, learning how to live a fuller life.

---

[164] Greer Garson (1904-1996) was nominated seven times for the Academy Award as Best Actress. She won the Best Actress Award in 1942 for *Mrs. Miniver*. She was also nominated for her role in *Madame Curie* (1943).

I'm awfully glad Simmy is going to Easthampton. That should take a big weight off your shoulders and for goodness sake take advantage of it.[165]

Yesterday I got letters from Bitz [Durand] and Amelie [Kane] dated in April & March something must have happened to them. I'll try to answer them soon but this is a very busy time of the year (half yearly reports etc) so if you see them explain I haven't heard a word of Peachy /p.5/ who was to arrive at Bombay Apr. 10th. I think she could have dropped me a card – she knew where I was and I don't know how to reach her.

Speaking of that Anne Carter Greene[166] dropped in last night. She's over here from Calcutta for several days. I tried to have her stay for dinner, lunch dinner today, lunch tomorrow but she would have none of it. She claims she's very busy with her work and hasn't a minute to spare which I guess is right. Anyway she did stop in and it was nice to see her.

On Sunday – we had some /p.6/ visitors who spent several days with us – a captain who had been in the Pacific and who had some interesting tales to tell and a lieutenant from Colombo who has given us more work to do at a very bad time. I played tennis with them on Monday, golf with Doris on Tuesday – the monthly mixed competition and tennis again last night.

Here's one for Warren see if he can get it
F U N E X   S U F X   F U N E M
S V F M   O R   I F   M N X

That's about all for now –
     God bless – cheers and
         Lots and lots & lots of love
            Al

----

6 July 1944, postmarked APO 886   From AWZ/APO to BSZ, Easthampton, Long Island, New York, U.S.A.   Rec'd July 14   ALS-I
[no header]

Wednesday July 5th

#73
Dearest Barb

Nothing from you since I last wrote. I sent you a package about a week ago (we can only send one a month now). It contained 2 pairs of sandals, 4 more elephants to make a dozen (if they don't match let me know) and some more Nepal jewelry to dispose of as you see fit. You might find something suitable for Anne Lincoln in the lot but I really got them for you and the girls especially the circular thing

---

[165] Simmy = The girls' English governess, who they appreciated.

[166] Anne Carter Greene = It is clear that Anne Carter Greene is indeed a member of High Society, and I suppose she was doing intelligence work in India, although she was in the Red Cross. She was the daughter of Louis Storrow Greene (1872-1947), and she had a brother Julian.

From *Washington Herald* (3 July 1910), "Society," p.2: "…Greene and Miss Anne Carter Greene have gone to Woodberry Forrest for the summer. Miss Lou Washington was the week-end guest …" Leading paragraphs on this page describe the activities of two cabinet officers and of Rep. and Mrs. Nicholas Longworth. The Longworths are to visit her parents, President and Mrs. Theodore Roosevelt, at Oyster Bay and then go on to the Summer White House in Massachusetts as guests of President and Mrs. Taft. This trip and the tension between TR and Taft that Alice and Nicholas were attempting to negotiate is discussed in Edmund Morris' new book, *Colonel Roosevelt* (2010).

From *Washington Post* (2 March 1913), "Alexandria Society" (pg. B10): "Dr. and Mrs. Louis S. GREENE, Master Julian GREENE, and little Miss Anne Carter GREENE, of Washington, were the week-end guests of Mrs. GREENE's parents, Mr. and Mrs. Julian T. BURKE, at their home in Prince street." And in the Papers of Mantle Fielding, in the Winterthur Library, Winterthur, Del., there is a letter from her, viz.: ".419 Letter, to Fielding, from Anne-Carter Greene, Washington DC, March 25, 1934 / Re: Washington portrait, possibly a Stuart."

with goddess of jade (I think). That, I thought, would be nice made into a hand bag for you. It would be the gadjet that opens and shuts it.

I'm glad you have some golf balls on the way. I hope you have some to play with at Easthampton. /p.2/ What happened to the ones in the wine closet? Also I understand you can take old ones back and have them exchanged for remade ones. You could use all my old practise [sic] balls for that.

I'm sending this to Easthampton but am worried for fear it needs a more definite address which you haven't furnished me. I hope it gets to you all right. How in the world did you move? Were you able to take the car? Did you all drive over and have your things sent by express or what. I hope you are very comfortable there and you all have a very pleasant healthy summer. Please don't worry about expense. I've been able to /p.3/ accumulate some dollars here that you never knew you had. I wish Stan would send me a survey of how our investments are doing every once in a while. I imagine they are doing quite well as our papers indicate stocks are doing quite well since the invasion [of France]. Also O&Z [Ott and Zimmermann, his wool broker partnership with Jack Ott] seem to be off to another good year. So don't worry!

Here is a new alphabet with a British twang

A for asses
B eef or mutton
C for Highlanders
D eferential
E ffervessence
F florescence
G for Polo (what they play Polo on)
H---for Hitler
I falutin
J afa oranges (in Palestine)
K Fraancis
L for leather
M phasis
N for eggs
O for a woman
P for relief
Q for theatres
R fa mo
S for me
T for two
U for me
V for victory
W for quits
X for breakfast
Y for mistress
Z ephyr winds

[Editor's note: On 5 July 1944, AWZ sent to his senior officer, JICA/CBI, the transcript of a lecture which he said he received from J. Harris, Esq., Central Intelligence Officer Karachi. The lecture was given to British officers on about 1 May 1944 by a person who he could not name. The 9-page typescript is shown in Appendix D. Zimmermann said in his cover message that this "probably presents the best British opinion on the Indian political situation as it exists today." Many other copies of the lecture were forwarded to others, including the State Department, although none of these have been found. The lecture was considered to be so sensitive that it was not declassified until 12/8/11, at my request through FOIA]

/p.4/ We celebrated the fourth in regal fashion. The night of the 3rd the vice consuls had a party at their flat with most of the diplomatic representatives there (Afghanis, Iranis, Iraqis etc) a pretty good party and last night our two Army Generals and Mr & Mrs Macy had a party that included about everybody important in town including the governor aur billi (Hindustani).[167]

It's still hot as the dickens and quite enervating and lots of work seems to have piled up on us so I'll close and try to get some sleep in this heat.

I miss you horribly dearest and hope the Allied successes in Europe mean my homecoming will be that much nearer.

   Oodles & oodles of love,
    Al

----

10 July 1944, postmarked APO 886 From AWZ/APO to BSZ, Easthampton, Long Island, New York, U.S.A.   Rec'd July 17   ALS-I
         [no header]

                     Sunday July 9th

#74
Dearest Barb

Your letter of June 29th arrived in 10 days – pretty good time. You sound so exhausted and worn out. I hope you're well settled in E.H. and can get some well deserved rest. For goodness sake, when you get back, don't take on a lot of extra work. You have enough to do running a family a house and two estates.

Who were the three boys that let Babs down at the last minute? I think that was despicable and would like to give them a good thrashing.

I'm sorry Letty is not working out. Isn't there anyone you can /p.2/ get? The war plants must be laying off a lot of people which should improve the help situation.

Too bad about Perry – too bad he didn't listen to a few of his friends. Joe Eastwick certainly tried to steer him away at the time among many others.

My mouth waters for all the delicious things you write about putting up in the deep freeze. I could certainly do with some good roast beef and real vegetables.

I loved the pictures you sent. You are getting very good as a photographer. How about keeping them coming and getting someone to take some of you. /p.3/ I miss you so much – pictures do help. Now don't be shy and refuse to have your picture taken.

I'm enclosing some with I took with our Navy camera a week ago at the War Services Exhibition. The explanation will be on the back.

I wrote you just a couple days ago so there isn't much new. Thursday night Howard [Voorhees] & I went over to Minora Island to "HMIS" Bahadur – an Indian Navy training school to a party Com. Carmarkar (see picture) was having to celebrate his being made a Commander. A rather dull evening but one we had to /p.4/ attend. We went over with the NOIC (Naval Officer in charge) Capt Coupe in his launch. After supper they played Bingo which you know thrills me no end.

Friday, Lt. Mallin (RIN Intelligence Officer), who is departing for England had a drinks party another must (had to call off having dinner with Buzz Sloan) and last night I played bridge at Doris's [Pettigrew] with the Blackwells. We had our first rain since January. It really came down for a while and [you] should have seen how excited the natives were.

Today is my duty day so I am able to catch up with a bit of writing.

---

[167] Brigadier Generals John A. Warden and Julian B. Haddon, USA; Hon. Clarence E. Macy; and "aur billi," presumably H.E., Governor Sir Hugh Dow – although this is not clear. I don't know the translation from Hindustani to English. However, two pages of Google references to these two words suggest that they may now have an unsavory or even pornographic meaning.

Good bye, dearest, don't let these [switches to margin here] gals worry you. You're the only one for me. I'll come home any time they let me.

Loads & loads of love   Al

----

13 July 1944, postmarked APO 886   From AWZ/APO to BSZ, c/o Parsons Cottage, Davids Road, Easthampton, Long Island, New York, U.S.A.   Rec'd July 20   ALS-I
[no header]

Tuesday July 11th

#75

Dearest Barb

I've just written you two days ago but as I'm leaving for Colombo tomorrow I thought I'd better write and explain a lapse that will occur in my mail. There is no sense in posting any letters en route. All mail has to come thru here, you see, and I'd probably beat it back.

Two letters from you came yesterday (June 16th & 21st). They were written before the one I received July 8 so it makes it very confusing.

I will have the sandals made and will send them on in about a week or so.

I'm glad to hear Warren is doing so well at tennis. He seems awfully /p.2/ small to be playing with Johnny Ott. I hope the association doesn't make him take on some of his characteristics. I think he's lazy and I'd hate to see Warren get as fat as he is.

I guess by now you are well settled at E.H. It must have been an awful job getting there but well worth it.

I'm thrilled with Lanie doing so well in school and being such a help to you at home. I hope everybody pitches in and takes the troubles off your shoulders. It would be a relief to know the children would do what they're told to do. It would be so much easier for you. How are they behaving /p.3/

There can't be much news from this end. Sunday night I had dinner with the Clees (Charlie & Mary). He is revenue commissioner, the job next in line for the governorship. A Mr. Tennyson was there none other than the great grandson of the poet.

Last night I did nothing and tonight the same. So there you have it.

I leave tomorrow at 6.30 on a flying boat that reaches Ceylon next morning. I am going to attend a meeting of the Aluslos in this area. I'm coming back by way of Agra and might get up to Delhi for a visit, just to have a look see.[168]

I should be "home" in about a week so it will be a week between this letter and the next you'll get.

We had some more rain today which of course is very welcome in this dry place. The trees, veg. gardens and the city water supply can certainly use it.

All my love dearest and lots & lots of hugs & kisses which I wish I could deliver personally.[169]

---

[168] ALUSLO = U.S. Naval Liaison Officer. This acronym is in *U.S. Naval Administration in World War II: History of Convoy and Routing / Headquarters of the Commander in Chief, United States Fleet and Commander Tenth Fleet*. XI – Appendix A. "Abbreviations" (accessed via Google, 2/8/11). Also from Google, we find the title (although not the name) of the man to whom AWZ reports. He is the Senior U.S. Naval Liaison Officer, I-B Theater = United States Senior Naval Liaison Officer, New Delhi. His title and location appears in a transcribed message that begins: "UNITED STATES NAVAL LIAISON OFFICE / 6. CHURCH LANE / CALCUTTA, INDIA / EN3-11(CT) A8-21 / Ser: 01139 / 9 September, 1945
From: The U.S. Naval Liaison Officer, Calcutta, India.
To: The Director of Naval Intelligence.
Via: The Senior U.S. Naval Liaison Officer, I-B Theater.
Subject: USS *Houston* . . ."
[169] AZ did not mention that he received an invitation about this time from a "Dodie Knott" (10 July 1944) to dinner "with us" on 15 July, to meet "Admiral Godfrey." This would be Admiral J. H. Godfrey, formerly Director of Naval Intelligence for the Royal Navy and in 1944 the Commander of the Royal Indian Navy. The invitation is in his Scrapbook, pasted to page 25(R); AZ's response is not shown.

Al

[This letter and the next five were filed by BSZ after the letter postmarked 15 Aug 1944. I have left the letters in their envelope where she put them, but the transcriptions are placed here, in the correct chronological order. It is not clear if they were delivered out of order, but I think she misfiled them.]

15 July 1944, postmarked APO 886   From AWZ/APO to BSZ, c/o Parsons Cottage, Davids Lane, Easthampton, Long Island, New York, U.S.A. Rec'd July 19   ALS-I
[no header]

Friday July 14th

[unnumbered 76]
Dearest Barb

Haven't yet left for Colombo. I've been down to the Marine Airport twice, all ready to go, but the weather has been too bad. I'm going down tonight again and expect to get off this time.

While I'm there I'm going to try to get up to Kandy. It's very pretty, they say, and of course quite important right now.[170] Coming back I hope to get to see Delhi, so it should be quite an interesting trip. The only movie film I have is the one you sent for indoors. I couldn't find the right filter /p.2/ but I have one that doesn't fit that I'll use, and hope for the best. I'll be able to take a lot of black & white anyway.

No mail since I last wrote and no news to tell you. Oh yes, there was a Lt Com. Groman that came thru on his way home from China from Bethlehem & He's married a girl that lives on Fischers [sic] Road (and still does) by the name of Richardson.[171] Too bad you're not home as he could probably very easily stop in to see you. He picked up a bit of amoebic dysentery and is going home to get over it.

That's about all, dearest. I wish I were on my way home.
Heaps & heaps of love,
Al

----

18 July 1944, postmarked APO 886   From AWZ/APO to WZ, c/o Parsons Cottage, Davids Lane, Easthampton, Long Island, New York, U.S.A.   ALS-I
[no header]

Tuesday July 11th

Dear Warren
Thank you very much for remembering my birthday and writing me a letter on it.

Your penmanship has changed quite a lot since you last wrote. I hardly recognize it.

Mummy tells me you are doing very well in tennis. Do you suppose you'll be able to beat me when I get home? We'll have a match to see who wins.

I'm afraid from what I see our Philadelphia baseball teams aren't doing very well this year. It's too bad. It makes it so much more fun to have your home team win.

I wonder what you'll do /p.2/ at Easthampton – play baseball or tennis? I hope Babs & Lanie will help you get acquainted so you meet friends you can play games with. Perhaps you, Babs, Lanie and Mummy can make up a nice doubles game at tennis. Anyway I hope you get some good healthful exercise and have some fun.

I'm very proud you got promoted with honors. That's swell.

---

[170] Kandy was in the interior of Ceylon, where the Headquarters of South-East Asia Command (SEAC) was located. The Commander-in-Chief of SEAC was Admiral Lord Louis Mountbatten. OSS HQ for India-Burma was also located in Kandy. The friend of AWZ and BSZ, nick-named "Peachy," who appears in the photograph that AWZ sent home after this trip, was stationed in Kandy as a civilian. She was in the OSS. Julia McWilliams (later Julia Child) was with the OSS in Kandy.

[171] Fisher's Road in Bryn Mawr, Pa., is only a short distance from the Baldwin School.

Be a good boy and help Mummy as much as you can.

Lots of love,
Daddy

P.S. How did you like the puzzle. I'll repeat it in case you didn't get it.
FUNEX, SVFX, FUNEM, SVFM, OK IFMNX

---

24 July 1944, postmarked APO 886   From AWZ/APO to BSZ, Parsons Cottage, Davids Lane, Easthampton, Long Island, New York, U.S.A.   ALS-I

[no header]

Saturday July 22nd

#77

Dearest Barb

    Back again after covering a lot of territory and feeling pretty weary. Going from one end of India to the other in less than a week is a lot of travelling.

    We were due to take off Wednesday the 12th from here but due to the weather and engine trouble we didn't leave till Friday evening. The weather was still not so good and we had rain most of the night but arrived in Ceylon just about on time early the next morning [Saturday, 15 July 1944].

    The two hour motor trip into Colombo was one of the prettiest I have ever taken. I took some /p.2/ movies on the way (the roll for indoor pictures you sent, using a yellow filter). I hope they came out all right. You go through beautiful groves of coconut trees skirting the picturesque coast line, rock bound with waves breaking high. The people are very interesting, a lot cleaner and more intelligent looking than Indians. The yellow robed individuals with their inevitable umbrellas are buddist priests (in the movies).

    On my arrival in Colombo whom should I find but Hampy Barnes. He's a Lt. Com. now and has seen a lot of action in the Pacific. I stayed at the /p.3/ [sheet 2] Officers Mess in Colombo and shared a room with Hampy who is really quite an amusing guy.

    After lunch we had a meeting and then went to Mt. Lavinia, a bathing beach, south of Colombo (I took some colored movies there the last time).

    That night, Hampy, the Aluslo from Bombay, Lt Com Curren and I went to the Galle F'ace Hotel for dinner and watched the dancing.

    The next day was spent mostly in meetings. I did get a bit of tennis in in the late afternoon at a very beautiful club. Really Ceylon is one of the most /p.4/ lovely parts of the world, anything grows, and right now surprisingly, the climate is quite delightful – quite a change from the hot, sticky weather we've been having.

    Sunday [16 July] night we had dinner at the mess (there are about 15 naval officers of various descriptions) and then played poker until about 11.30 – then to bed.

    One of the officers was a survivor who had some hair-raising stories of Jap atrocities that you probably will hear about later. Machine-gunning life boats would be considered humane compared to the further things they did. /p.5/ [sheet 3]

    The next day [Monday, 17 July 1944] we motored to Kandy – again a beautiful drive, once in a while seeing an elephant or two (tame) walking along the road. (Just before I left here I was able to get another roll of film so I took pictures along the way – the first part is of the Officers Mess at Colombo).

    Kandy is very delightful with a beautiful lake in the middle of town. We had a good look at headquarters, personally conducted by Com. Linaweaver our senior naval officer there. I forgot to mention on this trip was Lt Com. Curren, and Lt. Com. Dawson, Aluslo[172] Calcutta. /p.6/ We had lunch with Linaweaver and then stopped at O.S.S. headquarters and saw <u>Peachy</u>. I hadn't known she was in that part of the world till the night before when Hampy happened to mention 12 beautiful Am. girls were in Kandy connected with O.S.S. and sure enough one was Peachy. She looks very well and healthy

---

[172] Aluslo = ALUSLO (Naval Liaison Officer).

although she's had a touch of dengue. I'll write Bitz and tell her about my visit which necessarily had to be very short. [photo taken at OSS headquarters, *infra*]

That evening we arrived in time to go out to dinner with Hampy who had lined up a /p.7/ [sheet 4] couple of Wrens or at least one was a Wren and the other a Fanny (I'm not being rude). We went to the Silver Prawn, a night club effect, where we danced and had dinner. Don't worry, the gals were about as good looking as most of our waves but it was pleasant to have a bit of femininity along.[173]

The next morning [Tuesday, 18 July] I got up at five to take the plane for Delhi. We took off at eight and arrived at Delhi about eleven thirty [p.m.] with an hour for lunch at Bangalore and three hours for dinner at Agra. I saw the Taj Mahal from the air but no closer. I thought I'd be coming back the same way /p.8/ and would get a better opportunity at a more opportune time to see it but unfortunately I came from Delhi directly here. However perhaps sometime later I will see the "Taj."

Delhi and New Delhi were very interesting – nothing much to write about. I took lots of pictures (black & whites) which I will send on when developed. I also took other stills in Colombo all of which I hope you'll find interesting.

On my arrival home [21 July] (here unfortunately) I was very pleased to have your letters of June 15, July 5th & 8th awaiting me, the former containing the boys' reports. They certainly must be /p.9/ [sheet 5] getting big by the way their measurements are shooting up and very clever.

I'm sorry you had such a time getting to E.H. [East Hampton] with Babs sick and the car not preforming [sic] properly. Too bad Letty has turned out so badly. You seem to have had your share of things not going right. I hope by now everything is going smoothly and you are having a healthful pleasant summer. May be the next guy will get Hitler and I'll be back before you know it, taking my share of the responsibility of raising the family. I'm ready any time. Life here is interesting and some- /p.10/ times exciting but I want to be in my real home with <u>you</u>.

I'm supposed to be – right now – in entertaining some people who have come down from NWFP on <u>leave</u>. They called up a little while ago and told me of their presence so there was nothing to do but invite them over.

Thank God the news is getting better by the minute. I hope it won't be long now.
        All my love, dear,
            As always,
                Al

---

[173] Wren = Women's Royal Naval Service (WRNS; popularly and officially known as the Wrens).
waves = Women Accepted for Volunteer Emergency Service (USA), known as "WAVES."
Fanny = FANY = First Aid Nursing Yeomanry (UK)

"The Auck teaching a sepoy how to shoot" (AWZ's note)
From AWZ's trip to Delhi, showing Field Marshal Archibald

"Kandy – Entrance to S.E.A.C. Headquarters / [stamped] A. Zimmermann"

This remarkable image is the last photo that BSZ selected for the Photo Album.
Not because of its place in the chronology, but because of its historical significance.

---

26 July 1944, postmarked APO 886　　From AWZ/APO to WZ, Parsons Cottage, Davids Lane, Easthampton, Long Island, New York, U.S.A.　　　　　　　　　　　　　　　　　　　ALS-I

IN FLIGHT WITH "QANTAS"

Sunday July 23rd

Dear Warren

　　This paper came off the plane that took me on the first part of my recent trip. It was a "Cat."[174] I tried writing on the plane but it was too bumpy. We went through a storm most of the way. Besides /p.2/ it was at night and the one little light wasn't very strong.

　　Mummy sent me your reports and your physical measurements. I'm very glad you did so well in school and got promoted with honors. I'm very proud of you. Also I see you're getting bigger. I bet I won't recognize you when /p.3/ I get home.

　　Thank you for your letter of July 4th. It was very nice of you to write on such a busy day. Did you see any fireworks? I guess they were pretty scarce because of the war.

　　It looks as though Hitler was on the run. Maybe the war will be over /p.4/ soon and I'll be back with all of you again. My! that will be a happy day.

　　Well I guess I'd better close and write to Babs. Two very nice letters from her were waiting for me when I got back from my trip.

　　　　　　Lots of love,
　　　　　　　　Daddy

----

27 July 1944, postmarked APO 886　　From AWZ/APO to BSZ, Parsons Cottage, Davids Lane, Easthampton, Long Island, New York, U.S.A.　Rec'd Aug 7　　　　　　　ALS-I

[no header]

Wednesday July 26th

78

Dearest Barb,

　　I thought I'd have the pictures of my trip to send on, but they won't be ready until Saturday. I sent a box to you at Easthampton day before yesterday – mostly shoes, some more Nepalese jewelry and a novel cigarette case. I thought Lanie might have the set – like Babs' for doing so well in school this year, but if they particularly appeal to you, you keep them and give Lanie something else. Anyway you decide. To my mind the belt buckles are pretty puckka, just about the smartest that I've ever seen. You never say you want more of anything. I don't /p.2/ want to keep sending things if you don't like them and I will send more of the things you like. You could use them as Xmas gifts etc.

　　I haven't heard from you since my last letter so there are no questions to answer. I hope everything is going well at E.H.

　　Gordon Enders has evidently written an article on Afghanistan for the Sat. Evening Post.[175] I haven't seen it yet. I imagine it is one of the May issues or perhaps early June. When I see it I'll give you the exact issue. Prehaps [sic] he'll write one about our trip thru the NWFP, maybe use my pictures. I

---

[174] "Cat" = Consolidated PBY. The Catalina was an American flying boat of the 1930s and 1940s produced by Consolidated Aircraft. It was one of the most widely used multi-role aircraft of World War II. PBYs served with every branch of the US military and in the air forces and navies of many other nations. During World War II, PBYs were used in anti-submarine warfare, patrol bombing, convoy escorts, search and rescue missions (especially air-sea rescue), and cargo transport. The PBY was the most successful aircraft of its kind; no other flying boat was produced in greater numbers. The PBY remained in service in the U.S. Navy until the 1980s and it is still used in some parts of the world.

[175] This is the first reference I have seen to anything published by Gordon Enders since his last book (1942). His nephew and niece did not mention it when I spoke with them, and I believe they did not know of it. Both agreed that he stopped writing and lecturing after the war. There is nothing in the files of the *Saturday Evening Post*, so apparently the article was never published. He published, six years later, an article about Afghanistan, and this may be what AWZ is referring to. Gordon B. Enders, "The Nomad Woman," *Liberty* (7 October 1950), 16, 62ff.

kind of suspected he would, that is why I've never been too enthusiastic about doing it /p.3/ [sheet 2] myself. It thought by the time I got around to it he would have an article already published. Besides that's what he normally lives on anyway. He is now back in the states, having contracted a bad dose of malaria.

It is now raining again – such rain as we've been having is unheard of for here. We must be up to 14-16 inches for the year against 2 for last year. It throws the place into utter confusion. Nobody has a raincoat or umbrella and the streets are not designed to take off the water so that puddles of water lie around for days.

Three VIPs landed on us today /p.4/ but are leaving very early in the morning – two are Captains and one a Commander. A marine corps Colonel who was naval attaché at Chunking for about two years left today and said he would call you up when he got back to the states. A lot of these fellows suggest doing that and every once in a while I let them so you can get first hand information.

That's about all for the present. It's now 1230 so I'd better turn in.

I keep on missing you dreadfully and hope (I hope not against hope) I'll be seeing you soon.

      Heaps & heaps of love   Al

---

24 July 1944, postmarked APO 886   From AWZ/APO to BWZ, Box 50, Easthampton, Long Island, New York, U.S.A.
      ALS-I

IN FLIGHT WITH "QANTAS"

      Sunday July 23rd

Dear Babs,

Two very nice letters one dated June 18th and the other July 7th were waiting for me from you when I got back from my trip.

I hope you will get my telegram wishing you a happy birthday. I sent it two days ago so you should get it in time. /p.2/ Do you know what Qantas means? It stands for Queensland and Northern Territories Air Service. It's the longest air hop in the world 26½ hours. I didn't take it, of course, but the plane I was in did. I thought you might like to have a letter written on its stationery. I'm afraid it's a bit crumpled and soiled but what do you expect after /p.3/ coming from such a distance.

You sound like a very busy gal with all your doings. I guess you and Janet are pretty fast friends. You must be seeing so much of her as you do.

Your trip to Northfield must have been very interesting as well as educational. I'm afraid I'm inclined to agree with the /p.4/ person who thinks India should not have her freedom now. I think there would be chaos. The average Indian thinks only of himself, is dishonest and ruthless but I'll wait till I get home before going into the matter any further.

Easthampton sounds like a lot of fun. I'm glad to hear you're playing lots of tennis and are taking lessons. /p.5/ We'll have to have a lot of tennis when I get home. I think it's a swell game and one we can all play. How do you and Warren play? Is he able to give you a good game and may he beat you?

I'm sending along some pictures I took around town [not enclosed in the envelope]. I didn't put them in /p.6/ Mummy's letter as it was pretty fat already. There's nothing much to say about them other than what I've noted on the back. They pretty much speak for themselves. One thing you don't see are the millions of ants and flies that are around. Since the recent rains flies have come in hordes. I sit at my desk with a /p.7/ swatter and kill at least 150 a day.

Having written Mummy about my trip there isn't very much to say. In about a week I'll have some more pictures to send – those taken on the trip.

Be good, help Mummy as much as you can.

      With lots & lots of love,
          Daddy

---

2 Aug 1944, postmarked APO 886   From AWZ/APO to BSZ, Box 50, Easthampton, Long Island,
New York, U.S.A.   ALS-I
                                            [no header]
                                                                                            Monday July 31st

[unnumbered 79]
Darling Barb

  I suppose you've seen the enclosed [not preserved in this envelope]. I thought it quite funny.

  Dick Kelley dropped in today – on his way home. He's been here since March '42 – so it's certainly well deserved. He looks quite well not carrying any excess weight or anything like that, but has good color. I asked him about his girl and he replied he was going to take it slowly when he got back – a lot can happen in 2 ½ years. So maybe he isn't as ardent as he was. He reports to New Mexico on his return, but might get to see his family before, then he gets 30 days leave /p.2/ I'll try to send some pictures with him – of the NWFP and of my last trip.[176]

  Saturday night I had rather a large party here – the Cullens, Harrises, Macys, Cleis, Raymond Cougland, Doris Pettigrew and a couple odd men. I was repaying debts and collected quite an assorted crowd, but it worked out and everyone seemed to have a good time. After dinner we went to the Gymkhana to dance and listen to the Duke of Acosta's band that was captured intact in Abyssinia. Most of them are good looking Italians glad to be out of the war. /p.3/

  I stopped this letter at about 5.30 to play tennis and when I came back Dick had a fever of 102. I got a doctor who looked him over but couldn't tell what he had at this stage. He gave him some pills and said if he wasn't better in the morning he would arrange to have him sent to the hospital, not that it might be serious but they can take care of him so much better and can make various tests. He had malaria about 2 years ago and it might be a recurrence.

  Your letter of July 13th arrived this afternoon. Too bad Babs isn't meeting more girls who would see that she is having a good time. I guess some girls are /p.4/ like that.

  Yes, I remember Pamela Grout. I thought she was very attractive. I'm distressed to hear about her losing her husband. That's a bad break – also about Angus. What was he a paratrooper or was he in a plane? I must be awful for Indie.[177]

  I'm glad everybody in our family are well. I think of all of you often and wonder how everybody is. I hope you are all careful on your bicycles and when you go in swimming. The seas here are pretty rough due to the monsoons. Several people have lost their lives. So all of you be <u>very</u> careful.

  Good bye dearest. I still miss you terribly. There's nobody I've met can hold a candle to you, either. I want to be back with you and <u>soon</u>.
      Heaps & lots of love,
       Al

Tuesday A.M. Dick's fever has gone down so it might be just a passing upset. He says not to tell his family he's flying home.

---

9 Aug 1944, postmarked APO 886   From AWZ/APO to BSZ, Box 50, Easthampton, Long Island,
New York, U.S.A.   ALS-I
                                            [no header]
                                                                                            Monday Aug 7th

---

[176] Dick Kelley = Richard Carlyle Kelley Jr. (1919-2000), AWZ's nephew. He was a captain in the U.S. Army Ordnance Corps. He married (1) on 7 September 1944, Jeanne Hope Adams. They had five children. She had a stroke in 1995. He then married (2), as her second husband, Janet Best, who survived him. He had a successful career in the securities business and was involved with many charitable organizations in the Philadelphia area.
[177] Angus and Indie ____ have not been identified.

#80
Dearest Barb

A letter from both you and Babs came yesterday. It's pretty good getting a reply to a letter in three weeks.

I forgot to write down when I sent you a letter last week. I think it was Wednesday. Well Dick K went to the hospital next day with a fever of 104 which now is diagnosed as malaria. He had it about a year and a half ago so this probably is a new infection. I went out to see him Saturday, his temperature was normal and he was feeling pretty well. The previous day his fever had dropped to 99½ so I guess he is well on his way to recovery. They say the treatment ends next Saturday and if everything is all right then he will proceed home. I haven't written Anna [AWZ's sister, and Dick's mother] /p.2/ about his being here or his troubles as Dick didn't want me to. He wants to come home and surprise them so don't you say anything.

A courier just passed through – left last night – who kindly consented to take the pictures I took on the Colombo trip and the NWFP. Some day I'll have the latter enlarged to post card size but over here it's too expensive 1 rupee each (30¢). The negatives of the Colombo trip turned out so well I thought I'd splurge. I hope the box arrives safely and you like the pictures.

I've managed to get two more Kodachrome [rolls] for the movie camera so will be taking more movies when I get a suitable subject.

We've had an awfull lot of rain the last three weeks. Ten times as much as all last year when it rained /p.3/ [sheet 2] only two inches. Everthing [sic] is thrown into confusion around here when it rains. Most of the city is under high tide so the water doesn't drain off, air ports are isolated. Almost everybody's roof leaks – so you can imagine the mess. It's as much rain as they've had in the last 5 years put together and has broken all known records.

Life here is still the usual. Saturday the Grants had a party at their house and then to the Gym[khana] afterwards. John Grant is the Sind Government engineer and his wife Joy is a beautiful blonde. It was a fairly nice party but they are all pretty much the same.

Went to the movies last night and Thursday. Saw Fred Astair [sic] /p.4/ in one and Carmen Maranda [sic] in the other. I thought the Fred Astair one rather good with a very nice song in it something about "one hour."[178]

Your letter sounds a lot more cheerful about the Easthampton situation for which I'm very glad. You seem to be having a lot of visitors. Isn't that rather trying?

I heard about our ship and meant to say something to you about it. Too bad.

The news looks awfully good. Everybody sounds very optimistic. Boy, I certainly hope it means I'll be home soon. I just can't wait.

Good by dearest, lots of love. I hope I'll soon be able to give it in person.

Hugs & kisses & stuff,
Al

----

10 Aug 1944, postmarked APO 886   From AWZ/APO to BWZ, Box 50, Easthampton, Long Island,
New York, U.S.A.   ALS-I

[no header]

Monday Aug 9th

#22
Dear Babs

---

[178] We may never know what Fred Astaire movie AWZ is referring to. The only movie that I can find in which Astaire appeared in 1943-44 was "One for My Baby." The signature song contains the famous lines, "Make it one for my baby / And one more for the road." He made "Ziegfield Follies" in 1944, but that movie was apparently not released until 1946. I can't find his song with "one hour" in the lyrics; perhaps AWZ meant to write "one more."

Thank you for your letter begun July 18th and finished the 22nd. I'm glad I have one daughter that writes. I'm beginning to think I have only one as I haven't heard from the other in a dog's age.

Your life at Easthampton sounds very full. I hope you're enjoying it. At first Mummy thought you wouldn't have many friends to pal around with but from what I gather you're very busy.

Last week your cousin Dick Kelley stopped by on his way home. He was hardly here a minute before he contracted a fever that turned out to be malaria. Fortunately it was a very mild case, caught in time, and now he is practically well. He has to stay in the hospital till Saturday for his last treatment and then he'll be on his way home. He will have /p.2/ been away 2½ years. That's an awfully long time.

So you think Warren's pretty good at baseball. I'm glad to hear that. I'd hoped he'd be good. It's nice that some older "men" have let him play with them. I bet he is thrilled.

How's Albie getting along these days? Has he learned to ride a bicycle yet? Does he play ball too? Have you a piano in the house so he can practice? As far as I can see he's about the only one of the bunch that really knuckles down to practicing – that is besides Lanie.

There isn't much more I can say. Writing Mummy so often doesn't leave much new for anyone else. I do love getting your letters though, so keep 'em coming.

      Lots & lots of love,
       Daddy

----

15 Aug 1944, postmarked APO 886 From AWZ/APO to BSZ, Box 50, Easthampton, Long Island,
New York, U.S.A.
                        ALS-I
          [no header]
                        Sunday Aug 13th

#81
Dearest Barb

  Nothing has come in since your last letter and nothing is new here so there isn't really much to say.

  The first part of the week was spent horizontal with some unknown fever – nothing at all serious but just a fever. I got up on Wednesday feeling a little groggy but felt fine on Thursday and have since. Of course when you start one of these sieges you think you are getting malaria, cholera or dengue. You get a bit worried about yourself but this was just a fever from no known cause [Ed.: FUO].

  Friday I went out for dinner at Doris's and played bridge with /p.2/ Mollie Blackwell Wright and General Hinds (the "Fortress Commander"). Mollie is a daughter of the Blackwells who married an Army man and has come here to have her first baby. Her father as I've probably told you before is #1 at Burmah Shell. The General is a genial old bird and the evening was quite enjoyable.

  Yesterday it started to rain again and the prediction was that it would continue for 36 hours. With nothing to do outside in the way of exercise we took to bridge – started yesterday afternoon at five – played till eleven – started again at eleven today and finished about an /p.3/ hour ago – nine. It started off at Government House at the ADC's – continued at the Blackwells – I had them here for this morning and then we continued at Bill[?] Kings this afternoon until eight. It involved in all about eight people so it wasn't the orgy that you're thinking – we cut in and out. There really isn't a thing to do here when it rains and you know how I like bridge so don't be too hard on me.

  Dick Kelley doesn't get out of the hospital until Tuesday. He was supposed to get out today but last night his stomach was upset, he claims from ptomaine poisoning so he has to stay a /p.4/ couple more days.

  I enclose this picture of "Peachy's place" at Ceylon to complete the set I sent you. The one that was in the set I sent to Bitz. Don't you think Peachy is having a hard time? It really is a beautiful place.

  I'm sorry this is such a dumb letter but there just isn't the material to work with.

  Bye, bye, darling. I hope it won't be much longer.
      Heaps & heaps of love,
        Al

"Kandy –Lt. Com. Curren, Peachie,[179] Lt. Com. Dawson / [stamped] A. Zimmermann"
The photo, taken by AWZ on 17 July 1944, is in Photo Album (Sheet 7, #14). It was not enclosed with this letter, but it was instead probably in the box of photos that AZ shipped home much later.

----

18 Aug 1944, postmarked APO 886 New York, U.S.A.

From AWZ/APO to BSZ, Box 50, Easthampton, Long Island, ALS-I

[no header]

Thursday Aug 17th

#82

Dearest darling,

What do you think of the news? It sounds pretty good to us out here. I don't see how the Germans can go on. They must be dizzy trying to figure out where the next blow is going to fall.[180]

Jack Kunz arrived day before yesterday – brought me some peanuts and two bottles of whiskey – a gift from Powell. I was very much surprised he didn't bring any pyjamas from you. Didn't you say you'd sent them (2 pr) down to Brownie and some golf balls? Anyway they hadn't arrived on Aug 1st.

I reciprocated to Powell by sending home 2 bottles of gin via Jack. /p.2/

No word from you in quite a while. Have you given up writing or is it just the mails?

Dick Kelley is still here. He got out of the hospital on Tuesday. His treatment for malaria was finished on Saturday but unfortunately he ate something that didn't agree with him so he had to stay several more days. He's sort of a moody guy, quite independent and a bit stubborn. I would say that he

---

[179] Virginia "Peachie (aka Peachy) Durand, daughter of the Zimmermann's friend "Bitz" Durand, from Chatham, Mass. She was one of the first nine OSS girls who arrived at Kandy. Another was Julia McWilliams, who became well known as Julia Child when she married former OSS officer Paul Child.

[180] AWZ is doubtless referring to reports of Allied landings in Southern France on 15 August, so-called "Operation Dragoon."

has broadened quite a bit – is not adverse to drinking scotch etc and smokes as much as I do.[181] I would think he would be rather difficult to live with but I guess I'm feeling my generation.

Played bridge again on Monday /p.3/ as it rained all day again – or still – and there was no tennis or golf so Doris, Roger & Nell King continued our bout of the day before starting at six and finishing at ten for an early evening.

Last night and the night before I had planned to get to bed early but Tuesday Jack brought his crew over and last night he came over again and stayed till twelve. So you needn't feel badly about loading him up with stuff to bring out. He's been very well entertained out here.

I really can't think of anything else to say except I miss you terribly and hope this horrible /p.4/ mess is fast drawing to an end.

I think I would win some money from Mr. Strauss and a friend of his, if the war in Europe is over this year which I firmly believe.[182] I wrote the bets on a card and put them in the top of my chifferobe. So when you get home you might give them to Jack to collect.

You haven't said when you are leaving Easthampton. I hope you will let me know so I can send mail to the right place.

Good bye, dearest. Heaps and heaps and all my love,
    Al

---

24 Aug 1944, postmarked APO 886.   From AWZ/APO to BSZ, Box 50, Easthampton, Long Island, New York, U.S.A.   ALS-I

[no header]

Tuesday Aug 22

#83

Dearest darling,

Three letters from you came all the same day, dated Jul 28th, Aug 2nd & 7th. It was awfully nice to get them. There was quite a gap since your last letter.

It's dreadful news about Angus. It's just one of the horrors of this war – all so un-necessary.

Jim Winsor must be having some pretty exciting times these days. I hope he gets through all right. I've had several letters from him but none since he arrived in France.

You seem to be having a whirl with the Haynes brothers. I'm glad you have somebody to take you around but don't forget I'm still among the living.

I don't know who Hampy Barnes /p.2/ married but we met him years ago at Skinny Wheelers and I have seen him several times since. He went to Yale and knows Mac Aldrich[183] quite well although he's quite a bit younger. He now lives in New York.

---

[181] This may be an oblique reference to the fact that Dick Kelley and his father were raised in the faith of the Reorganized Church of Latter Day Saints (a branch of the Mormon Church). Richard Kelley Sr. was the son of Edward Levi Kelley, who was a Bishop of that church. Kelly Sr.'s wife Anna was the daughter of John Zimmermann Sr., who was also a Bishop of that church. Richard Kelley Sr. and Jr. later became Presbyterians. AWZ, too, was raised in the LDS (Reorganized), too. BSZ was raised in the Swedenborgian Church of the New Jerusalem and they were married by a Swedenborgian minister. AWZ and BSZ became Episcopalians, a denomination which is more tolerant of alcohol and tobacco than the LDS, Presbyterians, and Swedenborgians.

[182] AWZ lost this bet. Progress was not as rapid as had been hoped for, and by December the end was not yet in sight. And then, in that month, the Germans pushed back in what was known as the "Battle of the Bulge," and victory in Europe was not achieved until May 1945.

[183] Mac Aldrich = "Malcolm Aldrich, 86; Headed a Foundation" Obituary, *New York Times* (3 August 1986): "Malcolm Pratt Aldrich, the former president and chairman of the Commonwealth Fund, a national philanthropic foundation, died of pneumonia Thursday at Southampton (L.I.) Hospital. He was 86 years old and lived in East Hampton, L.I. Mr. Aldrich, a native of Fall River, Mass., graduated from Yale University in 1922, where he was captain of the football team. After college, he joined the staff of Edward S. Harkness, the philanthropist and president of the Commonwealth Fund in New York City. In 1940, after Mr. Harkness died, Mr. Aldrich became the

Glad to hear the sandals, jewelry etc arrived safely. You don't ever say whether you would like more of any particular item. I wish you would let me know. I'm about at the end of attractive things to buy and anything additional must be repeats so let me know what you particularly want.

I'm up to my ears in work just now. Howard Voorhees went to the hospital on Sunday with some skin infection. This means I'm /p.3/ all alone. Phil Haller [sic: later spelled Halla], the Coast Guard Officer is here but he can't do certain things but does help where he can.

The weather has been pretty miserable – more rain and it still stays pretty hot. The wind is from the land most of the time. If it would only turn from the sea it would make a lot of difference. Last year wasn't bad at all at this time. I'm covered with prickly heat which is a very common complaint. It's very annoying – there's nothing much you can do about it. If only it would cool off at night it would help – but it doesn't.

Saturday we had a party here for about 25 people – the nicest here /p.4/ It was quite successful and almost approached some of our singing evenings at home. We borrowed a pukka victrola expecting to dance but it was too hot so we sat around and passed the "call" as we do at Orpheus parties. This produced some quite amusing stories and songs. A siren by the name of Mrs. Hirst, wife of a British Army officer, here to cool off from the heat of Muscat produced some amusing stories, a bit risqué but funny.

Sunday I went to Sandspit with Doris and the kids – the first time anyone's been able to go for over a month. It was quite pleasant but awfully rough due to the monsoons.

Got to go now and do some [shifts to margin] work. Don't get too gay with those Haynes boys. Don't forget you're *mine*. Good bye sweetheart. With oodles of love, Al

---

29 Aug 1944, postmarked APO 886　　　　　　　From AWZ/APO to BSZ, Box 50, Easthampton, etc.
Forwarded 5 Sep 1944 to Rose Lane, Haverford, Pa.　　　　　　　　　　　　　　　ALS-I
[no header]

Sunday Aug 27th

#84
Dearest darling,

If you want to know when the war will end I refer you to the enclosed clipping from today's paper [no enclosure in this envelope]. By the way things are popping the old lama may be a bit on the pessimistic side. Today we get the news Bulgaria has asked for peace terms. Festung Europa seems to be losing a bit of its fest. I'm an eternal optimist but I can't see how Hitler is going to keep his house of cards together very much longer. If it completely falls tomorrow it won't make me mad. I'm ready to come home at any time.

So you think those ivory images I sent home "C'est impossible." Well here's another clipping to disprove that. This is a great country and "anything can happen here." /p.2/

Your letter of Aug. 12th arrived yesterday. You haven't mentioned when you intend to go back to Haverford but I suppose it will be around the first week in Sept., so this will be the last letter I'll send to E.H. unless a subsequent letter of yours says otherwise.

---

executor of his estate and the second president of the fund. … During World War II, Mr. Aldrich served in the Navy as a special assistant to the Assistant Secretary of the Navy." Malcolm Aldrich was a member of Skull and Bones. This letter shows that the Zimmermanns knew him before the war, and they continued their friendship with him at East Hampton after the war. His job in the Navy Department is not specified in the obituary, but it is not unlikely that he was involved in the appointment of officers, such as AWZ, for Naval Intelligence. His cousin Winthrop Aldrich was a member of "The Room," FDR's secret pre-war intelligence committee in New York City. The Room was chaired by Vincent Astor, who was called to active duty. FDR directed the Director of Naval Intelligence to provide orders to Astor to vet civilian recruits for the Office of Naval Intelligence on the East Coast.

I was thrilled to hear about how well everybody is doing. I'd love to see Warren play tennis and baseball. I'm glad to hear Lanie is alive. I haven't had a letter from her in <u>months</u>. I suppose with all the friends you've made you'll be wanting to spend future summers there at E.H. I guess it's as nice a place as any.

Things are pretty dull here. We haven't had any visitors for a week – that is other than the usual ships. /p.3/ The weather has been a bit better. The "rainy season" is apparently finished. They have opened the golf course so I think I'll try it out this evening. It's been unplayable for a month.

Today at noon I went to the Boat Club with the Blackwells. It is the usual place to go on a nice Sunday morning. The club itself is quite nice but it overlooks a mangrove swamp and is on a rather dirty backwater. They have a few small sail boats and sculls. Some people go in swimming – but not me. It's too near the Hindu bathing ghat. If you bring your own beer you can have beer. Otherwise you can have gin or maybe both. Gin out here seems to go with everything and surprisingly /p.4/ you don't get more than the usual aftermath. Besides the beer gin sequence they (and I too) drink whiskey and then gin right after. I never thought it possible. Last night I saw "Ali Baba" a beautiful picture but I'm afraid not a very accurate portrayal of the people of the time. From what I saw in Iraq and what I see here I'm sure the people of Ali's day were not as neat and clean as in the picture. [Editor's note: "Ali Baba and the Forty Thieves," starring Maria Montez and Jon Hall, was released in 1944]

I'm enclosing some more photos with the explanation on the back [not enclosed in this envelope]. I hope I'll be seeing you soon.

Loads & heaps of love,
Al

---

2 Sep 1944, postmarked APO 886   From AWZ/APO to BSZ, Box 50, Easthampton, etc.
ALS-I

[no header]

Friday Sept 1st

[unnumbered 85]

Dearest Barb,

Tomorrow I'm off on a jaunt, a trek, an expedition or something. Anyway I've got to go about thirty miles from here as the crow flies which means about fifty as I fly.

A cryptic message came by carrier pigeon that an American boat had washed ashore north west of here and it falls upon us (me) to see what it's all about. You can't get there by [illegible] or jeep, due to its being the tail end of a heavy rainy season so the only thing left is a camel. Communication is practically nil except by one way carrier pigeon (to here) so we can't reserve a suite of rooms ahead but will have to make our own arrangements as we go along.

We start from here at day break and drive for fifteen miles and then /p.2/ we will be in the hands of the most ancient mode of transportation. It shouldn't take more than 3 or 4 days and I'll be telling you about it when I get back.

The last few days has been gala, as far as mail from home is concerned. I've heard from Warren, Babs and you and also from my long lost daughter Lanie. Her hand writing has even changed perceptibly since last she wrote.

I'm glad Easthampton has worked out so well and a good time is being had by all. I sort of wish Warren would write more about what he is doing, what he likes etc. instead of what his baseball idylls are doing, in which my interest is only secondary. I'd rather hear about what his batting average is, who /p.3/ he's beaten in tennis etc. Not that I'm complaining but I'd like to hear more about himself.

I'm sorry about the Whitehead's son. That's something that must be pretty hard to take.

You are very sweet about the pictures. I thought they were pretty good but I don't think Life would be interested. They are enlargements, of course, done quite well locally. In one or two they missed the framing I had carefully composed. I had several done over and am enclosing one [not enclosed in this envelope] which I think is much better than the original although it's on different paper. Of course I'd be both surprised & pleased if Life showed any interest and I could send on the negatives which they could manipulate as they saw fit.

Howard is still in the hospital – expects to be out Tuesday. Phil Haller [Halla] has been very co-operative and we've been able /p.4/ to take care of what has come along. Speaking of that we have now staying with us a Catholic chaplain who has been in this part of the world for seven years. He's quite a nice guy and very easy to get along with.

That's about all there is and I had better tuck in for the night. I don't think tomorrow is going to be a very easy day. Camel riding besides being conductive to sea sickness, brings into play about every muscle of your body so my life is not going to be very comfortable for the next several days.

Good bye, dear, God bless and I hope it won't be too long now.
Heaps & heaps of love,
Al

[See Bibliography for the title of an official report from AWZ to CNO (DNI) on 1 September 1944]

6 Sep 1944, postmarked APO 886   From AWZ/APO to BSZ, Box 50, Easthampton, etc.
ALS-I
[no header]
Tuesday Sept 5th

[unnumbered 86]
Dearest Barb

I don't know how your letter of Aug 27th postmarked the 28th got here on Saturday Sept 2nd but it did and must have broken all records. The trouble is that now I'll have to wait a long time, even if you are very regular, for your next will probably take the usual 15 to 20 days.

I'm awfully glad you're having such a pleasant time at E.H. and your life is so much fuller with meeting people, golf, dancing and tennis. Your spirits sound so much higher than in previous letters.

Now, as regards to the pictures. You are very nice to be so enthusiastic and to go to all that trouble in /p.2/ peddling them. I'm afraid your enthusiasm is biased by the fact your boy took them. Probably by now you will have found they are unacceptable anyway, which would settle the matter. But if there is a glimmer of hope still left, there could be something done about it.

I'm not awfully proud of the NWF ones. I had to take about 50 pictures with a strange camera using filters etc before I ever saw the results of my work. Then, they are contact prints done by photographic shops along the way. With the negatives in hand I'm sure experts could do more with them in enlarging.

As to why I was there: it was a very nice gesture on British Intelligence's part (with whom we have the finest cooperation) to invite one of our officers /p.3/ [sheet 2] to make the trip that had been already instigated by the Military Attaché to Kabul (Maj. Enders) to give him an opportunity to see what was on the other side of the fence. Then again I think the British are anxious to let her ally have a look at some of her problems in India to counteract the crack pots who came over here and after six weeks write a book about the oppression of the Indian people and how they should unquestionably have their independence.[184]

Anyway I took the trip and took the pictures and there are very few people and I might say only two Americans[185] that have ever taken such a comprehensive trip of the Tribal Territory. /p.4/

---

[184] This is perhaps the only place where AWZ describes what he believes was the purpose of the trip, in amplification of what was in the message from IB Quetta in October 1943 and in his Orders from the Navy Department in November 1943. The relationship of the trip to the independence movement in India is new here; it is not mentioned at all in any of his notes or in the trip narrative. He also states unequivocally that Enders "instigated" the trip, although he doesn't say anything about why Enders wanted to make the trip.

[185] two Americans = AWZ and Gordon Bandy Enders.

As far as peddling them to a magazine is concerned I'm sure I should have Sir Benjamin's permission and also the Navy Dept. I've written them both today – just in case. As far a story goes – I started one some time ago – a bit more in detail than anything I've written in letters to you – but with the changes that have taken place here I gave it up more or less because I thought it a forlorn hope. Anyway pictures and story would have to be cleared before publication but, if what they have in hand, Life or Town & Country want to go ahead, I'm game. Hampy Barnes, when I saw him in Colombo and mentioned the subject, told me I was crazy not to go ahead as he had had several things /p.5/ [sheet 3] accepted by magazines and they were hungry for material and paid well. He also said the Navy Dept. approved of this sort of thing when routed through the proper channels.[186]

Now the Ceylon (and Delhi) pictures I'm quite proud of. I took them with more knowledge of what I was doing. True they haven't the romantic story behind them but it's a place that is going to hold more interest as the war moves this way. Did you show these to anybody or do they just not appeal? I'm sure it wouldn't take much to get them approved – but maybe that's the trouble with them.

The trip in Baluchistan was very interesting. I took lots of pictures.[187] I saw the negatives today /p.6/ and selected about 40 to be enlarged. I'll weave a story around them in the spritely manner of Life and send them to the Navy Dept (care of Curt Winsor). If they approve, I'll ask them to send the story and pictures to you so you can peddle them as you like.

Incidentally when I arrived home from said expedition I was so stiff and tired from bouncing around on a camel I took to my bed. Just for the fun of it I took my temperature and found I had a fever of 102. That was yesterday. The doctor gave me aspirin and codene [sic] and I felt fine this morning with temperature normal but this evening the damn thing's gone up again. Another one of these FUO's again I guess.

Good bye, dearest, heaps of love and kisses,
Al.

AWZ-Photo Album, Sheet #5, 3 (labeled 44).

---

[186] Curt Winsor provided official permission from the Office of Naval Intelligence for AWZ to submit his photos for publication, and BSZ communicated with *Life* magazine, *Reader's Digest*, and *National Geographic* in re his photos and stories about his experiences in India. Bromhead also gave permission for AWZ to use the photos he took on the NWFP trip (See Appendices for Winsor and Bromhead papers).

[187] For descriptions of the photos and details of the trip, see the file on Baluchistan Trip (now in Special Collections, U.S. Naval Institute). One photo is shown here as an example: it is "the thing" (a life raft from a Liberty ship) that washed ashore and was the object that AZ and his companions rode out on camels to examine. The trip report, 2-3 September 1944, was sent to CNO (DNI) on 13 October 1944; details are in the Bibliography, citing RG 38.

The legend in the Baluchistan Trip story is: "44 Examining the raft for whatever story it may tell." The legends on the preceding photos (not shown here) are: "Quite a crowd has gathered by the time the end of the trail is reached. . . . The "thing" turns out to be a life raft made in California." The life raft was from the Liberty ship *Sampan* (ex-*William I. Kip*). *Kip* had been transferred by Lend-Lease to the British and was renamed *Sampan* for the duration. It was transferred back to the U.S. after the war. The life raft must have gone overboard by accident.

----

12 Sep 1944, postmarked APO 886   From AWZ/APO to BSZ, Box 50, Easthampton, etc.

ALS-I

[no header]

Monday Sept 11th

[unnumbered, should be 87, but is an extra letter]
Dearest Barb

    I'm horizontal again. After the fever I had, which I suppose was a mild case of dengue, I got up not feeling too well but I had to as I was alone. I think it must be the reaction from the medication I took as my tummy is a bit upset. Anyway Howard is back so I went to see the doctor got some medicine and took to my bed to relax and get over whatever it is.

    The pictures of my camel expedition are back. They are pretty good. I haven't had time to do much on them but will get after them later.

    I don't want to raise your hopes but it looks as though I might be home in Nov. or Dec either for a visit (courier trip) or a change in /p.2/ assignment. An officer has just come through from home and he indicates one or the other. The way things are shaping up if I do get home on a change of assignment, my work having been completed here, or rather my time having been served, I might ask to resign as I don't see the further need for men of my age and qualifications. I understand quite a few of the older ones are resigning as the war draws to a close. There is a big pool of officers at home waiting to be assigned to a post like this that probably are better off in the Navy than anywhere else and are anxious for foreign duty. Why not let them take over? What do you think?

    I would like to make Lt. Com before I get out but that looks pretty hopeless. They are not giving them /p.3/ away these days and in our particular branch they are overloaded with them.

    An admiral came through on Friday and stayed till Saturday night. He had a Com. travelling with him. This was one of the reasons I got up. As far as I know it's the first admiral that ever visited Karachi.[188]

---

[188] The "admiral" is not named in this letter of 11 Sep 1944, and the word "admiral" does not appear again in AZ's letters to BSZ. Possibilities include (1) a U.S. Navy admiral, although I know of none who came through Karachi before Rear Admiral C. Julian Wheeler, who passed through Karachi on 26-27 October 1944; (2) Commodore Milton Miles (later RADM), who met AZ at some time and referred to meeting "Z and V" [i.e. Voorhees]; and (3) Vice Admiral J. H. Godfrey, who became Admiral-in-Charge of the Royal Indian Navy in late 1943 or early 1944, and who visited Karachi on 15 July 1944. AZ was invited to dine with him then – the invitation is in the Scrapbook, page 25(R). AZ's correspondence files show that at some time he met Commodore Miles and Admiral Wheeler, and although he doesn't say so explicitly, it can be assumed that he met Admiral Godfrey at some time, too. Voorhees was in the hospital when this "admiral" passed through Karachi, so it is unlikely that the admiral was Miles; it was probably Wheeler.

    Admiral Wheeler, in a letter of 7 November 1944, thanked AZ for hosting him on his way to his new post, and he refers to a "big Navy Day cocktail party" that took place on the day he arrived at his new post. This was probably on 27 October, so Wheeler must have come through Karachi in late October, say 25-26 October. The "Navy Day" party at the NLO was held on 27 October 1944 (see AZ's letter of 28 October, *infra* and invitations in the Scrapbook). Wheeler's letter is in the Scrapbook and a copy is in AZ's folder for Navy Record Documents. The party that Wheeler refers to is where he was on 7 November (probably Delhi, because he asks AZ to "look me up when you come this way again"). Since Wheeler arrived at his new post just in time for the Navy Day party, he

They no sooner left than 3 other people arrived. I was glad to have Howard get back yesterday from the hospital and Phil Haller [Halla] last night from Bombay to take over the obligations of playing host.

Your arrangements for this month sound complicated. I hop you don't wear yourself out. I think you're very wise tough to stay at the shore as long as possible.

Loads & loads of love,
Al

---

12 Sep 1944, postmarked APO 886.   From AWZ/APO to WZ, Box 50, Easthampton, etc.
ALS-I
[no header]

Tuesday Sept 11th

[unnumbered]
Dear Warren

I'm sorry to have taken so long in answering your letter but I have been very busy and right now I don't feel so well so I'm afraid this will have to be quite short. Thanks a lot for the lineup of the All Star game. I suppose you listened to it on the radio.

I'm awfully glad you are having such a nice time at East Hampton. It sounds as though there's lots to do to keep you occupied. It's nice they have a pool. You can swim and dive and have a great time. I'd love to see you do all the various activities you're interested in now.

Last week end I took a long camel ride 6 hours to see /p.2/ something that had been washed up on the beach north of here. It was quite a lot of fun. I'll tell you about it sometime.

I guess by the time you get this you'll be about back in school. What will it be the fifth or sixth grade?

Tell Babs & Lanie I'll be writing them soon. Oh I forgot I guess neither of them will be home. Why doesn't Albie write to me more often. I wish he would.

Lots of love,
Daddy

---

18 Sep 1944, postmarked APO 886.   From AWZ/APO to BSZ, Box 50, Easthampton, etc.
Forwarded on 11 Oct to N. Rose Lane, Haverford, Pa.   ALS-I
[no header]

Saturday Sept 17th

#87
Dearest Barb,

Your letter of Sept 3rd arrived several days ago and also two from Warren. What's gotten into him all of a sudden – writing three letters in about two weeks.

---

must have arrived at Delhi on 27 October. He probably came through Karachi on 8-9 September, and he came back again in October. In his letter of 28 October, AZ says two generals and the Governor were at the party, but he does not mention an admiral. Wheeler's letter is in the Scrapbook, page 27(R) and a carbon copy is in the AWZ file/folder for Navy Record Documents.

Admiral Godfrey may have returned to Karachi and AZ could have met him over the weekend of 8-10 Sep 1944. Godfrey's biography says that he came to Karachi several times, but he gives no dates for his visits. Admiral Godfrey was Director of Naval Intelligence (DNI) in London prior to being assigned to India, and while there he was the counterpart in the British Joint Intelligence Committee of Brigadier Stuart Menzies, head of GC&CS, better known as MI-6.

I am now up and about again but feeling a little woozy. I'll be all right in a day or two. There seems to be a lot of sickness around these days. Mr. Macy, the consul is in the hospital with dysentery and tape worm. Another friend has pneumonia, Tommy Weston. I'll certainly be glad to get out of this place.

Geoffrey Pike's family left this week for England. He'll be interested to hear from his cousins in America. He's a nice guy, loves bridge, about the best player here, but is frightfully British. I'm to play bridge with him on Wednesday /p.2/ of next week.

When you get home, you can tell Louisa Stuble that an old boy friend of hers has been staying with us several days, a Lt. Com. David Anon (sp?) was in the African cruise with them (Martha) some seventeen years ago.

Tonight I am going to the Clees for dinner and then to one of the few dances the Sind Club has. It should be quite a good party.[189]

A good friend of Freeman's & Clarence's [Freeman Lincoln and Clarence Lewis] came thru last week – a fellow by the name of Fischer from Pittsburgh. He wasn't here for long – just about a day.

Well that's about all the news there is except that I miss you terribly, which isn't news, and I can't wait till I'm on my way home.

Lots of love, dearest, I'll be seeing you one of these days,
Al

----

23 Sep 1944, postmarked APO 886   From AWZ/APO to BSZ, Box 50, Easthampton, etc.
Forwarded on 3 Oct to N. Rose Lane, Haverford, Pa.   ALS-I
[no header]

Thursday Sept 21st

[unnumbered 88]
Dearest Barb,

Haven't heard from you in a helluva while. I suppose I must blame it on the mails.

This week has been very dull so there isn't much to write about. Life seems to be very dull at the moment.

I played bridge last night and the night before but outside of that there is nothing to report. Night before last I held beautiful cards and won fifteen rupees ($4.80) playing for 4 annas a hundred. This is equal to about a twelfth of a cent a point. Last night I lost 8 annas or 16¢. I'm quite a bit ahead of the game at the moment and people are gunning for me. It suits me all right as I like to play and it is quite a /p.2/ diversion – it keeps your mind off other things.

Got a letter from Jim Winsor today. I don't envy him his job. It seems to have much to do about decaying cadavers. I don't think I'd like that. I don't suppose he does either.

Sheila and Suzanne Kane wrote me two very nice letters which I received about three weeks ago but I'm afraid I'm not a very good corespondent [sic] as I haven't gotten around to answering them. I just haven't felt like writing sprightly letters to little girls.[190]

I'm feeling a lot better than since I last wrote. I felt pretty lousy for a while but now have regained my vim, vigor and vitality.

---

[189] Supplement to the *London* Gazette (1 January 1946), page 7: Companion of the Order of the Star of India was awarded by the King to "Charles Beaupre Bell CLEE, Esq., C.I.E., Indian Civil Service Revenue Commissioner for Sind and Secretary to Government, Revenue Department, Sind." The same page of honors shows The Right Honourable Winston Leonard Spencer CHURCHILL, C.H., F.R.S., M.P., was appointed to the Order of Merit. Clee's wife was Mary.

[190] Sheila and Suzanne Kane = daughters of the late CDR (sel) Jack Kane and his wife Amelie. Their letters are in the folder of Correspondence with Family and Civilian Friends.

I'm sorry this is such a foul letter but you insist on hearing from me even when there's nothing to say [shifts to margin] so here it is. Can't wait till I'll be heading your way. I hope everything works out.
Lots & heaps of love   Al
[AWZ's penmanship in this letter is unusually sloppy and he had to correct the spelling of several simple words. He must have been ill or distracted when he was writing.]

----

12 Sep 1944, postmarked APO 886   From AWZ/APO to WZ, Haverford   ALS-I
[no header]

Saturday Sept. 17th

[unnumbered]
Dear Warren,
    You've about deluged me with letters – three in about two weeks. Thank you very much. It certainly is nice to hear from you.
    I'm glad to hear you say you liked Easthampton so much. It sounds as though you had a very healthy and enjoyable summer. I'm dying to see how well you play tennis. I wonder if you'll be able to beat me.
    By now you must be back in Haverford starting school. I guess you'll be glad to get back to see your old friends and make some new ones. You were fortunate this summer in having good pals as Freddy Thorne etc. Maybe he can come and visit you sometime. What grade are you in now the sixth or seventh? /p.2/
    There isn't much new with me. I've been a little sick for the last week but feel better now. This is a very easy place to get sick in, mostly because it's so dirty.
    Did Mummy tell you that I expect to be home in another couple of months. Boy, will I be glad. I'll bet you and Albie will be four inches taller since I last saw you. I'll hardly recognize you.
    Good by Warren, be good.
        Lots of love,
        Daddy

----

29 Sep 1944 postmarked APO   From AWZ to BSZ, Haverford, Pa. [This envelope was originally located after the envelope postmarked 25 Sep 1943. It was placed there in error by BSZ and I have relocated it to its proper place in 1944.]   ALS-I
[no header]

Friday Sept 29th

89
Dearest darling,
    It's very hard to write when I get no letters from you. I haven't had any mail from home in almost three weeks. Something must have happened along the way. The APO claims they are changing the route. Anyway I don't like it at all.
    The biggest news with me is that I'm pretty definitely leaving on the 5th of Nov. I don't know whether I'll end up in N.Y. or Miami but it will probably be the former. It should take about three days with the priority I will have. I will proceed immediately to Washington to get rid of what I'm carrying and get some leave. I don't know how much /p.2/ I'll get or whether I'll get it right away but the last fellow got it immediately and got 18 days and was able to get a few additional weekends while he was in Wash.
    It would be swell if you could meet me in N.Y. and ride down to Wash. with me – but I suppose the most sensible thing is for you to stay put and I'll telephone you as soon as I set foot on good old U.S.A. and figure out the best way to get together.

It's too bad I won't be home for Christmas but after a couple more months out here I should be home permanently.

That's about all I know now. Maybe the whole situation will change by the time I set off.

Life here is about as usual. /p.3/ I spent the day Sunday at Hawk's Bay with Arlo. He is going home in Dec. for a long leave. Monday all of us went to see Pride & Prejudice [Greer Garson and Lawrence Olivier, 1940] which I thought one of the best movies I'd ever seen. Tuesday I played bridge at Geoffrey Pike's with Pauline Blackwell and Doris Pettigrew – won about six rupees. Last night and the night before I had dinner here, entertaining transients.

Not hearing from you has worried me quite a lot. We've been hearing terrible tales of the hurricane that seemed to be at its worst at eastern Long Island but I guess plenty of warning was /p.4/ given and you would have cabled if anything serious had happened. Nevertheless I'll feel much better when I hear everything is all right.

Benjie Bromhead replied to my letter by telegram saying everything was in order for me to use to [sic] NWFP pictures and story. How is the situation? Is anybody really interested or should it just be dropped? Curt Winsor also replied that he would get a release from the Navy Dept.[191]

It will be wonderful seeing you again. I just can't wait until Nov.

        All my love and more,
           Al

----

5 Oct 1944 postmarked APO.   From AWZ to BSZ, Haverford, Pa.   ALS-I
                                 [no header]

                                                                          Wednesday Oct 4th

#90
Dearest darling,

One month from now I hope I'll be ready to leave. I should know definitely by the middle of this month. I guess the best plan would be for you to go to Washington & stay with Ginny [Mrs. J. Freeman Lincoln]. No matter where I land I probably will fly directly there. It should take about 65 hours. I'll send a cable with "all my love" the hour I leave. I just can't wait.[192]

Since I last wrote, three letters from you have come dated the 10th 12th and 20th. I won't go into the financial business now – that can wait till I get home – but if BS & R. & Stan [Brundage Story & Rose, his investment counselors] have any strong ideas for goodness sake carry them out. The thing's in their hands and I'm perfectly willing to reduce my percentage /p.2/ of common stocks.

You must have been having a busy time getting kids off to school and moving back to Haverford. I hope everything worked out as planned and you all had a wonderful summer.

I was quite worried about where you were when the hurricane struck and how you fared. You mention it very casually in your letter of the 20th so I guess everyone is all right and no damage done to our place. We have gotten very meagre reports but eastern Long Island was supposed to be right in its line.

Howard Voorhees is in the hospital again. This time with a fever. They thought it might /p.3/ be malaria but the tests so far have been negative. It probably is dengue or sand fly. It puts a lot of work on me but I don't mind. I'll be leaving here soon. They might send out another man while I'm away. I guess I've told you that when I get back here I expect to stay only about a month and then be relieved, for my 18 months will have been served. Then I'll go home, maybe by ship, and by that time I hope things will be such they won't need my particular services anymore.

---

[191] Permissions from Bromhead and Winsor were received. Bromhead's telegram of 24 September 1944 is in the Scrapbook, page 28(R); and Winsor's letter of 12 September 1944 is in his folder in the AWZ Wartime Papers. Both are in Correspondence with Others, Chapter 6, in this book.
[192] AWZ's hopes and tentative plans to return in November 1944 were based on correspondence from LT Curt Winsor (in the Winsor Correspondence). His hopes were, however, dashed.

There isn't much to say about what's going on here. Last night I saw Jane Eyre – a sad but good picture [Orson Welles, Joan Fontaine, Margaret O'Brien, 1943]. Today they are staging an exhibition baseball /p.4/ game and the general has invited me to go so that's practically an order.[193]

I'm afraid we'll have to throw a party on the 27th (Navy Day). We'll have to invite all the important people and it will mean a lot of work and preparation. It will be sort of a farewell party for me I hope.

I must go and get dressed so good bye dearest,
   With all my love,
    Al

---

10 Oct 1944 postmarked APO   From AWZ to BSZ, Haverford, Pa.   ALS-I
         [no header]

                         Tuesday Oct. 9th

[unnumbered 91]
Dearest darling,
  A letter from you, 2 from Babs and one from Lanie came today – a pretty good haul.

  The girls both seem to be enjoying their school and their work. Lanie must have had a terrible time getting to CH [Chatham Hall School] with all the wash outs etc. Too bad neither of them like their room-mates. I thought Babs was to room with Janet Hartwell, the girl she roomed with at CH. They both write very sweet and affectionate letters and I'm very proud of them.

  Jack Kane's yeoman came through here the other day on his way to report for duty at Bombay. He told me about the last days and something about the financial situation. Evidently Jack had borrowed up to the hilt on his life insurance and /p.2/ what was left went to pay his heavy indebtedness. Evidently there will be a monthly check from the government worked out on his gov insurance (on which you can't borrow) and his pay. This will go on at a certain rate until the twins are 18 and then will be cut. It sounds like a good arrangement in doling out what money there is. If it were paid in a lump sum I'm afraid it would be out the window in no time.

  I'm writing this letter with one foot in a hot water bowl with Epsom salts. I've picked up somewhere, goodness knows where, a case of athlete's foot. I've had it for 3 days and it is very stubborn to healing. I can't understand how I got it. I've been very careful and the only one to use /p.3/ my bathroom is myself. India can give you about everything with very little urging.

  Howard is back from the hospital looking very thin. He had bacillary dysentery. One of our men goes tomorrow for observation. They think he may have amoebic dysentery – so it goes.

  My trip home, I think, has been knocked into a cocked hat. After I was nominated out here, they sent for my orders and the reply came back negative. However it might work just as well if not better. They say we're only supposed to be here 18 months. I understand there are plenty of replacements in Wash. So I'm asking to be transferred, to get home for Christmas, get some leave. I think they'll do it, as /p.4/ there are plenty in Wash. sitting around doing nothing. When I get home I'll be able to appraise the situation. Maybe the war in Europe will be over and they won't need me any more. Anyway I'm doing all I can to get home – and by Christmas.

  I don't know where you get my letters being infrequent and impersonal, that we are becoming remote and that I _used_ to say you were my one and only. It seems I spend most of my spare time writing – I haven't read a book in a year. I try to be personal – at least it's that way on this end even if you don't read it that way. _You_ are still the one and only – need I keep repeating it. I never was a hand for handing out a lot of slush.

---

[193] One or more photos of this event are probably in the AWZ-Photo Album, undated.

I enjoyed the Kodachromes a lot. It's too bad I haven't a magnifier or projector. You can't see much [shifts to margin] in this size. I'm sorry I was premature on the trip. I thought it was all set. I'm dreadfully dissapointed [sic] that seeing you will be that much further off, if only a month.
    Loads & heaps of love   Al

----

17 Oct 1944 postmarked APO   From AWZ to BSZ, Haverford, Pa.   ALS-1
           [no header]
                           Monday Oct 16th
92
Dearest Barb

 A month old letter from you arrived today. It evidently got stuck right here. The A.P.O. has taken on a lot of new native help and I guess don't know their way around. If you add to the address USNLO this may not happen again.

 I'm very proud of Warren's feat in winning his first tennis tournament. You didn't go into details very much. How many matches did he play? etc. I'm glad he's interested enough to practice. He will be quite a star athlete with all the sports he's interested in. I'm dying to see him play.

 The hurricane must have been hair raising. I wondered why you hadn't mentioned it. I was quite worried when I learned that East /p.2/ Hampton or at least eastern Long Island was in the path of it.

 Did you read the article on the Khyber Pass in the Sept 30th New Yorker? He makes it sound like a very fearsome place. I thought it quite tame, I was rather disappointed. It's not much different than a drive through Del. Water Gap without the water. It didn't hold a candle to the many other places we visited in the NWFP but is shows what a guy with imagination can do.[194]

 This week has been about the same as any other. On Tuesday I went to the Harris's to dinner and then to the Boat Club to dance which didn't do my foot any good. Since then I've stayed off it as much as possible till yesterday when I tried nine holes of golf.

 Wednesday and Thursday, I stayed /p.3/ here and Friday and Saturday I played bridge – Friday Doris and I took on a Greek couple – the Dendrinos. They were quite elated as Athens had been just liberated. It's interesting to meet people you never dreamed of meeting before. She is quite attractive in spite of a gold tooth but he is like though southern Europeans we saw at Lido Venice. In fact the 8 or 10 Greeks here are all about the same – a pretty greasy lot.

 Saturday night I went to the Blackwells again with Doris. They are grandparents for the first time. Their daughter who lives in another part of India, came here to have a baby a 5 pound girl. They come rather small out here.

 My golf wasn't so bad yesterday. I had a 41. It's crazy /p.4/ golf – playing on a desert but it is something to do that's different.

 A guy came through from our desk in Wash. but he had left before my letter to Curt had arrived. So he didn't know the story in back of why the complete about face on letting us go back for a courier trip. I'm burnt up about it but if the other works out before Christmas I'll be more than compensated. In fact I prefer it but I thought the other a bird in the hand.

 I'm sending another package – mostly Kashmir scarfs. I thought you could use them for Christmas presents. The necklaces also came from Kashmir. They are semi precious stones – the red – carmeline but don't know what the green is. You take your pick (The silver things are swizzle sticks) [shifts to margin] and let the girls have the other two. You might give the Kane twins a plain white scarf each. Sheila thoughtfully has sent me a chess & checker game. They have both written me.
   Heaps & heaps of love   Al

---

[194] Ernest O. Hauser, "A Reporter at Large: Pathans Behind the Rocks" *New Yorker* 90 (30 September 1944), 169-222.

----

23 Oct 1944 postmarked APO   From AWZ to BSZ, Haverford, Pa.   ALS-I
                                     [no header]

Monday Oct 23rd

93
Dearest Barb

    Your letter (of Oct 9) and Albie's letter arrived the same day. It's been a long, long time since I've heard from him. He's getting to be quite a big boy writing his own letters. It was very nice one too.

    I see you've plunged right into a lot of work again – with the canteen and window box. Why do you do it? You'll just wear yourself out again. With servants as they are, you have enough to keep you busy at home taking care of children etc. There must be somebody else that could take over.

    Too bad Perry's marriage didn't work out. He must be pretty disallusioned [sic].

    Anna [(Zimmermann) Kelley] wrote me about Dick's [his nephew, Richard Kelley Jr.] marriage. /p.2/ I guess she's resigned to it but didn't sound overly enthusiastic. I thought when he was here that they would wait and not get married right away but I guess after what he's been through she looked pretty good to him.

    What did you do about football tickets this year? I hope you took them yourself. It sounds as though Penn had a pretty good team and I'm sure Warren will get a great kick in seeing the games, if you didn't want to go.

    I didn't know that Warren had even started playing football. It's certainly perfectly all right with me if he prefers soccer. Has Albie started playing anything at all? It's about time he did. We don't want a bench warmer in the /p.3/ family. I'm thrilled that everybody is so fond of the boys. I fill with pride. They are great guys and I hope I'll be seeing them soon.

    Speaking of that, there is no news. I've written requesting to get home by Christmas but it's too early to get a reply. In fact it will probably take some time to work out and when and if it comes it will be without warning. It <u>should</u> work out but of course you can't count on anything. I'll be frightfully disappointed if it doesn't – especially since the courier trip fell through. I had a secret hope that – getting home on the courier trip – they might decide to keep me there. The only thing to do now is to sit back and wait, anxiously opening each batch of mail / p.4/ that comes in. I've done all I can do at this end. It's now up to them.

    This week has been one and the same with the rest – played bridge Friday with Gen. Hind and Vere Birdwood and Doris P. The General is a nice old bloke in command of the British Army forces here. Vere is a daughter-in-law of the former C in C of India – the job Auchinleck has now.[195] Saturday I went to the races and yesterday to Hawkes Bay.[196] These special engagements are all very empty to me. My main desire is to get home to <u>you</u> and life here is meaningless. I miss you so much. Sixteen months is a helluva long time. I hope my best laid plans won't completely fall through. Don't worry about sending Xmas presents – even if I [switches to margin] don't make it by then it won't be long thereafter.

        Heaps & heaps of love   Al

---

[195] Vere Birdwood's mother-in-law was the aunt of AWZ's companion on the NEFP trip, Sir Benjamin Gonville Bromhead; and her husband, Christopher Birdwood, was in the Waziristan Campaign with Bromhead in 1919-1920 (see footnote with AWZ's letter of 4 June 1944).

    Auchinleck = Field-Marshal Sir Claude Auchinleck, GCB, GCIE, CSI, DSO, OBE, LLD. AWZ saw him, and perhaps met him, at some time. His photo, which he labeled "the Auck teaching a sepoy how to shoot," is one of the loose photos in a 7x10½ in. envelope labeled "India WW II" by BSZ (*supra*).

[196] Hawkes Bay = A beachfront site in Karachi, near Sandspit, and 22 km from Ingle Road, where the U.S. Naval Liaison Office was located. (From Google Maps, accessed 2/10/11).

28 Oct 1944 postmarked APO   From AWZ to BSZ, Haverford, Pa.   ALS-I
                                                                        Saturday Oct. 28th

[unnumbered 94]
Dearest Barb

An officer arrived yesterday bearing a note from Curt [Winsor] saying he had seen you and told you the bad news about my coming home – how you had taken it so bravely. I guess he really did work hard to get me home on a courier trip and I appreciate his efforts. It's too bad he couldn't convince the front office that another officer could be very easily drawn from one of the other posts to be here in my absence. Well anyway that's that. I hope my next idea won't meet the same fate.

An officer who should have the right dope passed thru here a couple of months ago from Wash. and said any of us who had been here for 18 months could on request be relieved, there being about 35 officers in the foreign pool awaiting assignment. Curt in his letter talks about mid-winter and mid-summer for my return. He either doesn't realize I've been here as long as I have or it's another bit of scuttlebut gone phut. My request is in, as I've told you, but I don't think Curt knew about it when he talked to you. It's too early to get any reaction so we'll just have to sweat it out waiting to see what happens. I've done all I can.

Last night we had our Navy Day party and quite an affair it was. We had H.E. the Governor of Sind and his wife Lady Dow, the two American generals [John A.] Warden & [Julian B.] Haddon, our Consul [Clarence Macy] etc etc in fact about all the pucca people in town altogether about ninety. It was a cocktail party that lasted till about 3 AM with /p.3/ the group captain in charge of transport planes etc promising me any sort of a plane I wanted when my time came to go home. As a matter of fact I think it would save time and be a lot more interesting to go by way of U.K. Maybe it will work out.

Day before yesterday [26 October 1944] I attended an "At Home" at the Iran consul's. It was quite interesting and the first of this sort of thing I've done. It was in the late afternoon and they served tea and cakes. The governor and his wife were there as well as the rest of the burrah sahibs and memsahibs about town.[197]

That's about all there is to write about. I haven't heard from /p.4/ you since I last wrote.

I'm awfully disappointed the trip didn't work out – it means my seeing you will be that much further away and believe me that has number one priority in all my desires. However it might be a blessing in disguise if the other works out on time.

                    All my love dearest,
                        Al

"Iranian Consul's 'At Home' / John Blackwell on left / Roger Collet on right, ADC to HE the governor / Consul with back toward camera" [photo is detached]

---

[197] A photo of AWZ at the Iranian consulate is in the loose photos envelope and in the Photo Album (Sheet 1, #15), and the invitation to the event is in the Scrapbook.

[See Bibliography for the title of an official report from AWZ to CNO (DNI) on 31 October 1944, on Karachi Port facilities. The report was sent via his superior officer, JICA/CBI]

4 Nov 1944 postmarked APO   From AWZ to BSZ, Haverford, Pa.   ALS-I
[no header]

Thursday Nov 2nd

95

Dearest darling,

    You are so sweet being so enthusiastic about my coming home – reserving rooms in New York and Washington – getting new dresses etc. all so everything will be perfect when I get home. You are a darling and I'm thrilled the flame is still burning so furiously.

    There is still no news, a bit early yet. I suppose no news is good news. If they have a mind to turn down my request, I would think they would do it right away. If they comply, they probably will wait until the change is to be made and not give me any advance notice. Anyway I'm standing by waiting patiently to see what my fate is. It won't take me long to /p.2/ get ready and believe me, I'll be a very happy boy when I'm finally on my way.

    I wrote Curt the other day, giving him a bit of my mind – not that I blame him in the least – I think he did all he could – but I was pretty browned off when I heard why I wasn't to make the trip and I wanted to point out a few exceptions to top-side's reasoning.

    Two letters from you came together yesterday (Oct 15th & 20th). Your Easthampton friends sound very nice and important. I'm afraid the possibility of my going to Bombay again is very remote so I don't think I'll be able to do much investigative work for Justice Haynes. /p.3/

    It was quite funny your meeting Batdorf at the Racquet Club. Tuesday, at the Boat Club, the night before I got your letter, a guy came up to me, introduced himself and it was Batdorf and the next day I got your letter. I don't think he seems very pre-possessing but it was nice to talk to somebody from home.

    Don't worry about Christmas presents. Even if I don't get home for Xmas it should be shortly thereafter and anyway it would be very uncertain getting anything out to me now. Powell's method does not seem to bring results – I haven't received the viewer, pyjamas, golf balls etc. you've mentioned as having been sent to him* – and remember last year – sending them via ONI (by ship)

* You might recall any stuff he has unless they are already on the way /p.4/ things didn't arrive till months later.

    The party at the Boat Club was thrown by Arlo for the Markleys. Ruth Markley has just arrived from U.S.A. Wives of civilians are now permitted to come out but it takes a long time. It was a nice party – dinner first at the Sind Club.

    Monday the Blackwells had a bridge party – two tables. They are very nice and I hope someday you'll be able to meet them and other of my friends. Some think they might come to America after the war.

    Wednesday I played bridge at Dixon Edwards house. He is the senior vice consul – more or less of a drip.

    That's about all, dearest, until our glorious reunion.

        All my love,

            Al

[Enclosed: Newspaper clipping: "India Shocks Him / Poverty and Misery Everywhere / London, Oct. 28," on the margin of which AWZ wrote, "The pot calling the kettle black"; and "Navy Day Programme / 21st October 1944 / In commemoration of Trafalgar Day," in both English and Urdu, on which AWZ wrote "simple language this"]

5 Nov 1944 postmarked APO   From AWZ to BSZ, Haverford, Pa.   TD [typed document]

This envelope contains five typewritten pages, on thin paper, 13x8¼ in., without comment, joined with a paper clip. The text occupies a maximum of about 11 vertical inches on each page. The pages are clean carbon copies. They span the entire NWFP trip of AWZ, Enders, and Bromhead, from the travelers' departure to Shabkadar on 15 November until the last day in Quetta, 12 December 1943.

The header on p.1 is:

Notes on Chitral Trip
Nov.15 -23, 1943.

Page 1: Nov 15, 16, 17
Page 2: Nov. 18, 19, 20, 21, 22, 23, and 27 [no notes for 24-26 Nov.]
Page 3: Nov. 27 (continued), 29, 30, Dec 1, 2 [no notes for 28 Nov.]
Page 4: Dec. 2 (continued), 3, 4, 5
Page 5: Dec. 5 (continued), 6, 7, 8, 9, 10, 11, 12

Images of the 5 pages are shown below. They were scanned from a copy in the AWZ Wartime Papers. Information about the period 24-26 November in Peshawar, which is missing here (including the Governor's garden party and meeting the Viceroy) is in AWZ's letter to BSZ and in his notes and drafts of this final report (see above and excerpted in the next chapter).
    In this report AWZ mentions several reasons for conducting this trip, and for his being invited to be on it – in addition to what he has previously stated in his correspondence (6 September, *supra*). In the report, he mentions looking for "Bolshies" in the Chitral portion; of Axis attempts to foment trouble with Hindus in India and using Ipi (who is also known as the Faqir of Ipi); several other places where Ipi is mentioned as a major trouble-maker along the Border; mines and tank traps that were laid in anticipation of German/Nazi invasion in 1942; and unrest characterized by shelling and bombing by the R.A.F. in the Waziristans. He also comments on the characteristics of the various tribes as he observed them, or reports that he received about them. The five pages are well-typed, with only a few typos and few signs of erasures. I suppose that they were typed by one of the yeomen in Karachi. The original was probably retained at NLO Karachi, where it was destroyed along with most of the other files when the office was closed in 1945. A copy of the report was probably sent to his superiors in India (Chief of ALUSLOs in India; Naval HQ IBT; and JICA/IBT or JICA/SEAC), and in the U.S. (ONI), but all of these copies have disappeared. None are in the U.S. Navy's files in NARA II.

Notes on Chitral Trip
Nov. 15 -23, 1943.

Nov. 15th: Picked up Sarwar Khan, Subhadar, Intelligence Liaison Officer to H.E. the Governor. Stopped at Shabkadar, picking up Mohammed Isacc, Asst. P.O. under Deputy Commissioner, Peshawar. At Shabkadar is an old fort with lists of dead more from disease, sunstroke and drowning than from fighting. Met a Malik who is enemy of the Usaf Kheyl Malik. Went up to Yusaf Kheyl and lunched with its Malik. Passed donkey caravans, bringing wheat into town for exchange into clothes. Yusaf Kheyls are a portion of the Lower Mohmand Tribe whose protection was guaranteed by the British. The Mohmands, however, misused this protection by taxing the Upper Mohmands who passed through their territory. Roads are now patrolled by Khassadars and illegal taxation is stopped, on threat of withdrawal of British protection. All villages fortified and concentrated. No schools to avoid fueds. The community is a surplus one, exporting products to India. The nearby Maliks are feuding, with the badi (blood feuds) widespread. Khassadars work alternate months. They are immune from blood feuds while on duty in order to avoid insulting government.

Nov. 16th: Trip to Swat. Country is same type as Lower Mohmand country, the route going up the Swat River. Passed the fort where, for six days, the British garrison was imprisoned and whence Gunga Din came down to the Swat River to get water for the troops. They could only heliograph from the fort at this seige. At Swat streets were lined by Levies. At Palace bugles were blown. Met by Attaullah, secretary to the Wali of Swat. Lunched with the Waliahad, Jehan Zeb. Swat was made a state in 1926 and is one of the newest. Peaceful. Wali suspected of pusing his brother over a cliff. Returning, inspected the headworks of the Malakand hydro-electric system. The Prime Minister, while hydro plant was discussed, said it was all very well but there is no evidence to show there is electricity in the Swat River. Many Buddhist towers remain along the road. Inspected a school, called a college, but of high school rank. Education is only hope to bring these people under peaceful government. (Sir Benjamin says Axis are conspiring with Hindus and Sikhs to boost the black markets throughout India, but educated Sikhs is of high type, used in the Army. Sikh merchants are lined up with the Hindus and Congress Party. Mohammedan mollahs are usually out for themselves rather than to further religion. Many mollahs are immoral and of bad type.) Dined that night with the Kenneth Packmans. He is P.A. Malakand and C.I.E., in charge of Dir, Swat and Chitral.

Nov. 17th: Held up at gate of Chakdarra, going into Tribal Territory of Dir State. Khassadars wanted to send an escort with us. We declined and Sir B said that we were American guests and what would we think if khassadars would not let us through. Dir and Swat are jealous of each other, Swat being partly taken out of Dir State. In Tribal Territory there are Scouts and Militia which are officered sometimes by British, sometimes by natives. Sir B. thinks it much better to use locals rather than bring in outsiders to foment trouble. So far Chitral and Kurram Scouts are mostly locals, but others are made up of locals and neighboring tribes. Chitral Scouts are officered by British as only relatives of the ruler may become officers. (Khazaks came down from Outer Mongolia to escape Bolshies in large numbers, with cattle and horses. They got to Kashmir, but lost all their belongings and now are wandering back to Outer Mongolia in small bands. We passed several of these groups.) The Dir Valley is very fertile, with terraced hillsides of rice (in summer) and wheat (in winter). The Nawab of Dir used to collect levies from the British Govt. for allowing troops to pass through his territories. To make it appear worthwhile, he occasionally shot up the troops the keep everyone convinced the levy was justified. The British troops were a battalion kept in Chitral prior to establishment of the Chitral Scouts. In tribal territory there is no written record of land causing many blood feuds, etc. as to ownership. (In South Waziristan coal was found and used by the Scouts. About 5 men laid claim to the land and a bad blood feud resulted, according to Sir B.). Mollah is tribal priest and Malik is Tribal leader. Propaganda against Ipi was that he tried to combine both offices in himself which was against the Koran. (Sugar cane is also raised in Lower Dir and Swat). (A sugar refinery is at Takht-i-Bhai). At Dir had tea with the Nawab and Waliahad, a Captain doctor. Nawab rather sickly, the Waliahad rather childish -- does not smoke in front of his father, etc. Waliahad is married (since 1941) and has one daughter. Dir is collecting point for antimony ore from Chitral and is roadhead. Dir city has electricity. -- hydro.

Nov. 18th: Left Dir at 9:00 a.m. People thought attempts to drive over Lawari crazy. Arrived at Gujar Levy Post (7,800 ft.) at 11:00 a.m. Reached top of Lawari at 12:30 -- mileage from Dir 14. Altitude of top -- 10,240 ft. Snow on top only; none on way up. Road narrow in spots, rough with rocks at upper end. On descent, ice and snow found in patches. Descent much steeper than ascent, with many awkward zig-zags. Arrived at Ziarat (7,500 ft.) at 2:30 and tead. Reached Ashret at 4:30 p.m. Ashret is roadhead on Chitral side. Mileage from top to Shret -- 8.8. Reached Drosh about 6:30 p.m. Drosh is 13 miles from Ashret. Stops on the way totalled at least 2 hours. Total running time about 7½ hours. Major White and Captain Hemming were at Scout HQ. Airfield at Drosh very small and L-shaped, cannot take modern planes -- dimensions about 1500 yds. by 200 yds. Marked with white circle in center; on right bank of Chitral river. Would be dangerous in cross-winds. Is served by an extension bridge to Drosh. Prevailing winds are up and down river valley and lengthwise of the field. Field about 1,500 feet below Drosh fort. Spet night in Drosh at Scouts Mess.

Nov. 19th: Left Drosh after lunch, reaching Chitral at 4:30 -- distance 26 miles. Met by Ghazi ud-Din, brother of the Mehtar, at the palace. Bugles and guard of honor. Had tea. Went out to polo game staged in our honor. Called on Thornburg, A.P.A. and his wife. (Thornburg sent telegram that ice, snow and mud on Lowari and advised against our attempt). Dined at palace with the Mehtar, the Prime Minister, Ghazi ud-Din, Dr. Kekyll, the Thornburgs and several officials who were relatives of the Mehtar. Chief of Bodyguard present. Saw Chitral and Hunza dances after dinner. State dining-reception room fantastic with chandeliers, artificial flowers, etc. Former Mehtar was pervert. Practically no Pathan seen wore glasses -- they do very little close reading and constantly watch distance. No women were in evidence in Chitral. Mehtar's name is Muzafar ul Mulkh.

Nov. 20th: Went hawking in a.m. using two perches and killing 8 chikhor. Mehtar's subjects kissed his hands, offered gifts, etc. State is feudal with Mehtar's powers reaching to confiscation of land, etc. Motored back to Drosh at 5:30. Called on the Mehtar's brother who is Governor of Drosh and heir to throne as Mehtar has no legitimate sons. Mehtar reported as about to marry a princess of Dir, the widow of his brother and predecessor as Mehtar. Spent the night at Drosh Fort Mess.

Nov. 21st: Left for Ziarat at 2:00 p.m. arriving at Ashret at 2:45 p.m. From Ashret to Ziarat (5.3 miles) in two hours, arriving Ziarat at 4:45 p.m. Spent the night at Ziarat, leaving peep at Scout Post.

Nov. 22nd: Left Ziarat at 11:00 a.m. in order to get sun on Lowari top. Made the 3.5 miles to top in 2 hours. Spent about 3/4 hours at one slick zig-zag turn; the rest were comparatively easy. The 3.9 miles from the top to Gujar were made in 25 minutes. The 10 miles from Gujar to Dir were made in 1 hour and 40 minutes. Total elapsed time 7 hours; running time from Ashret to Dir about 6 hrs., 5 minutes. Spetn night at Dir Rest House.

Nov. 23rd: Motored 160 miles from Dir to Peshawar. Ate late lunch at Malakand.

Gifts were given to us by the Mehtar of Chitral and the Nawab of Dir.

Nov. 27th: Trip up Khyber. Peshawar to Jamrud - 9 miles, entrance to Tribal Territory. Many blockhouses in Pass, tank traps, winding roads. To Landi Khotal -- 28 miles; 32 to Landi Khana. Forts and Shagai and Landi Khotal. Afridis on either side of the Pass as far as Landi Khotal, with Mullagaris (bastard Mohmand) a short distance to the North. RR goes through to Landi Khana. Shinwari tribe are the rest of the way along the Pass and into Afghanistan. Afridis a strong tribe; good fighters. In 1930 an Afridi Lushkar surrounded Peshawar. Since 1937 the Axis has caused unrest among Afridis. In 1938, parts of the Afridi tribe started a march on Jelalabad to loot it and embarrass the Afghan government. They were stopped by a blockade and effective bombing and made to pay a penalty in guns. Gun factory -- Nov. 28th -- Shahbadin, on the road to Kohat, has a small rifle and revolver factory, making about 10 revolvers and 10 rifles per month at a cost of Rs. 160 for a rifle (Rs. 140 for a sporting model); Rs. 120 for a pistol. Guns are good considering the crude machinery that turns them out. On the way back had tead with Subhadar Major (Honorary Captain, retired) and his son who is still in service. His name (the father's) is Tor Khan. Both father and son often decorated for valor and bravery. Water is found in good quantity at 10 to 20 foot levels below ground. Fruits, oranges, limes and tangerines may be grown here very successfully.

Nov. 29th: Left Peshawar at 8:00 a.m. - reached Kohat at 9:35 -- 41.9 miles. Left Kohat at 10:55 a.m. -- reached Thal at 1:15 p.m. -- 104.4 miles. Left Thal at 3:42 p.m., reached Parachinar at 6:15 p.m. - 160.3 miles. Saw the combination P.A. and D.C., Sheikh Sahib and his brother, Mir Ali for breakfast at Kohat. Saw Brigadier Barstow at Thal and lunched in Station Mess. Army insisted on giving us escort with armored carriers for 10 miles out of Thal. Tead and dined at Donald's (P.A., Kurram). Dinner guests were Major and Mrs. Boulter; Col. Francis and an Indian Captain. (P.A.'s full name is Kenneth Donald).

Nov. 30th: Climbed Peiwar Khotal on Afghan border with Denneth Donald, Lt. Col. Francis, Sir Benjamin and about 15 rifles of the Kurram Militia. Lunched on top of One Gun Hill - scene of an engagement between Afghans and Lord Roberts in 1879, and also involved in Third Afghan War in 1919. Distance from Parachinar to foot of Peiwar Khotal about 15 miles. Altitude of Peiwar about 10,000 ft. From Peiwar motored to Karlachi on Afghan border and point where Kurram River flows into India. Peiwar and Karlachi two main routes from Afghanistan in these parts. Had tea at KM post at foot of Peiwar (Teri Mangal). Returned to Parachinar for drinks with Capt. Echlin and dinner at KM Mess. Altitude of Parachinar, 5,800. Kurram Valley tribe are the Turis who are Shiahs and asked the British to come in to protect them from the surrounding Sunnis. Kurram Militia are recruited almost entirely from Turis and Militia has practically nothing but internal security problems. Country around Thal heavily filled with tank traps along roads and river-beds. This was defence against possible German threat during July 1942. Floods now washing away many from riverbeds. Total miles from Parachinar, via Peiwar Khotal, Karlachi and return -- 38.

Dec. 1st: Left Parachinar 9:40 a.m., arrived Thal 11:35 a.m. -- 54 miles. Here picked up Wazir Tribal Escort. Proceeded to Spinwam where we were met by Major Denning and thence went to Mir Ali. Met Lt. Col. Janson at Mir Ali and he accompanied us to Miram Shah. Arrived Miram Shah at 2:55 p.m. Stayed with Lowis, P.A. Tochi, and ate at Tochi Scouts Mess. Tochi Valley occupied by Daurs who are Sunnis and asked for British protection. Daurs are considered a bad lot, having conspired with Ipi, whose village is on the Tochi between Mir Ali and Kajourie on main road to Bannu. Ipi now on Afghan border west (and a little north) of Miram Shah. Two of his guns were reported moving on camels past Dosalli. Post at Sara Rogha and Ladha also warned to be on lookout for Ipi's guns which were reported in hands of 12 Mahsuds.

Gusht -- a maneuver or operation by Scouts or Militia	Naik -- Corporal
Tikala -- feast in the native style	Havildar -- Sergeant
Lushkar -- an enemy force of 50 or more men	Jemadar -- Viceroy's Commission 2nd Lieut.
Chigga -- a posse of Scouts or Militia	
Narai -- a hilltop or pass	
Subhadar Major -- Viceroy's Commission, Major.	Subhadar -- 1st Lieut & Captain
Serai -- Caravan resting place	

Rs. 2,000 is ordinary blood money for a murder and, once accepted, settles the murder. A murder can be bought for Rs. 50-100, which is less than the cost of a rifle. A British rifle brings about Rs. 1,200 -- a round of ammunition Rs. 1/2/-. Native round brings As. 6 to 8.

At Miram Shah heard that 10 Hindu women were kidnapped and taken into Afghanistan; also that a Bannu Hindu shopkeeper was kidnapped probably on information of his Hindu competitor. Ransom demanded for the shopkeeper -- Rs. 7,000 with negotiations under way. Huge Airdrome at Miram Shah.

Dec. 2nd: Left Miram Shah at 10:10 a.m. reached Dosalli at 11:20 a.m. -- 25.4 miles. Left Dosalli at 11:40 a.m. going to Iblanke Post. Saw Brigadier Mervin Hobbs enroute. He had just been shot up while out with his Brigade (from Razmak) on a practice maneuver. Arrived back at Dosalli at 1:00 p.m. -- 15 miles. Total miles for day -- 40.4. Tochi Scouts escort went with us to Dosalli. Brigadier Hobb's Brigade made up of 3rd Bn. Ghurkhas, a Bn of Dogras and a Bn. of Green Howards. All camped outside Dossali Scouts Post and were shot at at 7:15 p.m.

-4-

from near Iblanke Post -- 10 to 15 rounds from a light machine gun -- no reply although flashes were seen and a gun trained on spot in case of repeated shooting. At Dosalli was a 5.5 gun and truck which had fallen over khud killing 2, probably mortally wounding 1 and hurting 12 -- all natives. Had big tikala at Dosalli with guests (besides ourselves) a Major Taylor of the Dogras and a Captain from the Green Howards. At the Tikala were Cumbar Kheyl, Adam Kheyl and Kukhi Kheyl Afridis; Shiah Bangush (from near Kohat) and Khuttacks. These men command solid platoons of their own tribesmen, but brigades of mixed tribes are commanded by Subhadars without friction.

      re Tribes: All tribes above Kabul River quiet and easy to handle. Dir, Swat and Chitral tribesmen are ruled by autocrats under Mohmands and Afridis and British supervision and are easy to handle.
Mohmands and Afridis are democratic but fairly well handled by Maliks and Khans, therefore give little trouble.
Wazirs and Mahsuds take democracy to the point of anarchy; they will not listen to Maliks or Khans unless it suits them and are difficult to handle.
Tribes in Zhob area are democratic but settled and willing to heed maliks and therefore are no great problem.

Dec. 3rd: Left Dosalli at 10:15 a.m., reaching Miram Shah at 11:25 a.m. Went back to stay with Lowis. Total miles 23.4. Lowis had received word that a M.E.S. lorry had been held up near Datta Kheyl, a supervisor kidnapped and 2 men wounded. A dicker for the kidnapped man's release was reported by Assistant P.A. for Rs. 7,000. Two 500-lb bombs dropped on a gang of tribesmen who lived in caves near Miram Shah. Later reports said no casualties, but good moral results. "Kheyl" means a subsidiary of a tribe, for example, Adam Kheyl are a subsidiary clain in the Afridi tribe. "Zai" means "son of"; for example, Alizai means the sons of Ali. Notes: Grazing grounds are the keys to Tribal behavior. Also economics; for example, Rs. 1,000,000 worth of timber comes down the Tochi from both sides of the Durand Line just south and west of Miram Shah -- mostly brought by Madda Kheyl. If this were stopped the tribe would be hard hit, it would mean war. The job of controlling tribes, according to Benjy, is not a military but a Scouts one. For deserters a system of amnesties might be useful if it worked.

Dec. 4th: Left Miram Shah at 9:30 a.m. Had big trea at Mir Ali at 11:00 a.m. and took Major Denning to Bannu. Arrived at Tank at 2:30 p.m. and took the acting P.A. "Pat" Duncan in jeep to Jandola. Arrived Jandola at 3:35 p.m. Bannu is prosperous town compared to Tribal territories -- land is rich. Passed large groups of Gilzais migrating from Afghanistan for the winter. Generally speaking the whole Gilzai Tribe (including the Povindahs, or traders) and including the Suleiman Kheyl who refuse to send recruits to the Afghan Army, are friendly to the British. One small subdivision, known as the Dinar Kheyl (some 50 families), however, are sneak thieves and make trouble between Tanai and Gul Kutch.

Dec. 5th: Left Jandola 9:30 a.m. in Duncan's lorry and escorted by a Mahsud lorry in which Benjy and I rode. Had morning tea with Kushwakht (brother of Mehtar of Chitral) at Sora Rogha; lunched at Ladha (where tower was hit by 4.5 gun). Had afternoon tea at Tiarze. Arrived at Wana at 7:20 p.m. Ladha tower was shelled because its owner refused to turn over a murdered who had killed on post property. In this case the P.A. determines what action should be taken. Notes: Khassadars are organized by section of country, in companies of 40 to 100, with hereditary subhadars who are sometimes only boys (father to son). Companies can be moved around within tribal territory. Mahsuds have about 3,500 khassadars. They are responsible for safety on roads and can be suspended or demoted -- latter is very serious because subhadar's entire family is thus demoted. They can also be fined. Khassadar pay is Rs. 25, plus dearness allowances of Rs. 6/8 per month. Subhadars get Rs. 85 pr mo. Mahsuds steal from Wazirs and vice versa. Povindahs will fight if molested by either. 12 Mahsud guns, on Ipi's order were reported moving in this territory and PA gave orders to fire on 10 minutes notice. Tribal guarantees are mostly rifles (worth about Rs. 1,200 each). If fines are levied, tribal leaders are arrested and held until fines realized. Kaniguram is capital town of all three Mahsud tribes (Dre Mahsud) and has population of some 7,500. In it live Urmurs who speak no known language and are artisans (knives and guns) of Mahsuds; some now take to lorry driving. Urmurs feed jirga members on visits to Kaniguram gratis. This is their protection fee.

Roadworkers for M.E.S. in tribal territory are locals, who have not been yet fired on for fear of blood feuds. Ipi was born a Daur (Tochi Valley) but has turned Turi. Stays North of Datta Kheyl, near Ghorowekht (NW of Miram Shah). His chief followers are Madda Kheyl, but only a subdivision of them. He collects money from Congressites, from the Tochi Valley and from Bannu. His followers now number less than 400 and he is hard put to feeding and supporting them. At one time he had several thousands. His followers are deserters (like Mehr Dil who was in Tirah when we went through) and he attracts men who want loot -- particularly rifles. When he starts an action and is successful in an ambush or some such thing, his success often bring him support from locals who see an easy chance to loot. Census shows some 100,000 Mahsuds, but Duncan believes there are 150,000 with some 20,000 rifles. Janson estimates that Wazirs have more than the 30,000 rifles reputed to be held by them. Duncan reported a lorry shot up in Mahsud territory with an Afghan malik killed.

Dec. 6th: Loafed in Wana all day. Wana is a Brigade Hqrs. Abdullah Jan, an Afghan Brigadier, came into Wana to get British support because the Afghan Government had cut off his allowance. Nothing definite learned about British action but Duncan said he was a trouble maker and was probably stalling for time in an effort to get his Afghan allowances back again.

Dec. 7th: Left Wana at 9:30 a.m., reaching Tanai at 10:30 -- 10 miles. Picked up jeep. Left Tanai 10:55 a.m. crossing into Baluchistan at Gul Kutch at noon -- 26 miles. Met Peter Garrett and had tea at Gul Kutch. Left Gul Kutch at 12:45 arrived Sambaza at 1:30 p.m. Had lunch. Left Sambaza at 3:300, reaching picket of Tora Gharu at 3:30. Left Tora Gharu at 4:30 p.m., reaching Sandeman at 5:15 -- total miles -- 79. Dined with P.A. Searl and Brigadier Purvis. As Against NWFP rules, P.A. in Zhob does not attend jirgas but hears their decisions. If approved decisions are then carried out by Zhob Militia.

Dec. 8th: Went in Col. Keating's car and 2 escort trucks and a M.E.S. station wagon to see new road being made to Shawet which is big entry route for Gilzais, Suleiman Kheyl and other tribes. Visited Fort Shahigu -- 45.3 miles and alt. 6,695 ft., travelled old road across plain for 7.1 miles; new road to work camp (Zhob Militia) for 14.1 miles; went 5.5 miles beyond work camp for lunch. Total miles run -- 156. Left Sandeman at 10:00 a.m., returned at 8:00 p.m. Saw Nasra tribe at Shahigu, they are being chased by Suleiman Kheyl but escaped in time. This was mostly Suleiman Kheyl country, but much used by migrants such as Nasras. Water is somewhat scarce in Sandeman, some coming down from mountains and from wells which must be sunk 80 ft.

Dec. 9th: Loafed in Sandeman.

Dec. 10th: Motored to Quetta, leaving at 9:10 a.m., arriving at Residency at 5:05 pm. Saw large encampment of Karotis outside Hindu Bagh where we had tea in Zhob Militia Post. Karotis part of Gilzais and have special trade of Karez digging. Hindu Bagh is center of chrome mines (under Pop Wynn), producing 40,000 tons annually for U.S. of 52% ore. Wynn employs some 6,000 camels and donkeys to carry ore from mountains around Hindu Bagh and has one meter-gauge spur track for big working. He also gets some lease-lend trucks.

Dec. 11th: Saw Major General Money and Col. Bruce-Steer of Baluchistan Hdqr. Saw Father Wood, Major Platt and Alston.

Dec. 12th: Went for day beyond Hanna Valley for chikhor shoot with the Hays and Benjy. Also present was Major Woods-Ballard, P.A., Quetta.

10 Nov 1944 postmarked APO   From AWZ to BSZ, Haverford, Pa.   ALS-I
                                   [no header]
                                                    Thursday Nov 8th [sic: 8 November is Wednesday]
[unnumbered 96]
Dearest Barb

Still no news on my coming home. My heart thumps every time a pouch comes in. I don't suppose they are going to let me know in advance when and if they decide to relieve me.

No mail from you since my last letter. Yesterday I got a very nice letter from Babs. She seems to be getting along nicely and enjoying life at Smith. Her house and roommate didn't turn out so badly after all. Simmy [their children's governess] wrote me a very sweet letter about how she liked and enjoyed being with you this summer. She must be a very pleasant person – very thoughtful of her to write – always so nice to hear such nice things said about the ones you love and adore /p.2/

There really isn't very much to say – writing as I do every few days. I'm just patiently waiting around for something to happen. Our work here is now pretty much routine – we've done about all we can do as far as extra-curricular activities are concerned. I've played bridge twice this week with the same old people – went to a cocktail party at the Buslebys[?] on Monday night. He is the head of the Port Trust and lives in the nicest house here. It was quite a nice party with Gen. Warden and other burrah sahibs there. Tonight I am going out with Capt. Carrington the Army doctor who has taken care of all our sicknesses and innoculations [sic] (every 6 mo.). He wants to eat at one of the many Chinese Restaurants which doesn't appeal to me very much.

Good bye, my love, I hope it won't be long now.
          Heaps of love,   Al

----

16 Nov 1944 postmarked APO   From AWZ to BSZ, Haverford, Pa.   ALS-I
                                   [no header]
                                                                          Wednesday Nov. 15th
[unnumbered 97]
Dearest,

My letters must seem to get duller and duller but there isn't very much to write about. Life itself is dull. <u>I want to get home</u>. Still no news so I still have my hopes of getting home for Christmas. They at least haven't turned it down. Just heard today that Baker, who was here for a while then went to Bombay, has his orders to go home and he has been out less than a year. It's awfully to be indispensable.

Tomorrow should produce a bit of excitement. Ex-Ambassador Gauss[198] is arriving on his way home and Mr. Macy asked me to put him up. Com. Harrington /p.2/ is coming with him. He is the doctor who prescribed the trip to Brazil for the Madame. They are arriving in Gen Stratemeyer's own plane. I don't suppose we'll get much inside dope but it will be interesting to talk to them.[199]

---

[198] Gauss was Ambassador to China (*supra*). He was replaced by Patrick Hurley, who swooped into Delhi, Peshawar and then to Kabul in December 1943, shortly after AWZ and the others completed their trip in the NWFP. Enders made a fast trip back to Kabul to meet with Hurley. Hurley then flew on to Syria and sent a long message to FDR, and Enders returned to Delhi.

[199] Stratemeyer = Lieutenant General George Edward Stratemeyer (24 November 1890 – 11 August 1969). He was World War II chief of Air Staff. General Stratemeyer went to the China-Burma-India Theater in mid-1943, and was appointed Commanding General of the Army Air Forces in the theater and as Air Commander of the Allied Eastern Air Command. Although officially air advisor to General Joseph Stilwell, his status was comparable to that of Stilwell.... Part of Stratemeyer's command, the Tenth Air Force, had been integrated with the RAF Third Tactical Air Force in India in December 1943 and was operating under Mountbatten's South East Asia Command (SEAC).

Happened to get hold of a financial section of a New York Times dated Sept 10th. I noticed that all our investments are all doing quite well. I was especially surprised & pleased to see Warren Bros B at 42 and the C at 15.[200] This means my investment of $30,000 is now worth about $70,000 after looking awfully sour for a number of years. Also the common is paying a dividend of 50¢ a share which means the dividends on the preferred have been paid up to date. Money must be rolling into our /p.3/ coffers.

The weather here is getting quite delightful. The days are crystal clear with a warm sun shining uninterruptedly. The nights are warm at first, cooling to one blanket proportions in the early morning. We will be changing to khaki and blues in a few days.

That's about all there is except that I'm dying like hell to see you and wish my damn orders would come through.

               With all my love,
               Al

---

21 Nov 1944 postmarked APO   From AWZ to BSZ, Haverford, Pa.   ALS-I
                               [no header]

                                                                                                       Monday Nov. 25th

[unnumbered 98]
Dearest Barb

Our distinguished friends have at last departed. They left early yesterday morning. It was very interesting and really fun having them although a bit weary on one's constitution.

They arrived Thursday about seven having been reported as due to arrive at four. Mr. M. and I went out to meet them at the earlier hour so of course had to stand by for their actual arrival. We had a few people here for dinner including Mr. & Mrs. Macy and a newly arrived Wren. For their departure they were told the type of plane suitable to their station would not arrive for the next several days so feverish arrangements were made for their entertainment. Lunch was laid on for Friday at the consuls – a party that night at /p.2/ the vice consuls – an interview on Saturday with the Governor of Sind – a party that night at General Warden's the S.O.S. commanding general here.[201]

Well, everything went as planned up until Friday night when word came that a plane had come in and would take off at seven in the morning. So, after a successful evening of entertainment which included the burrah sahibs and memsahibs of this town we wended homeward to find the doctor in a relaxed and talkative mood. Phil, Howard and I stayed up till two thirty with well pulled ears over several beers.

    Don Lohbeck, *Patrick J. Hurley* (Chicago: Henry Regnery Company, 1956), 205: On 31 October, Hurley left for China, stopping at Baghdad, then Abadan, Karachi, and New Delhi, where he was met by MG George Stratemeyer, American Air Commander in India. He spent "almost a week" in New Delhi, meeting with Mountbatten, Wavell, Auchinleck, and American generals. On 7 November he flew to Assam and over the Himalayas to Kunming, China, where he boarded Stilwell's plane and few to Chungking and was met by Stilwell and Gen. Claire Chennault and had the first of several meetings with Chiang Kai-shek.

[200] Warren Bros = Warren Brothers Company, a paving and construction corporation that was founded by seven brothers of BSZ's mother, Mabel (Warren) Shoemaker. Seven of the eight sons of Herbert Marshall Warren formed the Warren Brothers Company in 1900, ten years after their father died. This company became the largest producer of asphalt paving in the United States. It developed subsidiaries in several other countries, including Canada, Cuba, Spain, Poland and Japan. The company's success was based on the discovery by Herbert's sixth son, Frederick (1866-1905), of "Bitulithic" pavement, a patented mixture of asphalt and stone, which was marketed as "Hot Mix" or "Sheet Asphalt." Warren Brothers' domestic subsidiaries included the Warren Brothers Roads Company, headed by a son of Mabel Warren, Jesse Warren Shoemaker (1898-1964).

[201] S.O.S. = Services of Supply. Brigadier General John A. Warden was Commanding General, Services of Supply, India, Burma Theater (SOS-IBT), 18 Dec 1944-10 Feb 1945. (From *Order of Battle of the U.S. Army Ground Forces in World War II*: Pacific Theater of Operations, p. 229).

At five-thirty we arose to have breakfast and to be at the airport at seven for the take off. We arrived at quarter to and after a bit of delay everyone was loaded onto the /p.3/ [sheet 2] plane, only to find something wrong with the first motor that was started. Mr. Macy and I waited until ten and gave our proper salaams as it was though the plane would take off shortly thereafter and no need for us to waste more time at the airport.

Arriving home and doing a job of work, I settled down to a bit of a snooze after lunch to catch up on sleep. But who should appear in the middle of it but our distinguished guests with the report that the plane wouldn't take off before late that evening or possibly the next morning. Having cancelled the general's party it was now on again with lots of telephoning.

I went back to the beckoning arms of Morpheus to be wakened at five-thirty with the news that the plane had been miraculously repaired and /p.4/ would take off at six-thirty – lots more telephoning. We got out to the airport – had a bite to eat and ten minutes before take off time a report came in about bad weather between here and Cairo – so the flight was scrubbed till the next morning if then. Back to town – more telephoning and off to the generals' party. It was a rather sticky affair with a private showing of Charles Laughton in "The Centerville Ghost" after dinner. Even a ghost story couldn't keep me awake so I caught up a bit on my sleeping. To bed at one thirty – up again at five and this time they did get off and I returned to the beach the rest of the day and I do mean returned.

There's a bit of news on my home coming. Howard received /p.5/ [sheet 3/ a note from the guy who assigns us jobs (who is a close friend of his) saying my request had been received and as soon as they selected the proper man he would be sent out to relieve me. No telling when this might come about but it is somewhat hopeful.

Your letter of Nov. 8th arrived day before yesterday. Awfully sorry to hear about Uncle Ralph[202] – he was a great guy. I'm glad, though, he lived to see his judgment vindicated and Warren Bros. in such good financial shape before he passed on.

Your trip down our hill sounds hair raising. I'm awfully glad it was no more serious. It could have been ghastly. Whatever happened to the brakes? That's one big disadvantage to hydraulic brakes. Make sure you /p.6/ have them [checked] then next time.

A new list of Lt. Comdr's arrived today. I wasn't on it – burns me up quite a bit but such is life out in the styx. The Washington chair bornes were much too busy thinking of their own promotions to worry about us. All of which makes me feel that the sooner I get home the better and as far as another job – let the highly honored ones carry on – as I understand it, they are falling all over themselves with nothing to do. I've done my bit and would continue to carry on if there were a scarcity in our line – but there isn't and I'm perfectly willing to step aside.

Good bye, dearest. I hope they make up their minds quickly and I'll be home to the best wife and family in the world.
   Heaps of love,
    Al

----

29 Nov 1944 postmarked APO   From AWZ to BSZ, Haverford, Pa.   ALS-I
           [no header]
                             Wednesday Nov. 29th
99
Dearest Barb,

I haven't had a letter from you in almost two weeks. Have you stopped writing or is it just the mails? Whatever it is I'm disappointed. I hate it when I don't hear from you regularly. It makes me think all is not well at home.

---

[202] BSZ's mother's brother Ralph Lambert Warren (b. 1874). Graduate of the University of Pennsylvania, B.S. (1895), M.E. (1896); a member of Phi Gamma Delta.

Yesterday I got a letter from Lt. Cdr. Off saying "as soon as I have an officer at hand whom I can recommend to take over in that area, we will start him on his way. I hope that it can be arranged that you may be back in this country by Christmas but frankly I cannot hold out much hope along these lines. Captain Baltazzi feels that the officer at Karachi should be the rank of lieutenant commander and at present /p.2/ we have no one at hand.["] Ironical to say the least that the reason I don't get home for Christmas is that they are waiting for a LtCdr to replace me. I'm pretty bitter about it. The last I heard there were 35 Lt Cdr's sitting around Wash. doing nothing. Well that's that. There's nothing much I can do about it.

Life here goes on in its same dull way. Thursday being Roosevelt Thanksgiving the Army staged a football game between a couple of GI units here. It wasn't a bad game – real tackling – the whiskey flasks and raccoon coats were replaced by coca cola bottles, shorts and helmets. /p.3/

Tomorrow we are going to celebrate Thanksgiving here. There will be eight of us – the Smythes, the Clees, Blossom Goldstraw, Doris Howard and I. Phil is in Bombay and won't be back in time. We have a turkey – got it off one of the ships and a ham. The assorted parcel of spirits I ordered months ago arrived. It contained a case of champagne so to make it a real celebration we're going to break it out. The way I feel now there won't be much left.

I played bridge a couple of nights in the last week – Monday at the Tambazzi's (Greeks) – Last night an Indian by the name of Thakirdas Gokalda gave a /p.4/ dinner for me at the "Karachi Club Annexe" – the pukka Indian boat club. He's in the wool businesses and is represented by Stuart Lamsbury, who had given me a letter of introduction to him. There were about 4 or 5 other men – Europeans – who also are interested in wool. It was quite nice, plenty of scotch and a fairly good dinner.

I suppose there still is a slight hope I'll be home for Christmas but if not by then it shouldn't be long afterwards.

Good bye, dearest, I'm dying to see you and hope the Navy will give me the only Christmas present I want – to be with you on that day.

    Heaps & heaps of love,
    Al

----

5 Dec 1944 postmarked APO   From AWZ to BSZ, Haverford, Pa.   ALS-I
    [no header]

Monday Dec 4th

#100
Dearest Barb,

You will note that this is the century mark in letter writing. One hundred is a lot of letters. Still no mail from you. I wonder what has gone wrong. I hope nothing at home.

On Thursday Delhi called and said they had a courier opening on the 31st of this month and offered it to me. I said I expected to be detached soon. They said why not let us request your orders for this courier duty – it might speed up the other. If they were willing to do this it was all right with me – if they agreed it would be a bird in hand – if they didn't it would smoke them out for the reason for not agreeing. The reply came back yesterday saying /p.2/ "negative – LT Zimmermann to be detached shortly" whatever that means but it's something a bit more tangible than I had before. My guess would be it will come thru between the middle and end of January. My hope that it will come through any day. Anyway I've done all there is to do in trying to get home. It now remains to see what happens.

Timoney sent me a Christmas card with a very nice letter. Good old Timoney – he's true blue, loyal and faithful.

Our Thanksgiving dinner went off as planned – not very exciting but nice – a token reminder of the good old days – but in a bit different atmosphere.

Saw a good movie on Friday night "The Life & Death of Colonel Blimp" /p.3/ an English picture in Technicolor. It doesn't come up to American standards as far as the mechanical part of the production but the acting & story are good.

Yesterday I took Doris and her two boys to Hawks [i.e., Hawke's] Bay. It was a beautiful day – the water the right temperature – good lunch. A pleasant time was had by all. I'm enclosing a picture I took of her baby, Susan, on her first birthday. Pretty cute don't you think?[203]

That about brings me up to date. Good bye, dearest, it shouldn't be long now. I just can't wait till it actually happens.

        All my love and more,
           Al

---

11 Dec 1944 postmarked APO.   From AWZ to BSZ, Haverford, Pa.   ALS-I
                              [no header]
                                                      Saturday Dec 9th

[unnumbered 101]
Dearest Barb,

I can't imagine what's happened to your letters or maybe you haven't written any. I haven't had a letter from you since Nov. 14th – one you wrote on the 8th – which means I haven't heard from you in over a month. I've heard from Babs, Hecker and Jack Ott in the meantime – so mail is getting through – but not a word from you. What has gone wrong anyway – I'm worried. If I don't hear tomorrow I'm going to send a cable.

This afternoon the Army is stageing a rodeo, of all things. I don't know what they are going to use for animals but they expect to have a reasonable facsimile of the real thing. Our presence is /p.2/ indicated – the general wrote requesting us to be there. We've invited nurses from the hospital to come in for lunch and go with us. Phil Halla is arriving back from Bombay bringing with him a W.S.A.[204] man and we've invited Dick Edwards, a vice consul so all in all we'll have about twelve customers for the gala event.

A new yeoman has just arrived to replace one of ours who has been in the hospital for two months with dysentery. The airport just called – we've been expecting him for over a week. We're glad he's here as we've been quite handicapped not having a disbursing yeoman.

No further news about my going home – I'm sick about it. I did so much want to get home to you [shifts to margin] by Christmas but it looks pretty hopeless. Good bye dear, I hope it's just the mails – and not that something has gone wrong at home All my love   Al

---

11 Dec 1944 postmarked APO.   From AWZ to BWZ, Haverford, Pa.   ALS-I
                              [no header]
                                      Sunday Dec 9th [sic: Sunday is 10 Dec]

[unnumbered]
Dear Babs,

I expect you'll be home by the time this letter gets to your side so I'll answer your letter of Nov 27th addressing this there.

You seem to be quite a traveller these days, hopping around to Farmington, Boston, Phila. etc. You sound very gay. I hope you're not wearing yourself out.

---

[203] There are many pictures of women and children in beach settings in AWZ's Photo Album. The photo referred to is in the Photo Album (Sheet 1, #22).
[204] W.S.A. = War Shipping Administration

It doesn't look as though I'll make home by Christmas and I'm terribly disappointed. It's been almost a year and a half since I've seen you and believe me that's an awfully long time. I wonder how much you've all changed.

I've just written to Mummy giving her the news so there isn't much /p.2/ I can write you – life is pretty dull here anyway – but I want you to know I'm thinking about all of you and longing for the day when we'll all be reunited.

There's been a big gap in letters from Mummy. I haven't had one in almost a month and I'm worried. It's not at all nice to be 12 000 miles away from home and not hear how things [are] going.

I hope you have a wonderful Christmas and I'm sunk about not being able to be there with you.
       Lots of love,
       Daddy.

---

16 Dec 1944 postmarked APO   From AWZ to BSZ, Haverford, Pa.   ALS-I
         [no header]

                      Friday Dec 15th

102
Dearest Barb,

I finally heard from you – two letters dated Nov 15 and 28th came on Tuesday. There's a mighty big gap between them – 13 days. As you've stopped numbering letters again I can't tell if one or more is missing in between. It took the one of the 15th long enough to get here – just about a month. Well anyway I'm glad to get them and know that all is well at home. In the same mail I got a letter from Babs which I will answer in due course. Of course Lanie and Warren have forgotten all about me.

In reply to your query about whether your letter of Nov 8 via air mail got through any quicker – it did taking only 9 days. I don't see that this information is going to do /p.2/ much good, as, by the time you get this, I hope you won't be sending many more. Nothing more definite has come through on my detachment but surely by that time something should happen.

Don't worry about what the Cadillac will cost – I'm so damn thankful you got through it without a scratch. Besides in reading the new tax law accidents of this kind can be deducted from income and in our bracket the government will be paying for about 85% of the bill. Make sure you call this to Bill Stuble's attention. Also his fee, Brundage Story & Rose's fee, I think are deductible and if doctor & hospital bills exceed a certain amount they are too. I've sent Bill a copy of income tax regulations as they apply to us. They are quite a bit different compared to a civilian. /p.3/ [sheet 2] I suppose he already realizes it but it does no harm in calling it to his attention.

Missing Thanksgiving and now Christmas makes me boil. It could have been worked very easily if the outfit I work for had a soul but I don't suppose in their comfortable seats they think much about those things.

You mean to say our Warren is neglecting his studies. That doesn't sound much like him. His first duty is his schooling with proper time allocated to sports.

You certainly are right when you gather that my letters sound impatient and restless. It's awful getting so near in promises and then not getting the orders that count. I'm frightfully disappointed I won't make it by Christmas but there's nothing much you can do /p.4/ about it. I've got a few things (couldn't find a thing for Warren & Albie) I had still hoped to take home personally by Christmas until it was too late to send them. Anyway it's safer to bring them myself – mail especially at this time of the year is very unreliable. Evidently you never got a box containing victrola records as you've never mentioned them. They were recordings of this beautiful Indian music we love so much.

The rodeo on Sunday was a great success, really pretty good – considering. They had quite a time making the animals ferocious but they did – some of them – and the performers, all local GIs, did right well.

Wednesday I had dinner at the Coughlans and last night we had 2 Red Cross gals and an Army nurse in for dinner. That's about all that's new with me.
>Always all my love,
>>Al

----

20 Dec 1944 postmarked APO	From AWZ to BSZ, Haverford, Pa.	ALS-I
	[no header]	

>>>Wednesday Dec 20th

103
Dearest Barb,

I suppose you've heard the wonderful news about my being here until Mar 1st. The billet doux brought by a passing officer a few days ago from Curt said that you had been "notified." This all makes a nice Christmas for me. My Christmas present from the Navy Dept has been a string of disappointments. Why didn't they let me take the courier trip in early November, why didn't they let me take the one at the end of this month, why didn't they make me a lieutenant commander? All these were promises that never materialized. It wouldn't have been so bad if they hadn't been. I guess I'm supposed to take comfort in their thought that I'm so valuable and doing such a good job.

I don't know whether Curt told you the program contemplated so here it is: /p.3/ I get 25 days leave and then report to New York to go to the Port Directors school for 9 weeks (live where I want to – not like the other one). It's supposed to be an honor, I guess, but once I get home I'm going to have a look around. The way I feel now – I don't want to deprive the boys in Wash the opportunity of getting some of these lush jobs. Well, that's that – I'm pretty browned off.

I sent two packages yesterday that only by a miracle will arrive by Christmas. They contain bits and pieces which I had planned to take home myself. My letter to Babs contains my ideas on distribution. I hope you'll like them. I've about reached the bottom of the barrel as far as unusual mementos of India are concerned that I think you'd like.

Don't worry about not sending me anything for Christmas. I've got about /p.3 [sheet 2] everything I need in the way of material things. I could stand a little of the other but you can't send that by mail.

We've had some pretty disagreeable weather in the last few days. It has turned cold with a high wind that blows around a lot of dust. These houses aren't built for that sort of thing. Of course there's no heating system but we are lucky to have a few oil stoves that we can put around that help but do not solve the situation.

No mail since I last wrote you. I suppose that can't be helped at this time of the year but it's just another item to add to life's little pleasantries.

I don't suppose it's made your Christmas more happy to hear of my additional sentence but it will /p.4/ make reunion that much sweeter when it does come. Might as well try to find some bright spot in an otherwise dull situation.

Good bye, dearest, I'm awfully sorry – I miss you terribly – just can't wait.
>>All my love
>>>Al

P.S. Send letters Air Mail

----

26 Dec 1944 postmarked APO	From AWZ to BSZ, Haverford, Pa.	ALS-I
	[no header]	

>>>Monday [sic: actually Tues.] Dec 26th

104
Dearest Barb

Well Christmas has come and gone. The Red Cross helped us out a bit. They supplied a lot of decorations, otherwise unobtainable, and Christmas stockings that were hung up on our dining room mantel piece. Phil's wife sent out a folding tree and some tinsil so what more could a man ask for.

The Grants had a cocktail party Xmas Eve which Howard and I attended. We came home feeling pretty high, had a couple of sandwiches and decided to go to church. Howard, being a good Catholic persuaded us to go to the Catholic Church – high pontifical mass. When we arrived there were thousands of people milling around trying to get in. Guess where we ended up – not standing on tip toes on the outside looking through /p.2/ a chink – not inside standing up or sitting down with the congregation but right in the sacristy – arm's length away from the monsignor who conducts the show, where we had a perfect view of the whole proceedings – practically helping to ring the bells and swing the incense. Shades of the Atlantic City Beauty Pageant.

Yesterday morning when I sat down to write you a letter three naval officers[205] on an anti typhus mission dropped in who had just arrived from the States. One (a Commander Steel) was with the Rockefeller Foundation before the war and knows Harry Schroeder very well. Of course letter writing went out the window. After showing them around town for a bit we came back for Christmas dinner. We insisted they stay for it in spite /p.3/ [sheet 2] of their protestations that they were mooching in on our party. Besides these three we had two Army nurses come in from the hospital, an officer from a Liberty in port and a courier who was passing through. For cocktails before hand we had our enlisted men. We opened our stockings – the Red Cross really did a good job – and a good time was had by all. In the middle of the proceedings Mr. Chauda one of our Indian wool friends brought in a large basket of fruit and nuts, very touching.[206]

Last night I went to the Sorbys for dinner and then to the dance at the Sind Club. The Pettigrews were there – Bill is home for Xmas leave – the Griffiths, Geoffrey Pike and Bill's adjutant.

Tonight the Dendrinas[207] are having /p.4/ a party to celebrate the liberation of Greece – not very fortunate timing in the light of recent events. Tomorrow the Navy is having a party at HMIS Bahadur a training school near hear – and so it goes – much too gay a time to suit me. I could do with a lot less but these people ask you so far ahead it's very awkward to regret.[208]

Your letter of Dec 7th arrived on Christmas day – the one you wrote Dec 15th had arrived two days previously – both very welcome. I see you now know about as much of my future as I do. If you keep in touch with my relief's wife – Mrs. Henry Groman, 334 or 344 Fischers Road – you will probably know just when to expect me home. He will bring my orders with him. I'll have to stay a couple of days after /p.5/ [sheet 3] he arrives. Figuring travel time both ways I should be home roughly 17 days after he

---

[205] The three naval officers were LT(j.g.) C. Rollins Hanlon, MC, USNR; a "Dr. Woods"; and this CDR Steel. Hanlon's letter to AZ, written in Calcutta, 13 January 1945, is in AWZ Scrapbook, Page 24(V). On 4 March 2011, I spoke on the telephone with Dr. Hanlon. The Blalock-Hanlon Operation for congenital heart disease bears his name, along with that of Alfred Blalock, Chairman of Surgery at Johns Hopkins University. Hanlon later became the Executive Director of the American College of Surgeons. Now retired, age 96, he lives in Kenilworth, IL. He doesn't recall the names of the other two physicians on this trip, but he remembers Karachi, where his "batman was so helpful." But he does not remember the Christmas party there in 1944, or his host, AZ.

[206] Harold Schroeder and Dr. (CDR) Steel have not been identified. Liberty ships were U.S. government freighters in World War II. They were built rapidly, unarmored, and protected only by a few sailors on board with machine guns. Many were sunk by submarines. Among the major builders of Liberty ships, Henry Kaiser was the most famous. Coincidentally, Kaiser's construction company was once owned by Warren Brothers (*supra*) but later became an independent giant company on its own. Warren Brothers' name has been forgotten, while Kaiser is still remembered.

[207] Dendrinas = The invitation to this party is in AWZ Scrapbook, Page 25(V).

[208] HMIS Bahadur = now PNS Bahadur, the major training base for the Pakistan Navy. British troops entered Athens on 14 October and the Germans left mainland Greece on 4 November. On 3 December, clashes began in Athens between leftists and Greek government forces. On 16 December, the German counterattack known as the Battle of the Bulge began; it would not end until 26 December.

leaves U.S. I can't possibly see why all this should take till Mar 1st as indicated by Curt but maybe he knows best. I hope he's wrong.[209]

Darling, I can't wait to get back. It's been so awfully, awfully long since I've seen you. I hope the agony isn't prolonged any further than necessary. As I figure it I can come right to Haverford and will not have to report any place for 29 days – 25 days leave and 4 days "proceed time."

I'm not going to make any plans or try to get any jobs in Wash. etc until I get home. I'll have quite a while to look around. I still haven't given up the idea that maybe my services will not be further required. /p.6/ I would certainly think East Hampton would be a good idea this summer no matter what happens.

What about Walter Johnson – hasn't he been asked to join the Orpheus Club?

The mail going your way seems to be about as fowled up as my mail. I wrote two letters #98 & 99 between Nov. 15th and Dec 45h that you evidently did not receive.

That's about all for now. Good-bye dearest, until then just waiting and longing.

                With all my love,
                Al

----

31 Dec 1944 postmarked APO   From AWZ to BSZ, Haverford, Pa.   ALS-I
                                 [no header]

                                                              Sunday Dec 31st
105
Darling Barb,

Here it is – the day before New Year's. Tonight will be the third straight New Year's Eve without you. Pretty rough I call it. Anyway I'll be thinking of you more than ever when the new year rolls in. Howard, Phil and I are going to the Pettigrews for cocktails and then to the Gym for the big Karachi party.[210] It will be a mad house.

This week has been pretty full – out every night but one to some sort of a party or other. Tuesday – went to the Dendrinas[211] for a cocktail party to celebrate the "liberation" of Greece. Wednesday went to the "Himalaya" – the Indian Navy training school here – a large dance. Thursday was a blank – Friday to Geoffrey Pikes /p.2/ to play bridge and last night to the Ack-ack school for a cocktail party. Sounds awfully gay but I'd a thousand times rather be in my little house in Haverford with you, my dearest.

No letter from you since my last one. The mail situation is pretty lowsy these days.

I'm sending a package today containing all my photographs to date. I hope it doesn't get lost on the way. I'll keep the negatives and either send them later or bring them. The camel trip pictures are there along with a slight story which, if you think [might] have interest, you might try to sell. I won't be at all disappointed if you don't. If you do it should have the Navy's approval (through Curt). The rest are scenes in and about Karachi [shifts to margin]. Am enclosing 2 snaps that Buzz Sloan took the other day. Still just waiting with no further news. All my love as always.

                Al

---

[209] The Henry Groman residence at 344 Fishers Road, Bryn Mawr, Pa., was but 0.5 miles from the Zimmermann residence at 400 North Rose Lane, Haverford.
[210] In AWZ-Photo Album, p.1, #6 was probably taken at the Pettigrews.
[211] AZ mentioned this party in his letter of 26 December. The invitation is in the Scrapbook, page 25(V).

"Buzz Sloan took this on Thanksgiving Day" (in Photograph Album, Sheet 1, #12)

Editor's comment: AZ, Voorhees, and one who is unknown, outside the NLO (Photo Album, Sheet 1, #12).

"Lunch or rather Tiffin on the last day of the year – Jack Sorby, Bill Pettigrew, Pauline Blackwell, Guess who, Frances Coughlan, Geoffrey Pike, Allen Say, Fernandez, Mohammed, John Blackwell, Doris Pettigrew, Howard Voorhees, Stella Sorby, Raymond Coughlan, Phil Halla"[212]

1945

5 Jan 1945 postmarked APO   From AWZ to BSZ, Haverford, Pa.   ALS-I
[no header]

Friday Jan 5th

106
Dearest Barb,
    I've just received your letter of Dec. 10th which is going backwards but never the less is something. I can't understand why you haven't heard from me for 2 ½ weeks. I've written you every five or six days. Something must have gone awfully skrewy with the mails.
    In answer to your queries. I guess my orders suit me – they are about as good as I could expect. The amount of leave is as much as I've heard anybody getting. I would like to spend some of it away but I don't just know where. If it were a little later I would like Hot Springs. I don't want to go north – I'd like to take the cooling off process a bit gradually. What do you think? I thought if I came Miami way you could come down there for a few days and go up with me but I've just read in the New [p.2/] York

---

[212] This must have been 31 December 1944 because Voorhees and Halla were not assigned to Karachi until 1944, and they were all gone by 31 December 1945. The luncheon ("tiffin") is not mentioned in AZ's first letter home in 1945. Most if not all of the people in this photograph and many others in the Photo Album are identified by full name, title, and duties in the Notes to AZ's letters to his family.

Times winter travel south this year will again be terrific and it would be an awful trip for you. I guess the best plan is to come right home and we can decide then. I rather like the idea of the New York school. It's near home etc and you can be with me the whole time. After that – I vehemently protest against being sent away on foreign service for another stretch. But I suppose that's looking pretty far ahead. We'll just have to wait and see what happens.

Yesterday Cy Polley called up from the Airport and I invited him in for dinner. He's on his way to China. I got pretty well up to date on some of our mutual friends. He told me a rather amusing story – not so amusing to the one who told it to him though. Knowing he was headed this way Browning Clement told him to call a man he knew that had /p.3/ been here. He wasn't home but he talked to his wife. She had just been to a cocktail party where several people had heard of my coming home and expressed their pleasure (I have no idea who they might be). It didn't please her very much as her husband was coming out to relieve me.[213]

He tells me Bryce Blynn is in Italy and is just about to be made a full colonel. Good for Bryce. I guess most of his other news you already know. It was nice to see him and I suppose he enjoyed this oasis after a rather rough trip across North Africa.

New Years passed with the usual fan-fare. We went over to the Pettigrews for dinner and then to the Gymkhana where things were rolling along in good shape. Kisses were a dime a dozen. New Years day we went to Bob /p.4/ Markley's for egg-nog. Since then things have quieted down very considerably – thank goodness.

With the turn of the year I feel I'm on the last lap and am I glad. It will be wonderful to see you again. Whatever you plan for my leave will be all right with me but I don't want to be awfully gay and go on a lot of parties – I want to be with you – <u>alone</u>.

Good-bye, dearest it won't be too long now.
    With all my love and heaps of kisses
     Al

----

12 Jan 1945 postmarked APO From AWZ to BSZ, Haverford, Pa.   ALS-I
         [no header]
                          Friday Jan 12th

#107
Dearest Barb,
  Still nothing new on when I set off. We are expecting Capt. Davis any day now. He is coming out to head all our offices in India & Ceylon. He probably will know when Groman is due to arrive.[214]

---

[213] Cy Polley = Cyrus Polley. He and AWZ were members of the Merion Cricket Club. Polley was one of the best squash players in America; he would be the U.S. veterans champion in 1947. Merion club member Vic Seixas, a relative of Amelie (Seixas) Kane and tennis great, was three-times a U.S. veterans squash champion. (from James Zug, *Squash: A History of the Game* [New York: Simon & Schuster, 2003]), 102, accessed on Google Books, 2/11/11). A memorial card on the website of the late Henry Dorr Boenning Jr. provides additional information about Cyrus Polley: "In 1950, two guys named Henry Boenning Jr. and Cyrus Polley sold a computer to a General Leslie Groves who was president of Remington Rand at the time. The computer was the ENIAC." Boenning died in January 2011; he was a graduate of the Germantown Academy and of the Wharton Business School, in 1933; he was a major in the U.S. Army, at the Battle of the Bulge.

Cyrus Polley was apparently directed by Browning Clement to speak with Mrs. Henry Groman, who was not happy to hear that AWZ would soon be coming home.

[214] CAPT Davis = "Ransom Kirby Davis, Captain, United States Navy," whose business card is in AWZ Wartime Scrapbook, loose between sheets 17(L) and 18(R), where it would be out of chronological order. CAPT Davis became Senior U.S. Naval Liaison Officer, I-B Theater. This title appears in fn to AZ's letter of 11 July 1944. Davis was not in that position in July 1944, but Davis was probably the Senior ALUSLO who received the document in 1945 that was quoted in that fn. Davis is mentioned in several letters in the Winsor folder, but his

151

Bill Stuble wrote that there is a possibility of JZ Sons being sold and mentions a price of $100,000 for the real estate, buildings etc and inventory." If such is the price without the cash and investments, I think it a good one but I certainly don't if they are to be included. Furthermore if the cash & investments are to be declared as dividends and the rest sold that would be bad as surtax would take most of the dividends. Whereas if it were sold lock stock & barrel, the profit on the sale would be taxed at a much lower rate. However /p.2/ undoubtedly they are aware of this and will handle it in the proper way. I do think Bill S. should be in on every move you make as our interests, due mainly to large profits of O&Z, are different from the other members of the family – not that I would want to block any reasonable advantageous sale for them. The whole idea comes pretty much as a surprise. Who is in back of it? What will happen to the present management etc.? I suppose I'll have to wait [for] further developments for answers to these questions. Unless it is an opportue advantageous sale I would rather go along as is than have the present management out of jobs.

We've been having some unprecedented cold weather. It's been very uncomfortable /p.3/ as we haven't the right clothes and the proper heating facilities in our office and quarters. We are lucky though, compared to others as we have two small fire places and several oil stoves. Which reminds me I hope you have plenty of oil to keep warm.

Your letter of Dec 16th arrived day before yesterday – not as bad time as previously. I guess you're pretty glad to get Christmas over with. I'm sure the kids all enjoyed it though. Babs wrote and told me how thrilled she was. Evidently my package didn't arrive in time which too bad. I hope it arrived before the girls went back to school. Too bad Warren was sick. How is he now? I'm awfully worried. Even tougher, he was up Christmas [but] you didn't say he was completely better. I'm glad our old friends still stick to us and you had open house again.

This probably is your last chance to /p.4/ ask me to bring anything you can think of from here (if this gets home quickly & your reply gets over her quickly). I'm sending a lot of stuff by sea and I'm allowed 65 pounds so I'll have a lot of room. What is your heart's fondest desire? I'll even bring some lemons if you want them.

Nothing much to report on my recent activities. Time drags slowly by. Night before last I did go to the annual dance given by the Digh[?] Road R.A.F.'s. It was quite a party. I went stag and ended up at a table with Barbara Bennett, the Browns – mother father and two daughters 17 & 15 and a couple of RAF blokes. I was pretty lucky as they were about the most attractive people there. The girls were very good dancers and it prepared me for twirling around my two daughters.

That's about all dearest. Why doesn't [shift to margin] Groman hurry up. I want awfully to be on my way to your arms. Heaps and heaps of love as always   Al

----

18 Jan 1945 postmarked APO   From AWZ to BSZ, Haverford, Pa.   ALS-I
                                    [no header]

Thursday Jan 18th

[unnumbered 108]
Dearest Barb

This week has been one of the dullest so far. I had expected Capt. Davis, our new boss, to stop over here a few days but he went right up to Delhi after a one hour's stopover in the wee hours of the morning. Castle, a lieutenant with him, brought the report that Groman wouldn't get out here before Mar.

---

given name is seen only in this business card in the Scrapbook. He and AZ did not start off with a good relationship, and AZ's correspondence shows that this precarious relationship deteriorated. I doubt that AZ got a good fitness report from Davis when AZ was relieved and returned to the U.S. That would impede his promotion to LCDR, although AZ was later promoted, after the war was over. However, that is just speculation; AZ did not save his FitReps.

1st which discouraged me no end. I had high hopes he'd be showing up any day now. I don't know what in the world's keeping him.

I spent Sunday and Monday in bed. We've had lousy weather which produced a cold for me so I thought the wisest thing was to stay in about the only place where I'd be warm. I still have a bit of the /p.2/ cold but it's over its worst stages.

Saturday night the Priestleys had a party which I attended. It was given for Gen. Hind who is about to be relieved as Fortress Commander. He's a great guy and everybody is sorry to see him go.

We've had a Capt. Habecker staying with us the last few days awaiting a plane to Colombo to be on Mountbatten's staff. Peachie [Durand] is gathering quite a name for herself. I told him[215] to give her my salaams and he said he'd already been given her name to look up.

There's really nothing more to say other than I'm burnt up that our reunion is taking so long to come about. I miss you awfully, dearest, and can't wait till we'll be together again.

With all my love always,
                Al

You don't seem to like my newspaper enclosures but I thought I'd try again. I think this will amuse you as well as the pictures. [enclosures were not preserved in this envelope]

----

25 Jan 1945 postmarked APO   From AWZ to BSZ, Haverford, Pa.   ALS-I
                                  [no header]
                                                                    Wednesday Jan 24th
#109
Dearest Barb,

Nineteen months ago today. Nineteen long months since I last saw you. Thank goodness it's not going to be much longer.

Your letter of Jan 1st arrived a couple of days ago. You didn't mention the package I sent in early December so I guess it hadn't arrived. I hope nothing has happened to it as it was pretty valuable.

John[216] seems to be up to his old tricks again. He's tried it before to get someone with him on the opposition. He should have sense enough to know we wouldn't line up with him. The answers to my questions in a previous letter are /p.2/ now pretty well answered. 1,300,000 is a bit more like it. I think it's a good price if we can get it.

The last few days seem to have been ones of farewell parties. Two last week for John Petty, an Indian Civil service bloke. Last night the Sind Club had one for General Hind. Everybody seems to be changing jobs. Saturday afternoon Eric Layman, Gen Hind's A.D.C., was married. Howard and I were the only Americans invited which was quite a compliment. I managed to get some Kodachrome recently from Bombay so I took a few feet of the wedding. It was quite an event. Sunday night we had some British naval officers here for supper along with some other people. They are here only /p.3/ for a short time and I thought it would be nice to do something for them during their stay. They are all young and are very fond of singing so it was that kind of a party.[217]

Don't let the enclosed pictures of a sketch done of me worry you. They tell me I don't really look like that but I thought I'd send them along any way. The artist was here for a few days in the interests of the Coast Guard and Phil Halla had him do one of each of us. The pictures were taken by the V mail camera at our local A.P.O. He also did a water color of our office and mess which I enclose.[218]

---

[215] See Habecker to AWZ, 23 Jan 1945, in folder of "Winsor and Other ONIs," Chapter 6.

[216] John = John Zimmermann Jr. (b. 1892), AWZ's older brother.

[217] The movies that AWZ took at the wedding have not been found, but still photos of it are in the photographs in the Album: Sheet #1, 26, 27; Sheet #2, 1, 2, 3. The invitation is in the Scrapbook, pasted to page 25(V).

[218] The pictures are no longer in this envelope. Photos of the sketches done by the Coast Guard artist are elsewhere in the AWZ Wartime Papers. Some are in the Scrapbook and others are in the Loose Photos, including two Chief

153

Tell Warren and Albie their /p.4/ Christmas presents are being made. I guess I might as well bring them with me when I come rather than trust them to the mails.

That about brings me up to date so I'll close this dull letter. I'm sorry I'm not more brilliant but last night's party has me down a bit. Noic (the naval officer in charge) Capt. Coupe, insisted Howard and myself go up to his room for a night cap which was a bad idea.

Good bye, dearest, I hate this waiting. I want to go home and soon.

With all my love as always,

Al

Petty Officers, two USAAF officers, a U.S. Army Transportation Corps Officer, and a civilian. None are identified. AZ says little about the enlisted personnel at NLO Karachi, so these images of the CPOs are especially valuable. The water color of the officer has not been identified but a small copy of it may be in the Loose Photos File Folder.

The artist was Joseph Di Gemma (who printed his name "Di Gemma" on each drawing). One of his watercolors was offered for sale by Hake's Americana & Collectibles: "Original Watercolor by Coast Guard Combat Artist Joseph Di Gemma" (1944). Bidding ended without bids, on 30 August 2006. The item was described as follows: Auction 189, Part 1. 14x18" image in 19x23" wood frame. Illustration shows servicemen boarding interior of ship. Officer at right. Left foreground shows photographer taking photo of soldiers. Lower right shows "Log Book/USS Gen. G.M. Randall/AP 1 15". Upper left has artist's name and year 1945 (partially matted out). Attached to back of art board is typed text "Coming Aboard/Soldiers Coming Aboard a Coast Guard Manned Troop Ship at Norfolk, VA. En Route To the South Pacific. Watercolor by Coast Guard Combat Artist Joseph De Gemma, Brooklyn, N.Y. Virginia, 1944." Also, two typed stickers reading "Property of Public Relations Division U.S. Coast Guard Headquarters Washington, D.C./Government Property." Wonderful example of WWII military art. Exc. (I/J - $400 to $1000).

Di Gemma was a well-known artist before and after World War II, and there are many pages of citations to his works in Google. Another example: "Lot 562: JOSEPH P. Di GEMMA (American, 1910-) MAINE LIGHTHOUSE. Oil on canvas scene shows tall white lighthouse with connected buildings, rocky ledge with colorful sea-grasses. Signed lower left with etched signature "Joseph Di Gemma 1940 Maine". SIZE: 26" x 35". CO … James D. Julia: USA / Auction Date: 2003.

---

31 Jan 1945 postmarked APO From AWZ to BSZ, Haverford, Pa. ALS-I
  [no header]

Tuesday Jan 29th

#110
Darling Barb

Here it is our anniversary again and still 10,000 miles between us. I hope this will be the last one we're ever separated. I don't like it. I hope Caldwells and Albrechts did what I told them. Next year will be our 20th. We'll have to make up for the ones we missed.[219]

Your letters of the 10th and 16th have just arrived the latter enclosing Brad Story's summation of what's been going on in our accounts. I would say by the looks of it that we are doing pretty well. I really think the money spent with him has been well invested. Besides I believe you can deduct their fees, as a necessary evil of being rich, for income tax purposes. I guess Bill Stuble has that well in hand.

I'm awfully glad to hear what /p.2/ you say about the children being so nice and developing so well. I'm awfully awfully proud of them and can't wait till I see them. Twenty months is a long time and they must have changed considerably espeacially [sic] the boys.

Curt is in with a note containing several suggestions for my future which he said he talked to you about.[220] He asked me if I wanted to fill his shoes in Wash. while he's on a two months courier trip. This is all right with me as long as I get some leave. I certainly don't want to step right into a job as soon as I get there. Splitting my 25 days leave is all right too but a "bird in hand" you know. They might forget about the balance. I look forward to going to the school in New York. I understand it's a good school – not too confining /p.3/ and is close to Phila. and offers workable living conditions that should be easy to work out. To sum up I ended my letter to Curt by saying this is what I really would like considering everthing [sic] and let him work it out from there: two weeks leave on arrival – work at the India desk in Wash. for the two months Curt is away (I suppose we can live at the Hay Adams or some place) report to the Port Directors school in N.Y. for the 9 week course – take the rest of my leave (which should be sometime in the summer so I could go to E.H. I imagine week ends there could be arranged while I'm at school) then get a job in the Port Directors office in either Phila. or New York where I'd be /p.4/ near my family – in the meantime picking up an increase in rank. I'm told I missed being considered for the last promotion by 18 days. I don't think he can mind my not being keen on another foreign job espiacially not right away. I'm perfectly willing to let the boys who've been in Wash. get out in the field for a while. Things never work out as hoped in the Navy so we'll just have to wait and see what happens. Anyway by the way things are rolling along in Germany and the Pacific it might not be many months before this whole damn thing is over.

Same old stuff going on here – a party at the Cargills Sat. night, a cocktail party tonight at Bob Markley's (#1 at Standard Vac). You'd hardly know there's a war on.
[shifts to margin] That's about all at the minute, dearest. Time certainly travels slowly. Hurry that guy along if you can.

  Heaps & heaps of love as always, Al

---

5 Feb 1945 postmarked APO From AWZ to BSZ, Haverford, Pa. ALS-I
  [no header]

Sunday Feb. 4th

---

[219] Caldwells and Albrechts = J.E. Caldwell and Company is a prominent Philadelphia jewelry firm. There is no longer a firm known as Albrecht in Philadelphia.
[220] The suggested rotation and duties in the U.S. did not appear in Winsor's letters to AZ, nor did this visit of Winsor to Karachi appear in Winsor's letters (see CW to AZ of 19 Jan and 9 March 1945).

#111
Dearest Barb

What should appear, several days ago, but two pyjamas, three tennis balls and a half dozen golf balls – sent by Powell Browning from Miami to Mail & Dispatch Sect. They were a bit late but none the less welcome. I was just about out of tennis balls and my pyjamas are just about gone so they came in the nick of time. The golf balls I think I'll bring home as they look to be pretty good ones and I haven't been playing much golf recently.

We had Capt. Thornton U.S.N.R.[221] stay with us several days in the last week. He was in the Phila. office when Jack Kane was there and /p.2/ was very fond of him. Amelie probably knows him quite well. He's stationed in Cairo and was awaiting transportation back to there. He tells me that Pete Rosengarten's wife's family are very wealthy, own half of Egypt, entertain lavishly espiacially [sic] our Navy boys stationed there. He suggests that I stop over in Cairo and meet them. The idea doesn't appeal to me at all, as once I get released from here I aim to get home as fast as possible.

I had a nice letter from Babs this week. She's a very faithful corespondant [sic] and keeps me informed of all her doings. Dottie Boericke[222] also wrote to give me her address /p.2/ in Miami in case I went home that way which is a possibility but a rather small one. I rather think I'll be routed pretty much the same way as I came out. No letters from you since I last wrote. Hecker has been very thoughtful in sending me the weekly overseas edition of the Evening Bulletin and also sending a short newsy note with each. Sally Winsor has also been very thoughtful in writing once in a while. It's nice to have such nice friends.

Nothing much to report on last week's activities. Life is just damn dull sitting around waiting. This place is pretty much drying up. I'm afraid my successor is /p.4/ going to find it pretty hard to keep himself occupied.

Last night we went to see "Double Indemnity" – a pretty good picture I thought – just down my alley [Fred MacMurray, Dorothy Stanwyck, Edward G. Robinson, 1944]. Yesterday I went to the races, with Phil Halla, Doris and Geoffrey Pike, and had a pretty good day winning about 100 Rs. That's about all the excitement I've had.

By the looks of things maybe I'll be home just in time to celebrate VE day. Germany looks about done though she's looked that way before. Certainly the rope must be getting awfully short.

Good bye, dearest, keep your chin up. One more month should just about do it.
       Until then   all my love,
          Al

[See Bibliography for the title of an official report from AWZ to CNO (DNI) in February 1945, on Karachi Port facilities. The report was sent via his superior officer, JICA/CBI. It is the last of four reports about the Port that have been preserved in the NARA II archives.]

12 Feb 1945 postmarked APO   From AWZ to BSZ, Haverford, Pa.   ALS-I
                               [no header]

Monday Feb. 12th

#112
Dearest Barb,

No letter from you since Jan 29th. It's awfully hard to carry on a one sided correspondence. I suppose you've been writing and it's the mails again.

I haven't heard any more about when Groman is to leave. He certainly is taking his time about it. Capt. Davis our new boss is due to arrive here at the end of this week. It would make a lot more sense to have Groman here at the time. I have an idea he's going to find something wrong with himself at the last minute which will be just dandy.

---

[221] CAPT Thomas A. Thornton. See Note to AZ's letter of 7 July 1943.
[222] Dottie Boericke = Dorothy (aka "Dee"), daughter of BSZ's sister Dorothy (nee Shoemaker) Boericke.

No word as yet about the package I sent home before Christmas. I /p.2/ hope nobody's pinched it on the way. I can't imagine why it isn't there by now. I'll be frightfully disappointed if it doesn't get there. Several days ago I sent another package – a pair of chablis (sort of sandals) for each of you. I got a pucka shoemaker to make them and they are much better made with much better leather than the last ones. Babs had written asking me to get her a pair and I thought I might as well get them for all of you. I hope I have the sizes right. I allowed a full size bigger for Warren and Albie and a half size for Lanie and Babs. I also included some /p.3/ bracelets and necklaces from Tibet and Kashmir and a couple of buttons from China you might like to put on a dress. There is also a couple of weird highly colored Hindi table scarfs characteristic of this part of India.

Warren's Christmas present is still in the process of being made. Albie's is finished and looks quite smart. I'll bring them along when I come home.

From all I can gather the Central African route would be the best way of going home. This would put me in Miami where I'll immediately phone on my arrival. This route will be more comfortable, different, and your /p.4/ chances of being held up are a lot less. However there is a chance of getting a courier job with a #1 priority and going the northern route. Flying continuously day and night for two and a half days has its advantages but also its disadvantages. I would probably be a wreck by the time I arrived, judging from the stories of others who have done it that quickly.

Life here goes on in its dull way. Yesterday, however we went out to the Dalmia Cement Factory to watch the antics of Prof. Ramesh Chandra Arya which ended by his getting buried for 24 hours. I took some movies of the party which I'll send along.[223]

Bye, bye darling – Happy Birthday. I wish I were spending it with you.
       All my love and more,
          Al

---

26 Feb 1945 postmarked APO   From AWZ to BSZ, Haverford, Pa.   ALS-I
                               [no header]

                                                                                                Monday Feb. 26th

[unnumbered; probably 114, and 113 was lost]
Dearest Barb

Curt arrived yesterday and brought a letter from you. Also I have received two others since I last wrote – one dated Jan 25th and the other Feb. 9th. Jack [Ott]'s letters seem to get thru a lot quicker than yours – but it's rather late in the game to do anything about it.

I don't know whether I'll beat this home – not that I think I'll be home so quickly but you've said my letters have been taking so long recently.

It was good to see Curt and get the right dope on my future at least my immediate future.[224] I didn't have an awfully long time with him as he arrived early in /p.2/ the morning and left after breakfast. However he did manage to spare a few minutes although he was very tired. His arriving precluded any idea that I might relieve him in Wash. My orders will read as before – report to the school in N.Y. after 25 days leave. I might be able to arrange splitting the leave after I get home.

---

[223] The invitation for this event is in AWZ Scrapbook, loose between pp. 25-26. Many rather fantastic feats are promised.
[224] Winsor arrived "early in the morning" on 25 February and departed "after breakfast," presumably on the same day, because AWZ says they spent only "a few minutes" together. However, Winsor's letter of 9 March to Col. Bales at ONI, written while he was still in India (probably in Delhi), shows that Winsor and AWZ must have exchanged important information that AWZ does not mention here. It is very likely that Winsor showed AWZ photocopies that he brought with him of a two-page "Personal" letter from Winsor to AWZ (21 Feb 1944), which had been intercepted by the then-Commanding Officer at Karachi, LCDR F. Howard Smith; and a one-page letter from Commodore Milton Miles to RADM Schuirmann, Director of Naval Intelligence (22 Aug 1944), in which Miles suggested that Winsor ought to be fired.

Henry G. [Gorman] was supposed to arrive in Cairo today, although Curt and I both feel he's optimistic in getting there that quickly with a 3 priority not having left till after the 19th. He is going to spend several days in Cairo before coming on – why I have no idea. I don't think I'll have to spend more than a couple (3 or 4) of days with /p.3/ him which all figures out to my leaving between the middle and end of next week (Mar 7th to 10th) allowing for Henry being delayed en route. Until he arrives I don't know just how I'll be going home. The speed should be between 2 ½ and 6 days. If I'm able to get top priority it will be the former and I will be heading right for Wash. to be there – a question of [illegible]. The other way offers two alternatives – New York or Miami – either way I'll be heading straight home. I think the best bet is for you not to try to meet me anywhere. I will call by telephone as soon as I can and we can appraise the situation and decide what to do. The chances are I'll be heading home as soon as possible (not Wash.)

    I loved getting the pictures of you /p.4/ and Lanie – she looks completely different but you look exactly the same. Boy it certainly will be good to see the reall [sic] thing. I hope Lanie's diet will be able to take off a few pounds. She looks a bit plump.

    Our boss is due to arrive today which means I'll be pretty busy preparing for his visit so I'd better get going. I'll bring you up to date on my recent activities (which really haven't been very interesting anyway) when I get home which is very apt to be before this letter anyway. Nobody is any sorrier than I that this thing has strung out so long. Thank goodness it is finally shaping up.

    I'll cable you as soon as my trip is definite and try to give you some idea of time and route.

    All my love, dearest just can't wait

    Al

----

8 Mar 1945 postmarked APO	From AWZ to BSZ, Haverford, Pa.	ALS-I
	[no header]	

Tuesday Mar 6th

#115

Dearest Barb,

    I had gotten as far as the above when I had to go down town. On my return a letter awaited me from my dear friend Groman with the delightful news that he is not coming out to relieve me. This certainly was swell news after all the previous disappointments. I was wondering what in hell was keeping him. He was supposed to leave on the 19th etc etc. His story is – two days before he was about to set off "the question about his amoebic dysentery came up" and they decided not to send him. A fine time to decide that. He consoled me with the fact they were getting somebody started immediately. That probably would mean another three months judging by his case. /p.2/ Not a very happy prospect. However, I got busy and called Capt. Davis, our new boss in Delhi and he said he would nominate someone from out here. I hope this means I'll soon be on my way. Certainly it will take a lot less time than waiting for somebody to go thru the same riggermarole as Groman. If all goes well I may be out of her in two weeks. There's not much hope of moving before then.

    I haven't written in the last week as I was sure I would beat any letter home. Capt. Davis was here from Monday till Thursday. We had various people in to meet him – Gen. Warden (U.S. Army) and Colonel Shirley for lunch on Tuesday and a dinner party Wed. /p.3/ night with Capt. Watt (Naval Officer in Charge RIN) & wife, Mr. and Mrs. Macy and a couple of Wrens. I guess he was fairly satisfied with our work here. He certainly is not the kind of guy to give you a big pat on the back or weigh you down with praises. I wasn't sorry when he left – it was a bit of an ordeal having someone pry into your affairs using a fine tooth comb.

    The Van Dusens have finally arrived back after being months on the way mostly in Australia. It was good hearing from someone who had stayed at our house though it was a long time ago. They were glowing in their praise of you and the children. I can now understand /p.4/ why you didn't go into raptures about them. She is not exactly your line of country.

Clarence Lewis[225] came through on Friday. Unfortunately he didn't call me when he arrived (which was considerate as it was 4 AM) so they put him out in a stageing area miles from town. I was able to manoeuver to get him into our place but not till Sat afternoon, in time to go to the races. Sat. night we had dinner here and then went to the Gym where the Vice consuls were having a party – a farewell to Dixon Edwards who is going to Chungking as second secretary to the embassy. We finally got Clarence off at 4.30 AM after going to Joe Sparks bungalow (one of the VCs) and eggs here. We didn't know he was to go that night but /p.5/ the call came from the airport to get him out there immediately. It was good to see him and get the latest dope from home. He looks very well and everybody liked him espiacially Mr. Macy. The two got along like two old shoes, starting off on common ground – both being named Clarence.

Sunday I took Doris and her kids to the beach – a welcome restful day after a rather rough night. The beach is very pleasant now – the weather and temperature being just right.

Last night the RAF had a big dance at their officers' mess. It was quite an affair but I'm sick of all these parties. I want to be home with you. I'm dreadfully /p.6/ sorry about all this delay. I've done all I could but my best laid plans have gone awry. You must have been sunk when you heard this last bit of news. It's been just one thing after another. Keep your chin up dearest, it can't be long – something's got to happen and believe me the sooner the better.

I had a hunch about this latest bomb shell – I think I mentioned it in a letter to you. I can't help thinking there has been some skullduggery afoot.

Good bye dearest – I'll keep plugging – things are bound to break our way soon.

    All my love and loads more

        Al

----

13 Mar 1945 postmarked APO   From AWZ to BSZ, Haverford, Pa.   ALS-I
                                 [no header]

                                                        Tuesday Mar 13th

[unnumbered 116]

Dearest Barb,

Still no definite word on when I'm to leave. Sunday a dispatch came in from Wash. and I was sure it concerned me but I was wrong again. It contained Howard Voorhees' orders. Now wouldn't that burn you up. It looks as though he would get out of here sooner than I and he left the States about four months after I did.[226]

Jim Jarvis appeared Sunday evening on his way to Delhi. I asked him to try to find out what the score is up there (Capt. Davis's headquarters) but so far nothing has come through. I feel awfully discouraged and blue about the whole damn thing. Things have got to break my way some day but that day seems to be taking a helluva time getting here. /p.2/

Received your letter of Feb. 26th, written just after you'd received the delightful news from the Gromans. I know it must be awfully hard on you – making plans etc. I'm dreadfully sorry but there's nothing much I can do about it. I've done everything I can think of to do.

You say you hadn't received any letters between Jan 25 & Feb. 12th. I wrote two between those dates (You could tell by the numbers). I can't understand why you haven't received them. Everything seems to be fowled up these days.

It was refreshing to see Jim Jarvis. He seemed happy & cheerful after the rather arduous journey over here. We spent the evening together. I put him on his plane about 10 P.M. after a little liquid refreshment /p.3/ [sheet 2] and probably the best meal he's had in a week.

---

[225] Clarence Lewis, who lived next door to the Zimmermanns in Haverford. He was in the OSS as a USMC officer.
[226] Voorhees was, however, not relieved before AZ. He was still there on 8 April and signed AZ's orders on that date, authorizing per diem payment (see folder for Travel, Orders, and Receipts). Voorhees does not appear in any file or folder that I have seen after 8 April 1945 and I know nothing about his pre-war or post-war activities.

Lanie sent me a very nice letter from Chatham. She seems to be striving hard to get her grades up to B-. She tells me her Spring vacation ends the 3rd of April. I hope I'll be able to make home by then but I've given up trying to figure things out. She's a sweet girl. She specializes in quality rather than quantity in her letters. I don't think I've had more than two in the last six months.

Jack [Ott] has also written – urging me to try to get out of the Navy on my return. I must say the way I feel now the idea has lot of merit.

My activities since last writing /p.4/ home have not been very exciting. Thursday Jack Lintoll, a British Army Captain, had a so called singing evening. We went to the club first and then to his bungalow for a snack, after which he produced an old song book and led those present in some old bromides. They asked me to sing which I did – Mother Carey and All Black Magic – but the reception was luke warm. Maybe I'm losing my grip or my heart wasn't in it. Anyway the party was not up to the standard I'm used to at home. I long for the good old days with my good old friends.

Good bye, dearest, maybe I'll be able to give you some good news one of these days.

    Oodles of love as always,

        Al

---

19 Mar 1945 postmarked APO   From AWZ to BSZ, Haverford, Pa.   ALS-I
                        [no header]

                                                                                                             Sunday Mar 18th

#117
Dearest Barb,

I guess I opened my mouth and put my foot in it. After waiting ten days, after receipt of the news about Groman, I called Capt. Davis in Delhi to ask if there was anything new in my situation. I got promptly sat on for calling on such a "trivial" matter I wanted the information espiacially to pass on to you to give you some idea on when I'd be coming home as I heard he had asked to have a man in Calcutta transferred here and I wondered how long that would take. Needless to say I got little or no inkling on the subject. I hope he doesn't punish me by keeping me here longer /p.2/ for such a grievous error. It's all very discouraging and I'm sunk that things take so long to work out. I did hope to be home for at least part of Lanie's spring vacation. (I hope Babs will be able to come down any time for at least a weekend.)

Your letter of Mar 5th arrived several days ago. I know how you must feel. I'm dreadfully sorry. There's nothing much I can do about it as you can see from the above. Any additional move would only aggravate so I'll just be quiet as a mouse and await developments.

I'm sick about the McCown boy and Jan B's husband. This whole mess is so awfull and un-necessary. Why do they go on. It's ruining so many lives and causing so much heart-ache and grief. /p.3/

As you can well imagine my life here has been quite meaningless. It's not much fun doing anything and you can't plan ahead very far. Any invitation is accepted with "if I'm still here" and of course I've been able to fulfill them all much to my regret. Thursday Bill Cullen took Howard, Doris and myself to a play at Napier Barracks given by the Repertory Players – a group of talented citizens of this town. Its main attraction was that it was written by the same person [Terrence Rattigan] who wrote "French Without Tears" [play, starring Rex Harrison, 1936; movie, starring Ray Milland, 1939] – remember? – but not nearly as good.

Yesterday, Phil Halla and I went /p.4/ to the races. Had some thrills and disappointments – ending the day with a $6 loss. Betting here is on a very small scale. You can place as little as 60¢ on a horse. Howard, Phil and I are about to set off for Hawks Bay to spend the afternoon. It should be very pleasant as the weather is ideal. Arlo Bond has let us use his hut while he's in America.

I'm enclosing a recent picture [not retained in the envelope] to give you an idea of what you're to expect one of these days. Believe me, that day can't arrive too soon. Until then, Good Bye dearest and all my love,

        Al.

22 Mar 1945 postmarked APO	From AWZ to BSZ, Haverford, Pa.	ALS-I

[no header]

Wednesday Mar 21st

#118

Dearest Barb,

Here's a bit of news at last. Delhi called this morning to say my relief's orders were there. He is being sent over from Calcutta and should arrive here in about a week, which will allow me to leave April 1st. As it's set up now, at least tentatively, I'll be going home on top priority, reporting directly to Washington. I'll not be there long however – just a few hours and then I'll have my leave. The chances are very strong I'll land in New York and it should be about the 4th. I will of course phone you as soon as I can. Possibly I'll be going from N.Y. to Wash by train and you could ride down /p.2/ and back if you wanted to. That's about all I can figure out now. It still isn't very definite but it should work out <u>this</u> time. I'll try to let you know by cable if it all works out.

So, keep your fingers crossed and be looking for me somewhere about that time but don't worry if I don't appear on time as a trip like that is, of course, subject to delays even after you once get started.

This should be my last letter as I should beat any subsequent one. I won't know anything definite till just before I leave and will let you know by cable (probably through Jack) whether or not this schedule stands.

Bye, bye, dearest, till the 4th. All my love,

Al

---

2 Apr 1945 postmarked APO	From AWZ to BSZ, Haverford, Pa.	ALS-I

[no header]

Saturday Mar 31st

[unnumbered 119]

Dearest Barb,

Can you believe it – another disappointment. My relief came last night and brought the news I was not to be the courier (pity they couldn't tell me before this) – leaving this Sunday. That means of course all my plans in the last letter go the way the previous ones have. It's all so damn discouraging. My relief[227] (the dear boy) says I'm not to leave until he understands everything thoroughly. What that means I don't quite know but I have visions of that taking another week or two. The most baffling thing to him seems to be the mess about the mess my predecessor bequeathed to me. He's a nice enough guy, I suppose, and is probably taking his cue from the god /p.2/ that sits in the shadow of his Hindu brethern [sic] in Delhi. I have used all manner of pleas with him – espiacially the longing that I think my family has for me – which I hoped would speedily bring the light of day to his foggy mind. But being a bachelor – not appreciating the finer things of life – he has elected to go to the beach tomorrow and not burden himself with the problem of "understanding everything thoroughly" even though that might be uppermost in my mind.

Well anyway, I have a lot to be thankful for – you are all well and I'll be seeing you sometime – perhaps not in my normal mind – as these delays are driving me rapidly mad. I'll cable you tomorrow about the delay, so you won't be expecting me on the fourth.

From your burnt up but still longing & loving husband,

Al

---

[227] The new CO, LCDR Edward F. O'Connor relieved AZ. O'Connor signed AZ's orders on 9 April 1945, allowing him to depart for home.

Here ends the last letter in this box. AZ's modified orders to depart Karachi were endorsed on 8-9 April by the new commanding officer, LCDR Edward F. O'Connor and LT Howard Voorhees (who authorized his per diem). AZ departed Karachi on ATS (Air Transport System) on 10 April and he arrived in Miami on 17 April 1945. His duties during the rest of the year appear in other files and folders. After he returned to the U.S. he had 30 days of leave, at home. On 19 May he reported to HQ, 3rd Naval District, NYC, for an 8-week course of instruction at Port Director School.

While at this school he developed a recurrence of a peptic ulcer and was evaluated at St. Albans Naval Hospital from 23 July until a time in September. A Medical Board recommended on 11 September that he be released from active duty, and he was sent home on what became terminal leave. His active duty concluded at midnight on 30 December 1945.

Five Naval Officers at Naval Liaison Office, Karachi, India
LTs Howard Voorhees, in center, and Albert Zimmermann, 2d from L.
The other three are LTJGs, Baker, Callahan, and Reflord.
(winter of 1944-1945)

Mahan Kishin,
Zimmermann's "bearer"

"Our bungalow"
Naval Liaison Office Karachi
254 Engle Road

Undated but labeled group photo of the leaders of SOS (Services of Supply), Karachi, showing 2nd from L (J. Harris), 4th and 5th from L (BG Haddon, BG Warden). Second row, 3rd from L (A. Zimmermann). Photo Album, 29(R).

Undated, unlabeled photo of AZ and Anglo-American staff in India, probably Karachi. AZ is second from left in front row. He is the only U.S. Naval officer in the picture.

31(V) in photo album

LT Zimmermann: "Himself in the flesh on the banks of the Indus"

AWZ-WP, Loose Photos

Photographs of Naval Liaison Office, Karachi

LCDR F. Howard Smith, USN – CO, NLO, until March 1944

LCDR Smith and Hashim Ali Shah, head watchman

Note equipment used for coding and decoding messages

The Three Lieutenants
At the Naval Liaison Office, Karachi

LT Howard Voorhees, LT Albert Zimmermann, LTJG Phil Halla
About March 1944

Di Gemma

LTJG Phil Halla at NLO, Karachi

Wikipedia

Senator Richard Russell
Visited NLO Karachi
(see Zimmermann's letter,
23 August 1943)

Paintings by April Swayne-Thomas, in *Indian Summer*
Zimmermann was her guest at dinner on 18 February 1944
Her paintings are of locations mentioned in Zimmermann's letters

April Swayne-Thomas, from frontispiece

"A corner of Peshawar," p. 60

"Crags on the way to Ziarat," p. 104

"Malakand," p. 63

"Camels at Sandspit," p. 32

April Swayne-Thomas' paintings, continued

Clockwise, from upper L: "Chakdara Fort, Swat Valley," p. 63.
"Political Agent's bungalow, Malakand," p. 62.
"Quetta Hills," p. 105.

April Swayne-Thomas' paintings, continued

"Peshawar City," p. 61

"On Guard, Landi Khotal," p. 61 (L)
"Men of the Khyber Rifles," p. 61 (R)

Lady Vere Birdwood
"Abode of Peace"

Lady Vere Birdwood was his frequent bridge partner
(see Zimmermann's letter, 4 June 1944)
Her First Prize Photo, "Abode of Peace"
From *Illustrated Weekly of India*, 12 November 1944

Mentioned in Zimmermann's letters to his wife

From http://www.unc.edu/depts/diplomat/item/2006/1012/enge/engert_corn.html (L)
Harry S Truman Presidential Library

Cornelius Van EngertCharles Thayer
See his letter of 2 December 1943 for Engert and Thayer

https://nekropole.info/en/Gene-Markey
From http://www.thefridaytimes.com/tft/cunningham-and-the-tribesmen-1940s

Gene MarkeySir George Cunningham

See letters of 9 and 10 July 1943, 15 June 1944 for Markey
See letter of 2 December 1943 for Cunningham

Google Images
Ambassador Clarence E. Gauss, in Chungking, 1944
In letters of 15 September 1943, 23 November 1944

Di Gemma, USCG
BG John D. Warden
In letter of 24 October 1944

Google Images

Winston Churchill, after the Battle of Malakand
(mentioned in Zimmermann's letter of 2 December 1943) and after World War II

Khyber and Bolan Passes

Khyber Pass – Afghan Fort

Bolan Pass
(painting in 1838)

One of 21 passes on the route that Zimmermann's train used on his way from Quetta to Karachi in December 1943

Peshawar

Google Images

Dean's Hotel
It was the most famous hotel in Peshawar – destroyed in about 1979
Kipling, Curzon, Churchill, Lowell Thomas, and Gordon Enders stayed there
Zimmermann stayed at Dean's on both of his visits to Peshawar
He had dinner at Dean's on 15 and 26 November, and had lunch there on 24 November

The Peshawar Club

Google Images

Zimmermann attended a dance there on 27 November 1943 and had lunch there the next day

Peshawar

Google Images

Peshawar Services Club

Sir Benjamin and Lady Bromhead stayed at the Services Club with their children.
The facilities included hotel rooms for service members and a club-like atmosphere.
Zimmermann dined at Services on his first night in Peshawar, 14 November 1943.
He had dinner at Services with the Bromheads on 24 and 27 November.

Peshawar

Zimmermann met the Viceroy of India at a Garden Party at the Governor's House

See letter of 2 December 1945

The Governor's House

Google Images

The Governor's Garden

Commodore Milton Miles and Ambassador Patrick Hurley

Courtesy of family of Milton Miles

Group picture in China showing (L to R): RADM Milton "Mary" Miles, wearing one star, U.S. Consul General Walter Robertson, Ambassador Patrick Hurley, and Gen. Tai Li
RADM Miles was known as Commodore Miles when he visited Zimmermann in Karachi
Amb. Hurley was Gen. Hurley when he visited Enders in Kabul (Correspondence with Enders)

Field Marshals Montgomery, Wavell, and Auchinleck

Google Images

Bernard Law Montgomery, Archibald Wavell, and Claude Auchinleck
Zimmermann met Wavell at the Governor's Garden Party on 27 November 1943
He met Auchinleck in Delhi in July 1944 (see photo with letter of 22 July)

4

North-West Frontier Trip Along the Indian-Afghan Border

**Peshawar – Chitral – Quetta
Centered on the Khyber Pass
15 November – 15 December 1943**

**By
LT Albert W. Zimmermann, USNR
Major Gordon B. Enders, AUS
Major Sir Benjamin Bromhead, IA**

"People thought attempts to drive over Lowari crazy. [A] tortuous trip over the pass in Maj. Enders' jeep – the first time a car of any kind had attempted the feat"
(AWZ, 18 November and 25 November 1943)

*I was the first Navy man in Chitral,
and there had been only one other American there before us.*
(AWZ, 25 November 1943)

*The Viceroy had heard we'd been up the Lowari to Chitral in a jeep so came over
to where we were standing and asked all about it.*
(AWZ, 6 December 1943)

The trip in 1943 – It was all part of the Great Game.
(Col. H. R. A. "Tony" Streather, OBE, 20 January 2010)

First Official Notice of the Planned Trip

From IB Quetta to John R. Harris, Esq., Central Intelligence Officer, Karachi
26 October 1943

```
EXPRESS        SECRET & PERSONAL              No. 9432
                                         INTELLIGENCE BUREAU,
                                              QUETTA.
                                         26th Oct., 1943

Dear John

     I have just heard from Major Sir Benjamin Bromhead of the
N.W.F. Public Relations Bureau that with the blessing of the
Governor N.W.F.P. he is taking Major Enders, U.S. Military
Attaché, Kabul, on a personally conducted tour of the Frontier
and Baluchistan from Chitral to Quetta with the idea of making
it clear to the American Legation in Kabul what are our frontier
problems and our ideas and policy in dealing with them and the
Afghans.
     I promptly asked him whether he could also take one of the
American officers from the U.S. Naval Liaison Office if they
would like to send one. He replied in the affirmative subject
to the Governor's sanction which he said he thought would
certainly be forthcoming.
     Would you put the offer to Smith and ask him to telegraph
me a reply so as to reach me by 1st November, just saying
whether they would like to send an officer and if so whom.
Bromhead's dates are :-
     10.XI.43  leave Peshawar for the Kurram (Parachinar)
     15.XI.43  return Peshawar.
     18.XI.43  leave for Chitral.
     25.XI.43  return Peshawar.
     29.XI.43  leave for Waziristan.
     10.XII.43 finish tour in Quetta or Peshawar.
     As you can see, it means a month away from H.Qs.
     I don't know if Smith would be interested in this somewhat
unique opportunity of getting a first-class background for his
own office and Naval H.Qs. at Washington to use in connection
with any reports emanating from U.S. sources in Kabul or Delhi,
or whether he could spare an officer for so long. You will
readily appreciate the necessity for carefully picking the
officer so that he does not get hold of the wrong end of the
stick or miss important points.
     How Smith would explain to Enders and Engert the presence
of this officer would be Smith's headache and not ours!
     If Smith's wire contains an affirmative reply I will
wire Bromhead in Peshawar to get H.E.'s sanction and convey same
direct to Smith, including instructions regarding date and place
where the officer should report. After wiring Bromhead as above
I fade out of the picture and negotiations between Smith and
Bromhead are then direct.
     Smith will realise that the weather will be bitterly cold
with the possibility of snow in Waziristan and Chitral, so warm
clothes are essential.
     Bromhead's address in Peshawar is :-
          Major Sir Benjamin Bromhead, OBE, IA,
               Deputy Public Relations Officer,
                    Frontier Tribes, PESHAWAR. (N.W.F.P.)
     Will you please convey this message to Smith ?
                                        Yours sincerely,
                                             /s/

J.R. Harris, Esq., I.P.,
Central Intelligence Officer,
Karachi.
Copy to Major Sir Benjamin Bromhead, OBE, IA, in
continuation of our conversation of yesterday's date.
     The designation and address of the American Naval Liaison
Office in Karachi is :-
          United States Naval Liaison Office,
          254 Ingle Road, KARACHI,
and the telegraphic address is  ALUSLO, Karachi. The
Commanding Officer is Lt.Commander F.Howard Smith, U.S.N.
```

This letter to J. R. Harris from the anonymous person known only as "Intelligence Bureau Quetta" was transcribed and printed in *Proceed to Peshawar*, xix-xxi.

Orders for Lieutenant Albert W. Zimmermann
8 November 1943
"on or about 11 November 1943 . . . proceed . . . to Peshawar"

OFFICE OF THE UNITED STATES NAVAL LIAISON OFFICER
KARACHI, INDIA

Cable Address: ALUSLO
205065
8 November 1943.

EN3-11(KA)P16-4/00/A-1/jah
Serial: 558

From: The United States Naval Liaison Officer, Karachi.
To: Lieutenant Albert W. ZIMMERMANN, I-V(S), USNR.

Subject: ORDERS - Temporary additional duty.

Reference: (a) SecNav Ltr. to All Ships and Stations dated April 30, 1943.

1. Upon receipt of these orders and when directed by proper authority, on or about 11 November 1943, you will proceed via transportation furnished by the United States Army, to Peshawar, North West Frontier Province, India, and such other places as may be deemed necessary for the proper performance of the duties assigned you. Upon completion of this temporary duty you will return to this office and resume your regular duties.

2. Transportation to Peshawar, North West Frontier Province, India, is to be furnished by the United States Army and you are authorized to defray any additional travel, including transportation by military or commercial aircraft, subject to reimbursement by the government.

3. Per diem allowances while traveling in obedience to these orders is authorized in accordance with reference (a).

FRANCIS H. SMITH.

cc BuPers

FIRST ENDORSEMENT
EN3-11(KA)P16-4/00/A-1jah

U. S. NAVAL LIAISON OFFICER
KARACHI, INDIA
November 12, 1943.

From: The United States Naval Liaison Officer, Karachi.
To: Lieutenant Albert W. ZIMMERMANN, I-V(S), USNR.

1. You departed at 1540 this date.

FRANCIS H. SMITH.

SECOND ENDORSEMENT
EN3-11(KA)P16-4/00/jah

U. S. NAVAL LIAISON OFFICER
KARACHI, INDIA
December 15, 1943.

From: The U. S. Naval Liaison Officer, Karachi.
To: Lieutenant Albert W. ZIMMERMANN, I-V(S), USNR.

1. Reported and resumed regular duties at 1130 this date.

FRANCIS H. SMITH.

(Side margin notation: F. RICHARD SMITH, Lt. Cdr. USNR., SEO. U.S. NAVAL LIAISON OFFICER, KARACHI, INDIA, Dec. 24, 1943 Paid Per Diem allowance for period Nov. 12, 1943 to Dec. 15, 1943 on PV#197-44, thirty four days at $8.25 per day, total paid $280.50.)

The title of the book, *Proceed to Peshawar*, is taken from the words in these orders

The North-West Frontier Trip[228]

DAY 1 – Friday, 12 November 1943
From KARACHI en route to PESHAWAR

Departed Karachi 1540 (3:40 p.m.) by train

The first part to Lahore was in an air-conditioned car, sharing a compartment with a Col. Fagin, another American Army officer, and a British Army officer. Fagin knows Lanie Hunter quite well and could understand why he hated Karachi. He seemed to think he was a pretty pampered individual. The trip was quite pleasant . . . We were very lucky to be in an air conditioned car as the route took us over the Sind desert . . .

From Lahore it was a bit different – no luxuries whatever – not even a dining car. I had to fight my way onto the train even tho I had a reservation. I found later it's just another way of getting baksheesh . . . During the night there was all sorts of commotion . . . supposed to be in I class but it was nearer III . . . I finally would up with a London limey . . . and a British colonel and his wife as my bunk-mates. The limey's bearer was also there spending the night sitting on the floor. . . [AL]

*Attock – The gateway to Northwest India.
Here the North West RR crosses the River Indus over a strategic bridge* [AN]

---

[228] AN = autograph (handwritten) note by AWZ, on back of photo. Also, a few handwritten notes in his files.
AL = authograph (handwritten) letter by AWZ. See previous chapter for details of the letters.
TR = Typed report of trip, submitted by AWZ
TRd = typed report draft, by AWZ

DAY 2 – Saturday, 13 November 1943
From KARACHI en route to PESHAWAR (continued)

Crossing the Sind Desert –

On Indian trains you carry your bed with you in what's known as a bedding roll. If you don't own one you can rent one from Cooks. I borrowed Cam. Smith's. Your bearer (if you have one along – I didn't) unrolls your roll on one of the four long seats and then is supposed to spend the night in a bearer's compartment or in III class, to appear the next morning, help you dress and roll up your bed again . . . and your chances of picking up new traveling companions (very small ones) are thus minimized. . . . Too bad the colonel's wife wasn't a bit younger and more attractive. It would have added a bit more zest to the intimacy of our bed room. [AL]

Attock – Showing the characteristic terrain of northwest India [AN]

DAY 3 – Sunday, 14 November 1943

Arrival at PESHAWAR

I was met by Major Sir Benjamin Bromhead on my arrival and after a good hot bath at my hotel was wisked to the Peshawar Club for lunch. It being Sunday, lunch at the club was quite festive. The English colony was out in full force, drinks and lunch being served to the accompaniment of a grand Indian orchestra playing most of the good modern tunes . . . The city is divided into two sections – the native and the cantonement. The cantonement is surrounded by a barbed wire fence with gates on the entrance roads. This is a protection against the plundering tribes that inhabit this part of the world.

    The Northwest Frontier Provinces have a peculiar set-up for government. The Administered Area is governed with Indian laws. A boundary extends thru the provinces that separates the administered from the non-administered area or tribal area and then comes the boundary between India & Afghanistan known as the Durand Line, more or less arbitrary one set up around the turn of the century. The Tribal Area has its own local governments ruled by mehtirs, nawabs, Walis or mullahs according to where you are in the Provinces. [AL]

*Peshawar – Capital of Northwest Frontier Province – Railhead –
Although military RR extends 25 miles thru Khyber Pass* [AN]

Major Sir Benjamin Bromhead, 5th Baronet Bromhead, in uniform, meets Zimmermann

DAY 4 – Monday, 15 November 1943

From PESHAWAR into TRIBAL TERRITORY, to YUSEF KHEL via SHABKADAR

On Monday we made a trip into Tribal Territory to Yusef Khel. Major Enders . . . was supposed to be with us but didn't arrive till the next day. . . . we had lunch with the malik of the lower Mohammad Tribe (one of the peaceful ones). We started off having our hands washed by pouring water over them as we sat at the table. There were no implements. . . . An interesting time was had by all. I was more or less a curiosity being American and in the Navy.
    At the entrances . . . there is a gate for proper identification both ways and a place to park your shootin' irons when you are entering the land of law and order. Caravans of donkeys, mules and camels pass thru taking grains, poultry, etc., to the towns . . . They would never think of traveling thru most of the Tribal Territory without a gun.   Stopped at Shabkhar . . . an old fort . . . Met a malik [master] *who is enemy of the Yusef Khel malik ... All villages fortified . . . badi (blood feuds) widespread. Went up to Yusef Khel and lunched with its malik. . . Yusef Khels are a portion of the Lower Mohmand Tribe whose protection was guaranteed by the British. We returned in time for dinner at the Bromheads' hotel and were joined by Enders who had arrived from Kabul.* [AL]

Shabkadar – On the border of tribal territory. Old British fort on right [AN]

Shabkadar – Sawar Khan subhadar, Intelligence Liaison Officer to H E the Governor, two Indian army officers, Mohammed Isaac, assist. P.O. [Political Officer] *under deputy commissioner. Peshawar Fort in background* [AN]

DAY 4 – Monday, 15 November 1943 (continued)
From PESHAWAR into TRIBAL TERRITORY, to YUSEF KHEL via SHABKADAR

Shabkadar – bazaar where farmers from tribal territory come to trade their products for bolts of cloth and other manufactured products of India [AN]

Shabkadar – bazaar woman hiding her face – This place is between Peshawar and Malakand – on the border of Tribal Territory . . . for sale are grains, cloth, green tea . . . quite a few money changers [AN]

DAY 4 – Monday, 15 November 1943 (continued)
TRIBAL TERRITORY, to YUSEF KHEL via SHABKADAR and HALKI GANDAO

Shabkadar – end of bazaar street – coming out into sun – by this time we had collected quite a following
Major Sir Benjamin Bromhead is slightly to the right in front row

Halki Gandao – Tribal Territory. Khassadar on promontory guarding road [AN]

DAY 4 – Monday, 15 November 1943 (continued)

TRIBAL TERRITORY, to YUSEF KHEL via HALKI GANDAO and KASAI MUNIDI

Halki Gandao. Khassadars [AN]

Kasai Munidi. Reception Committee [AN]
Bromhead is leaning on the car

DAY 4 – Monday, 15 November 1943 (continued)
TRIBAL TERRITORY, to YUSEF KHEL via KASAI MUNIDI

The malik [master] *of Kasai Munidi. His beard is red showing he's been to Mecca* [AN]

Yusef Khel – Hill 19408 where Auchinleck fought in 1935 [AN]

DAY 4 – Monday, 15 November 1943 (continued)
From PESHAWAR into TRIBAL TERRITORY, to YUSEF KHEL

Yusef Khel – lunch with the malik and his right hand man. Shazada – the malik – in fore ground

Typical Pathan head dress. Lunch – roast chicken (very tuff) hard boiled eggs, many varieties of cuke (gourd), chapattis. All done with right hand. Asst. Political Agents left foreground and on right [AN]

DAY 4 – Monday, 15 November 1943 (continued)
From PESHAWAR into TRIBAL TERRITORY, to YUSEF KHEL

Yusef Khel – Shazada the malik of Y.K. [left in photo] and his right hand man [AN]

Shazada [center, lower row] *and his body guard – Yusef Khel* [AN]

DAY 5 – Tuesday, 16 November 1943
From PESHAWAR into MALAKAND. First stop: SWAT

Trip to Swat . . . up the Swat River. Passed the fort where, for six days, the British garrison was imprisoned and whence Gunga Din came down to the Swat River to get water for the troops. They could only heliograph from the fort at this siege [TRd]

Churchill as a sub-lieutenant fought in these mountains and Auchinleck in 1935 was stationed there [AL]

Lower Malakand country –taking farm products to market – Walled villages in background. Donkeys provide most of transportation [AN]

Malakand – P.A. [Political Agent] Packman. Wandering musicians [AN]

192

DAY 5 – Tuesday, 16 November 1943 (continued)
From PESHAWAR into MALAKAND. First stop: SWAT

Malakand – Valley of Swat River – Gunga Din came down from piquet on hill to get water from river

Malakand – Valley of Swat River – Another view of piquet with musicians in foreground [AN]

DAY 5 – Tuesday, 16 November 1943 (continued)
From PESHAWAR into MALAKAND. First stop: SWAT

The first day we had lunch with the Wali of Swat, a venerable man who is said to have pushed his brother over a cliff. We were greeted by saluting soldiers lining the streets, a blare of trumpets at the "palace" gate and the old Wali himself at the door, along with his son the Walihad and the "prime minister." [AL]

Saidu – Palace of Wali of Swat. Major Bromhead, Secretary to Wali, P.A. Packman, Walihad, Wali of Swat [AN]

Saidu – Band that greeted us [AN]

DAY 5 – Tuesday, 16 November 1943 (continued)
From PESHAWAR into MALAKAND. First stop: SWAT

His palace is astounding – Persian rugs, modern furniture, up to date plumbing, etc., etc., in the middle of mud huts and buffaloes. A very good lunch with all the courses and trimmings of India. [AL]

Saidu – very modern furnishings, bath rooms, etc. [AN]

Saidu – Swat License Plate #1 [AN]

DAY 5 – Tuesday, 16 November 1943 (continued)
From PESHAWAR into MALAKAND. First stop: SWAT

Zimmermann's notes for 16 November 1943

Swat – School children – slates and books on heads [AN]

Malakand – Fortified village - rugged country. Swat River running thru mid ground [AN]

DAY 5 – Tuesday, 16 November 1943 (continued)
From PESHAWAR to MALAKAND

*Headworks Malakand Hydroelectric – gates to Swat River Canal.
Bundles of straw being carried on heads* [AN]

Headworks Malakand Hydroelectric – gates to Swat River Canal [AN]

DAY 5 – Tuesday, 16 November 1943 (continued)
From PESHAWAR to MALAKAND

That night we spent with the Political Agent, Major Pachman, at Malakand (one of Churchill's first books was about Malakand – get it & read it – it's supposed to be good). He and his wife and two daughters live at the fort in a very nice house over-looking the Swat Valley. As with the other P.A.'s, they live a rather lonely existence, living so far from civilization and seeing very few white people for months. They have a baby ibex as a pet [AL]

Malakand – from P. A. Packman's house. Fort on hill [AN].

Malakand – P.A.'s house – pet ibex

DAY 5 – Tuesday, 16 November 1943
MALAKAND

Malakand – terrace of P.A.'s house – British Flag – "Darmanian over palm + spirit" [AN]

DAY 6 – Wednesday, 17 November 1943 MALAKAND to DIR
The next day we drove to Dir at the foot of the Lowari Pass.
We had dinner with the Nawab and Waliard (heir to the "throne"), spent the night [AL]

Jeep "Ma Kabul" ready to start for Dir state. Maj. Enders at wheel Political Agent Kenneth Packman, Bengie [Bromhead], Mrs. P. [AN]

DAY 6 – Wednesday, 17 November 1943 MALAKAND to DIR (continued)

Malakand Fort. P.A.'s house on top of hill [AN]

Bat Khel – near Malakand – stopped to buy rope. Residence of Political Agent in background [AN]

Held up at gate at Chakdarra, going into Tribal Territory of Dir State. Khassadars wanted to send an escort with us. Sir B said that we were American guests and what would we think if khassadars would not let us through. [TRd]

DAY 6 – Wednesday, 17 November 1943
MALAKAND to DIR (continued)

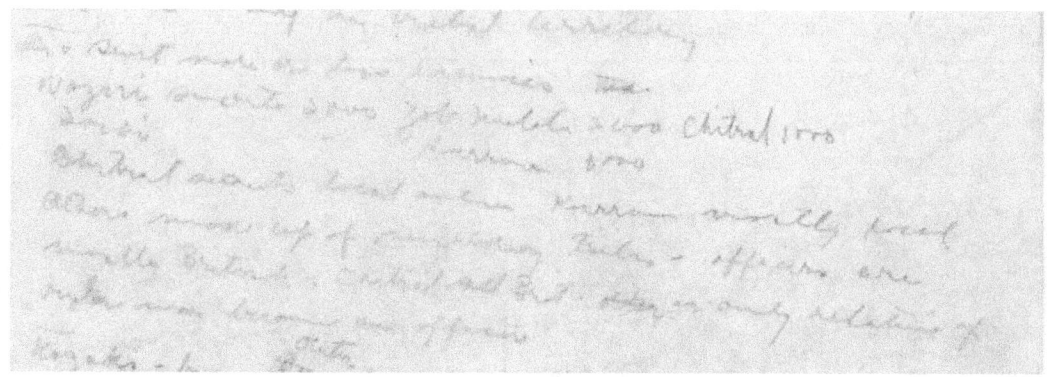

Notes, 17 Nov

Punjkara River Valley. Camel caravan on way to Dir [AN]

Punjkara River Valley on way to Dir. A falcon for H. E. [AN]

DAY 6 – Wednesday, 17 November 1943 MALAKAND to DIR (continued)

The Dir valley is very fertile, with terraced hillsides of rice (in summer) and wheat (winter). [TRd]

Dir state – Rice Patties [sic] – Rice and wheat are main crops [AN]

Dir – Rest House. Had tea here with H. E. the Nawab of Dir [AN]

DAY 6 – Wednesday, 17 November 1943
MALAKAND to DIR (continued)

At Dir had tea with the Nawab and the Waliahad, a Captain doctor. Nawab rather sickly, the Waliahad rather childish – does not smoke in front of his father, etc. [TRd]

Dir – Bazaar. Sewing machines on street [AN]

Dir – Serai (caravan resting place). Bags of antimony. Headdress on man left foreground rolls down like stocking. Typical of Dir and Chitral [AN]

DAY 7 – Thursday, 18 November 1943   DIR to CHITRAL
Over LOWARI PASS and night at DROSH

Started early next morning for the tortuous trip over the pass in Maj. Enders' jeep – the first time a car of any kind had attempted the feat. The trail is no more than a donkey caravan route, one of two, the other being thru Gilgit, that give the state of Chitral communication with the outside world. [AL]

Dir – ready to start for Chitral. Capt Behts of Chitral Scouts, the Nawab's doctor, Gordon Enders, the Waliahad of Dir [AL]

Dir – ready to start. Donkeys carried our kit [AN]

DAY 7 – Thursday, 18 November 1943 DIR to CHITRAL
Over LOWARI PASS and night at DRASH

Left Dir at 9:00 a.m. People thought attempts to drive over Lowari crazy. Arrived at Gujar Levy Post (7,800 ft.) at 11:00 a.m. Reached top of Lowari at 12:30 – mileage from Dir 14. Altitude of top – 10,240 ft. Snow on top only; none on way up. Road narrow in spots, rough with rocks at upper end. … [TRd]

Dir to Gujar. Passing donkey caravan. Trail is a bit narrow [AN]

Dir to Gujar [AN]

DAY 7 – Thursday, 18 November 1943   DIR to CHITRAL
Over LOWARI PASS and night at DROSH

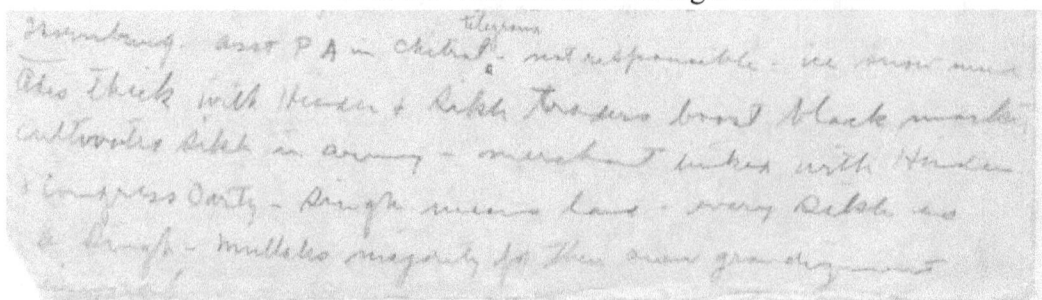

Notes, 18 Nov

Gujar Levy Post [AN]

Gujar Levy Post [AN]

DAY 7 – Thursday, 18 November 1943　　　　DIR to CHITRAL
Over LOWARI PASS and night at DROSH

Nearing top of Lowari Pass. Mountains are about 14,000' [AN]

Approaching top of pass [AN]

DAY 7 – Thursday, 18 November 1943　　　　DIR to CHITRAL
Over LOWARI PASS and night at DROSH

*Top of Lowari Pass. Hgt. 12,000' Chitral Scout, me, Benjie.
Hindu Kush Mountains in background* [AN]

LT Zimmermann (L), Chitral Scout, Maj. Enders
Photo by Major Sir Benjamin Bromhead

DAY 7 – Thursday, 18 November 1943   DIR to CHITRAL
Over LOWARI PASS and night at DROSH

Top of Lowari Pass. Benjie + Behtes [AN]

Descending from Lowari Pass. Enders driving, Zimmermann on R, facing camera
Photo by Maj. Sir Benjamin Bromhead – On cover of *Proceed to Peshawar*

DAY 7 – Thursday, 18 November 1943   DIR to CHITRAL
Over LOWARI PASS and night at DROSH

Descent much steeper than ascent, with many awkward zig-zags. Arrived at Ziarat (7,500 ft) at 2:30 and tead. Reached Ashret at 4:30 p.m. Ashret is roadhead on Chitral side. Mileage from to to Ashret 8.8. Reached Drosh about 6:30 p.m. . . . 13 miles from Ashret. . . . Total running time about 7 ½ hours. . . . Airfield at Drosh, very small . . . Would be dangerous in cross-winds. . . . Spent night in Drosh at Scouts Mess. [TRd]

Going down. Getting out to see what's ahead [AN]

Difficulty at a "zig-zag" on way down [AN]

DAY 7 – Thursday, 18 November 1943   DIR to CHITRAL
Over LOWARI PASS and night at DROSH

After squeezing between a stone wall on one side and the edge of the trail on the other, which sometimes dropped several hundred feet to the valley below, creeping over the icy and snowy places and coaxing the jeep around many zig-zags we arrived at Drosh about 7:30 that evening. We had gone from 4000 ft to 10,000 ft and down again to 4000 ft. Quite an exciting trip. There were four of us, having picked up an officer of the Chitral Scouts at Dir [AL]

Encountering ice [AN]

Tough going + the rope came in handy [AN]

DAY 7 – Thursday, 18 November 1943   DIR to CHITRAL
Descent from Lowari Pass – Photos by Major Sir Benjamin Bromhead

DAY 7 – Thursday, 18 November 1943　　　　DIR to CHITRAL
Descent from Lowari Pass – Photos by Major Sir Benjamin Bromhead

DAY 8 – Friday, 19 November 1943   DIR to CHITRAL (continued)

Ziarat Levy Post [AN]

Ziarat Levy Post [AN]

DAY 8 – Friday, 19 November 1943   DIR to CHITRAL (continued)

Ziarat – Yes, this is the road [AN]

Looking back at Ziarat [AN]

DAY 9 – Saturday, 20 November 1943   CHITRAL

We spent the night and most of the day at Drosh and went on to Chitral late in the afternoon. Chitral is surrounded by snow-capped mountains, the highest being Tirichmir, 25,600'. We were fortunate to see its top as it is usually in the clouds. But the 2 days we were there, the weather was perfectly clear, giving us wonderful mountain scenery to look at all the time. [AL]

Left Drosh after lunch, reaching Chitral at 4:30 – 26 miles. Met by Ghazi ud-Din, brother of the Mehtar. . . . Bugles and guard of honor. Had tea. Went out to polo games staged in our honor. Called on Thornburg, A.P.A. and his wife (Thornburg sent telegram that ice, snow and mud on Lowari and advised against our attempt). Dined at palace with the Mehtar, the Prime Minister . . . and several officials who were relatives of the Mehtar. . . . Saw Chitral and Hunza dances after dinner. State dining-reception-room fantastic with chandeliers, artificial flowers, etc. Former Mehtar was pervert. Practically no Pathan … wore glasses – they do very little close reading . . . No women were in evidence in Chitral. Mehtar's name is Muzafar ul Mulkh. [TRd]

This is where polo originated. After the game we went to the Assistant Political Agent's house for drinks and then to the "palace" for dinner. . . . All of the dancers were men, of course. In fact, we didn't see a woman except on the road, the whole trip. The women are kept secreted from curious eyes . . . Most of the time they were working hard carrying wood . . . mostly on their heads, moving their goods & chattels from the high lands to the low lands for the winter [AL]

"Tirichmir" at dawn. Height 25,500' [sic] (and of Fairies)
Breakfast, hawking lunch, chicor hunting at Chitral. Dinner & night Drash [AN]

Editor's note: Tirich Mir is 7,690 m = 25,230 ft. From Chitral to summit is 40.04 km = 24.9 miles

Tirich Mir

Google Images

The South Face of Tirich Mir,
View from Chitral

DAY 9 – Saturday, 20 November 1943   CHITRAL

Zimmermann's notes, 20 November 1943

Tirich Mir from the south, over Chitral polo field

Chitral Palace

(Still shots from 16mm movies taken by LT Zimmermann, 11 November 1943)

Editor's Note: LT Zimmermann did not take any additional still pictures until Monday, November 29 (Day 18), when he departed from Peshawar with Bromhead and Enders to proceed through the Waziristan Tribal Territories to Quetta, in Baluchistan. His account of what happened on the intervening days is shown in Chapter 3 of this book, in which his letters and trip reports appear. For purposes of continuity, I will summarize what happened on those days, before continuing with his photos.

DAY 9   CHITRAL to DROSH   Saturday, 20 November 1943

[W]e went horse back riding to watch "hawking." I drew a spirited ex-polo pony. I took a dim view of his thinking he was still on the polo field and it took all my strength to hold him down to a slow trot. . . . . Hawking consists of going to a high place where the river is narrow, getting about 200 men to beat the brush, rousing the birds to flight (chikhors – like partridge), releasing the hawks (falcons) as the chikhors fly by and watching the kill. Not my idea of real sport. . . . After lunch we went for a jeep ride about 13 miles north. As soon as we got back the mehtir had arranged a markor shoot. Markors are mountain deer that come to lower altitudes in the winter ... Major Enders killed one but didn't have time to see it as we were leaving for our journey home. It was all very interesting 7 exciting and something very few have had the opportunity to do. In fact, I was the first Navy man in Chitral, and there had been only one other American there before us. The trip back was much the same as going up. The mehtir of Chitral and the nawab of Dir gave us chagas (native coats) in honor of our visit. [AL]

Editor's note: Zimmermann took movies of the scenes described above in Chitral, but they were poorly preserved and I have not tried to enhance still shots from them. The original films are now in Special Collections, U.S. Naval Institute, Annapolis.

Motored back to Drosh at 5:30. Called on the Mehtar's brother who is Governor of Drosh and heir to throne as Mehtar has no legitimate sons. . . . Spent the night at Drosh Fort mess. [TRd]

DAY 10   DROSH to ZIRAT   Sunday, 21 November 1943
Breakfast and lunch Drash. Polo, drinks at Thornburgs. Dinner & night Ziarat. [AN]

Zimmermann took movies of the polo game. They are poorly preserved (see 20 November)

Left for Ziarat at 2:00 p.m., arriving at Ashret at 2:45 p.m. From Ashret to Ziarat (5.3 miles) in 2 hours, arriving Ziarat at 4:45 p.m. Spent the night at Ziarat, leaving jeep at Scout Post. [TRd]

DAY 11   ZIARAT to DIR   Monday, 22 November 1943
Left Ziarat at 11:00 a.m. in order to get sun on Lowari top. Made the 3.5 miles to top in 2 hours. Spent about 3/4 hours at one slick zig-zag turn; the rest were comparatively easy. The 3.9 miles from the top to Gujar were made in 25 minutes. The 10 miles from Gujar to Dir were made in 1 hour and 40 minutes. Total elapsed time 7 hours; running time from Ashret to Dir about 6 hours, 5 minutes. Spent the night at Dir rest house. [TRd]

DAY 12   DIR to MALAKAND and PESHAWAR   Tuesday, 23 November 1943
Motored 160 miles from Dir to Peshawar. Ate lunch at Malakand. Gifts were given to us by the Mehtar of Chitral and the Nawab of Dir. We arrived back Tuesday pretty well tired out. Enders is taking me up the Khyber Pass and Saturday the Viceroy of India is visiting Peshawar and we are going to attend the garden party to meet him. [AL]

Editor's note: Zimmermann took movies of a parade of Indian Army soldiers, probably changing the guard somewhere, perhaps at Dir, or possibly in Peshawar. The location cannot be identified, and the movies are poorly preserved (see above).

DAY 13   PESHAWAR   Wednesday, 24 November 1943
Signed HE the gov. book / Lunch at [illegible] - dinner at Service [Hotel] [AN]

DAY 14   PESHAWAR   Thursday, 25 November 1943
Sikenada / Dinner at Deputy Commissioner [AN]

Editor's note: Zimmermann wrote a 14-page letter to his wife from Peshawar, summarizing his trip up to this point. His AL was dated "Thursday, the 25th" and was passed by censor. Some of the captions on his photographs mention "see letter." The letter of 25 November was typed in December 1943 by Mrs. Zimmermann and it was shown to family and friends in the U.S. Her transcription was somewhat inaccurate, and I have used the original letter as my reference.

DAY 15   PESHAWAR   Friday, 26 November 1943
Dinner at Dhams / Bromheads / Carters Col. / Mrs. Liepor [actually, Leeper] [AN]

DAY 16   PESHAWAR to KHYBER PASS to PESHAWAR   Saturday, 27 November 1943

Trip up Khyber. Peshawar to Jamrud – 9 miles, entrance to Tribal Territory. Many blockhouses in Pass, tank traps, winding roads. To Landi Khotal – 28 miles; 32 to Landi Khana. Forts at Shagai and Landi Khotal. Afridis on either side of the Pass as far as Landi Khotal, with Mullargaris (bastard Mohmand) a short distance to the North. RR goes through to Landi Khana. Shingari tribe are the rest of the way along the Pass and into Afghanistan . . . In 1930 an Afridi Lushkar surrounded Peshawar. Since 1937 the Axis has caused unrest among Afridis. In 1938, parts of the Afridi tribe started a march on Jalalabad to loot it and embarrass the Afghan government. They were stopped by a blockade and effective bombing and made to pay a penalty in guns. . . . On way back had tea with Subhadar Major (Honorary Captain, retired) and his son who is still in service. . . . Father and son often decorated for valor. [TRd]

That afternoon the gov. of Northwest Provinces had a garden party to meet the Viceroy. Maj. Bromhead arranged for Enders and me to be invited. It was a grand affair in a beautiful setting on the government House Lawn. All the mulliks [maliks] (tribal leaders) from the surrounding country were there along with important Britishers of the community. The Viceroy had heard we'd been up the Lowari to Chitral in a jeep so came over to where we were standing and asked all about it. Enders did the talking as he has a flare for it and modesty is not one of his virtues. By the way – he has written three books. You might be interested in reading at least one of them. His first name is Gordon. I'm sorry not to know the titles. (Foreign Devil is one I think.) [TL, 6 Dec 1943]

DAY 16   PESHAWAR to KHYBER PASS to PESHAWAR   Saturday, 27 November 1943 (cont'd)

Khyber in morning / Garden Party for Viceroy aft./ Dinner at Services / Dance at Club [AN]

Ed.: Zimmermann's card shows that he was at the Club, but did not schedule any dances

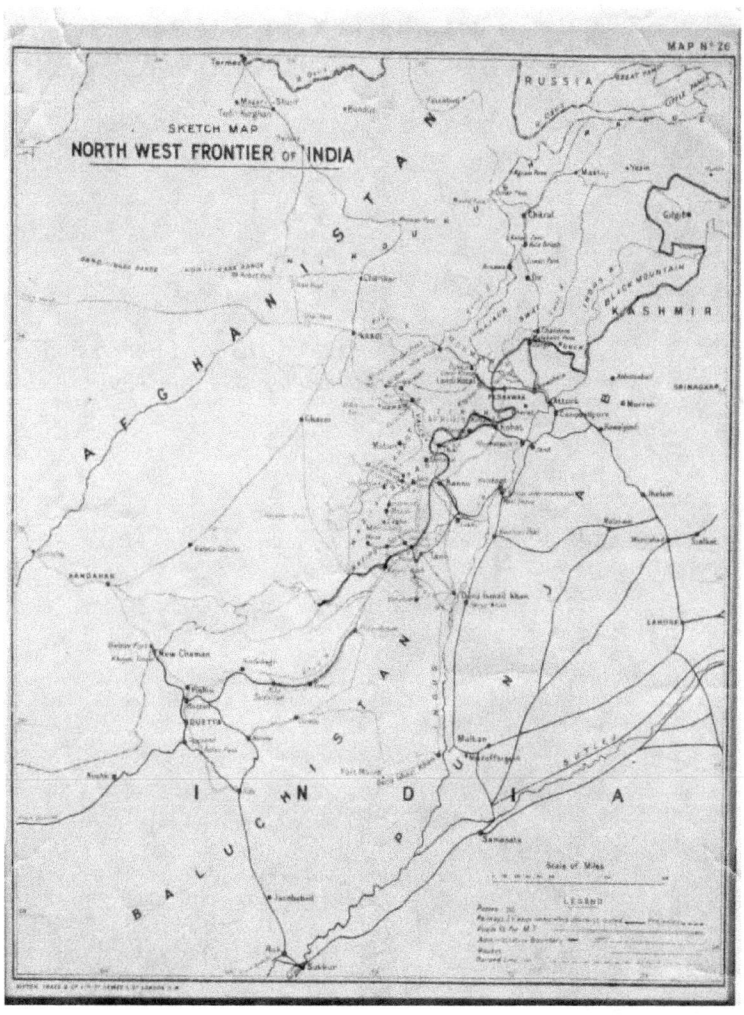

Editor's note: Zimmermann was in Peshawar from 23-29 November 1943. During this time, he worked on his Notes for the trip to Chitral. At this time, he also marked part of his route in ink on this map of the North West Frontier of India.

DAY 17   PESHAWAR   Sunday, 28 November 1943
Notes in morning / Lunch at Club / Gun factory in afternoon / Tea with Met. Maj. & son / Dinner at Comms. [AN]

Editor's note: Zimmermann took no photos or movies in Peshawar, perhaps he was concerned about security. Many British military bases were then located in the city and near it, including the Peshawar air field (where elements of the U.S. Tenth Air Force were also based). At least two groups of U.S. Navy personnel were stationed in Peshawar, probably at the air field: Navy radiomen provided official U.S. communications for the Army, Navy, and State Department, and some were located in Peshawar; and a U.S. Navy storage facility was located there, used to transport goods to Kabul and China. Little is known about these operations. Zimmermann would surely have visited them, but he wrote nothing about them.

DAY 18   PESHAWAR to KABAT, THAL, and PARACHINAR   Monday, 29 November 1943

Left Peshawar at 8:00 a.m. – reached Kohat at 9:35 – 41.9 miles. Left Kohat at 10:55 a.m. – reached Thal at 1:15 p.m. – 104.4 miles. Left Thal at 3:42 p.m., reached Parachinar at 6:15 p.m. – 160.3 miles. Saw the combination P.A. and D.C., Sheikh Sahib and his brother, Mir Ali, for breakfast at Kohat. Saw Brigadier Barstow at Thal and lunched in Station Mess. Army insisted on giving us escort with armored carriers for 10 miles out of Thal. Tead and dined at Donald's (P.A., Kurram). Dinner guests were Major and Mrs. Boulter; Col. Francis and Indian Captain. (P.A.'s full name is Kenneth Donald). [TRd]

Kohat – Enders, Deputy Commissioner Sheiku, his brother, Mir Ali [AN]

Kohat – D.C.'s House [AN]

DAY 18   PESHAWAR to KABAT, THAL, and PARACHINAR   Monday, 29 November 1943 (cont'd)

On the way from Kohat to Thal [AN]

Kohat to Thal [AN]

DAY 19   PARACHINAR to PEIWAR KHOTAL PASS   Tuesday, 30 November 1943

Climbed Peiwar Khotal on Afghan border with Kenneth Donald, Lt. Col. Francis, Sir Benjamin and about 15 rifles of the Kurram Militia. Lunched on top of One Gun Hill – scene of an engagement between Afghans and Lord Roberts in 1879, and also involved in Third Afghan War in 1919. Distance from Parachinar to foot of Peiwar Khotal about 15 miles. Altitude of Peiwar about 10,000 ft. From Peiwar motored to Karlachi on Afghan border and point where Kurram River flows into India. Peiwar and Karlachi two main routes from Afghanistan in these parts. Had tea at KM post at foot of Peiwar (Tari Mangal). Returned to Parchinar for drinks with Capt. Echlin and dinner at KM Mess. Altitude of Parachinar, 3,800. … Country around Thal heavily filled with tank traps … defense against possible German threat during July 1942 [TRd].

Peiwar Khotal. Afghanistan outpost in distance back of which Afghanistan's highest mt. about 16,000'. Kurram militiaman in foreground [AN]

The Tora Bora Mountains, where Osama bin Laden hid after 9/11, are in the background. Zimmermann wrote, prophetically, that this is the "*potentially most powerful position in the world*"

DAY 19   PARACHINAR to PEIWAR KHOTAL PASS Tuesday, 30 November 1943
(cont'd)

Peiwar Khotal. Afghans tried to invade India thru there in 1919 (also 1879) [AN]

Peiwar Khotal. Enders, Benjie, Col. Francis (C.O. Kurram militia),
Sergeant Thomas Nicolson, Kurram Militia – Afghanistan in background [AN]

DAY 19   PARACHINAR to PEIWAR KHOTAL PASS Tuesday, 30 November 1943

Peiwar Khotal. Tommy, Col. Francis, Benjie, Political Agent Kenneth Donald, myself [AN]

Same as before, Enders in my place [AN]

DAY 19   PARACHINAR to PEIWAR KHOTAL PASS Tuesday, 30 November 1943

*Parachinar Plain from Peiwar Khotal. Height 10,000'.
It was through this valley the Afghans tried to invade India in 1919.* [AN]

DAY 19   PARACHINAR to THAL and MIRAM SHAH   Wednesday, 1 December 1943

Left Parachinar 9:40 a.m., arrived Thal 11:35 a.m. – 54 miles. Here picked up Wazir Tribal Escort. Proceeded to Spinwam where we were met by Major Denning and thence went to Mir Ali. Met Lt. Col Janson at Mir Ali and he accompanied us to Miram Shah. Arrived . . . 2:55 p.m. Stayed with Lowis, P.A. Tochi, and ate at Tochi Scouts Mess. Tochi Valley occupied by Daurs who are Sunnis and asked for British protection. Daurs are considered a bad lot, having conspired with Ipi, whose village is on the Tochi . . . Ipi now on Afghan border west (and a little north) of Miram Shah. Two of his guns were reported moving on camels past Dosalli. [TRd]

Rs. 2,000 is ordinary blood money for a murder and, once accepted, settles the murder. A murder can be bought for Rs. 50-100, which is less than the cost of a rifle. A British rifle brings about Rs. 1,200 – a round of ammunition Rs. 1/2/-. Native round brings As. 6 to 7. [TR]

At Miram Shah heard that 10 Hindu women were kidnapped and taken into Afghanistan; also that a Bannu Hindu shopkeeper was kidnapped probably on information of his Hindu competitor. Ransom demanded for the shopkeeper – Rs 7,000, with negotiations under way. Huge airdrome at Miram Shah. [TRd].

DAY 19   PARACHINAR to THAL and MIRAM SHAH   Wednesday, 1 December 1943

Thal – Indian Army Post [AN]

Spinwam – Tochi Scout outpost. Armored trucks. Near here is airport, planes from which bomb troublesome Pathans in their hideouts [AN]

Since we've been here . . . a road engineer's lorry held up, the supervisor being held for ransom. At Dosali, . . . about twenty shots were fired from a hill into the army camp . . . Ten Hindu girls were kidnapped . . . the girls aren't considered worth ransoming. Another lorry was held up and three of the highwaymen were killed, one an Afghan. The R.A.F. bombed some outlaws who were hiding in caves. Such is life on the Northwest Frontier. It forms one of Britain's big problems and costs millions a year to keep in as good order as at present. [TL, 6 Dec 43]

DAY 19   PARACHINAR to THAL and MIRAM SHAH   Wednesday, 1 December 1943

*Dasali – Piquet on hill top – Tochi Scouts.
Near here a battalion of Indian Army was fired on from surrounding hills* [AN]

Dasali – Iblanke Piquet – inside. Tommy, P.A. Lawis, Benjie, Enders, Tochi Scouts [AN]

DAY 19   PARACHINAR to THAL and MIRAM SHAH   Wednesday, 1 December 1943

Dasali – in court of Tochi Scout outpost [AN]

Dasali – Old malik, Political Agent Lawis [AN]

DAY 19   PARACHINAR to THAL and MIRAM SHAH   Wednesday, 1 December 1943

Dasali – Gurkha at attention [AN]

Dasali – Members of Brig. Hobbs brigade at soccer game   [AN]

DAY 19   PARACHINAR to THAL and MIRAM SHAH   Wednesday, 1 December 1943

Dasali to Miram Shah – Body guard of Khassadars in rear [AN]

Most of the way we have been escorted by Khassadars (local policemen). The line the road on either side (about 1/2 mile apart). Others ride in lorries ahead and behind us. It makes one feel quite important having so much fuss made of you but it's the only way to travel in this country with comparative safety. Some of the local boys just don't like strangers and might decide to take a crack at them or try a little highway robbery or kidnapping. [TL, 6 Dec 43]

Dasali to Miram Shah – Defended villages [AN]

DAY 19   PARACHINAR to THAL and MIRAM SHAH   Wednesday, 1 December 1943

Miram Shah – the P.A.'s bodyguard [AN]

Miram Shah – The P.A.'s bodyguard.
This is headquarters of Tochi Scouts and the Political Agent [AN]

Editor's note: LT Zimmermann made handwritten notes and prepared a typed report that included events and observations on 2, 3, and 4 December, but took no photos on those days. His photography resumed on 5 December. He later mentioned that he had a bad cold, and he was suffering from it at this time.

DAY 21   DOSALI   Thursday, 2 December 1943

Left Miram Shah at 10:10 a.m., reached Dosali at 11:20 a.m. – 25.4 miles. Left Dosali at 11:40 a.m., going to Iblanke Post. Saw Brigadier Mervin Hobbs en route. He had just been shot up while out with his Brigade (from Razmak) on a practice maneuver. Arrived back at Dosali at 1:00 . . . Tochi Scouts escort went with us to Dosali . . . At Dosali was a 5.5 gun and truck which had fallen over killing 2, probably mortally wounding 1 and hurting 12 – all natives. Had big tikala at Dosali . . . All tribes above Kabul River quiet and easy to handle. . . . Wazirs and Mahsuds take democracy to the point of anarchy; they will not listen to Maliks or Khans unless it suits them and are difficult to handle. Tribes in Zhob area are democratic but settled and willing to heed maliks and therefore are no great problem. [TRd]

DAY 22   DOSALI to MIRAM SHAH   Friday, 3 December 1943

Left Dosali at 10:15 a.m., reaching Miram Shah at 11:25 a.m. Went back to stay with Lowis. Total miles 25.4. Lowis had received word that a M.E.S. lorry had been held up near Datta Kheyl, a supervisor kidnapped and 2 men wounded. A dicker for the kidnapped man's release was reported by Assistant P.A. for Rs 7,000. Two 500-lb bombs dropped on a gang of tribesmen who lived in caves near Miram Shah. Later reports said no casualties, but good moral results. . . . Grazing grounds are the keys to Tribal behavior ... The job of controlling tribes, according to Benjy, is not a military but a Scouts one. For deserters a system of amnesties might be useful if it worked. [TRd]

DAY 23   MIRAM SHAH to BANNU and JANDOLA   Saturday, 4 December 1943

Left Miram Shah at 9:30 a.m. Had big tea at Mir Ali at 11:00 a.m. and took Major Denning to Bannu. Arrived at Tank at 2:30 p.m. and took the Acting P.A. "Pat" Duncan in jeep . . . Arrived Jandola at 3:45 p.m. Bannu is prosperous town compared to Tribal territories – land is rich. Passed large groups of Gilzais migrating from Afghanistan for the winter. Generally speaking the whole Gilzai Tribe . . . are friendly to the British. One small subdivision, known as the Dinar Kheyl (some 50 families), however, are sneak thieves and make trouble between Tannai and Gul Kutch. [TRd]

Gilzai Tribes near Bannu – on the march (move with seasons) [AN]

DAY 23   MIRAM SHAH to BANNU and JANDOLA   Saturday, 4 December
1943 (continued)

Jandola – Hdqts South Waziristan Scouts. Enders, P.A. Duncan, Col. Janson (C.O. South Waziristan Scouts), Capt. Cole [AN]

Jandola – Armored truck [AN]

DAY 23   MIRAM SHAH to BANNU and JANDOLA   Saturday, 4 December 1943

Jandola [AN]

DAY 24   JANDOLA to KONIGURAM and WANA   Sunday, 5 December 1943

Left Jandola 9:30 a.m. in Duncan's lorry and escorted by a Mahsud lorry in which Benjy and I rode. Had morning tea with Kushwakht (brother of Mehtar of Chitral) at Sora Rogha; lunched at Ladha (where tower was hit by 4.5 gun). Had afternoon tea at Tiarze. Arrived at Wana at 7:20 p.m. Ladha tower was shelled because its owner refused to turn over a murderer who had killed on post property. . . . Kaniguram is capital town of all three Mahsud tribes . . . In it live Uramurs who speak no known language and are artisans (knives and guns) of Mahsuds; some now take to lorry driving. Uramurs feed jirga members on visits to Kaniguram gratis. This is their protection fee . . . Census shows some 100,000 Mahsuds, but Duncan believes they are 150,000 with some 20,000 rifles. Janson estimates that Wazirs have more than the 30,000 rifles reputed to be held by them. Duncan reported a lorry shot up in Mahsud territory with an Afghan malik killed. [TRd]

Village near Laaha [AN]

DAY 24   JANDOLA to KONIGURAM and WANA   Sunday, 5 December 1943

Jandola to Sora Rogha – tea – Ludha lunch – Tirwaza Wanna dinner [AN]

Laaha – outpost of South Waziristan Scouts. From here a neighboring village was shelled because they wouldn't give up a murderer who had killed on Post property [AN]

South Waziristan Khassadars – acting as body guards on escort truck [AN]

DAY 24   JANDOLA to KONIGURAM and WANA   Sunday, 5 December 1943

South Waziristan Khassadars – body guards on escort trucks [AN]

Koniguram – Is capital of Mahsud Tribes.
Pop 7500 – Inhabitants speak no known language. Make guns and knives [AN]

DAY 24   JANDOLA to KONIGURAM and WANA   Sunday, 5 December 1943

Koniguram, South Waziristan [AN]

South Waziristan [AN]

DAY 24 JANDOLA to KONIGURAM and WANA Sunday, 5 December 1943

Wana – Indian Army outpost [AN]

DAY 25 FORT SANDEMAN to WANA Monday, 6 December 1943

Mahsud Khassadar [AN]

DAY 25   FORT SANDEMAN to WANA   Monday, 6 December 1943 (continued)

Zhob Militia Post on border of Baluchistan [AN]

Sambaza – Zhob Militia Post [AN]

DAY 25   FORT SANDEMAN to WANA   Monday, 6 December 1943 (continued)

Sambaza [AN]

*Tora Gharu Piquet near Sambaza – manned by Zhob Militia.
This is relatively peaceful country, no escorts needed* [AN]

DAY 25   FORT SANDEMAN to WANA   Monday, 6 December 1943 (continued)

*Fort Sandeman. Political Agent Searle lives in the fort –
His family takes to hills in background in hot summer*   [AN]

Fort Sandeman – late afternoon. Headquarters of Zhob Militia   [AN]

DAY 25   FORT SANDEMAN to WANA   Monday, 6 December 1943 (continued)

Fort Sandeman fort on hill. Mrs. Searle, Maj. Peter Garrett (Zhob Militia), P.A. Searle   [AN]

Fort Sandeman fort in background. Peter & officer in Zhob Militia   [AN]

DAY 25   FORT SANDEMAN to WANA   Monday, 6 December 1943 (continued)

Fort Sandeman – Maj. Garrrett being saluted in Militia Lines   [AN]

Fort Sandeman – Militia Lines.
Their main food is chapattis – a flat wheat cake and rice. Big feast is known as a tikala. [AN]

DAY 26   WANA to JANAR and FORT SANDEMAN   Tuesday, 7 December 1943

Left Wana at 9:80 a.m., reaching Tanai at 10:30 – 10 miles. Picked up jeep. Left Tanai 10:55 a.m., crossing into Baluchistan at Gul Kutch at noon – 26 miles. Met Peter Garrett and had tea at Gul Kutch. Left Gul Kutch at 12:45, arrived Sambaza at 1:30 p.m. Had lunch. Left Sambaza at 3:00, reaching picket of Tora Gharu at 3:30. Left Tora Gharu at 4:30 p.m., reaching Sandeman at 5:15 – total miles – 79. Dined with P.A. Searle and Brigadier Purvis. All lived at Zhob Militia Mess; Benjy with Lt. Col. Geoffry Keating, and Tommy and I with Peter Garrett. [TRd]

Hindubagh – Tent of wandering Gulzai tribe from Afghanistan

Hindubagh Gulzai camp – see letter. Woman trying to hide her face.

DAY 26   WANA to JANAR and FORT SANDEMAN   Tuesday, 7 December 1943

Hindubagh Gulzai Camp [AN]

Hindubagh Gulzai Camp. The Afghan dress is more colorful than Indian [AN]

DAY 26   WANA to JANAR and FORT SANDEMAN Tuesday, 7 December 1943

Hindubagh Gulzais – famous for locating underground streams – see letter

Hindubagh Gulzai Children. They wouldn't have an idea what a school was

We are now at Fort Sandeman in Baluchistan, staying with the Zhob militia. Tomorrow we leave for Quetta . . . While it's been interesting it's been rather hectic, none of the places have modern plumbing ... Also it's been quite cold, because of the altitude, between 4 to 10,000 feet. It's not as bad here – it's only 4500. My place in Enders' car is in the back seat. It's an Army car, open of course, and the breezes and the dust make life far from comfortable . . . I finally picked up a cold, one of the good ones. Today we rest. I hope it will do some good. Most of this country is pretty God-forsaken. You marvel that anyone can scratch a living out of it. That's partly the trouble. Some cannot, so they take to plundering and pillaging. [TL, 9 Dec 43]

DAY 27   SANDEMAN   Wednesday, 8 December 1943

Went in Col Keating's car and 2 escort trucks and a M.E.S. station wagon to see new road being made to Shawet which is big entry route for Gilzais, Suleiman Kheyl and other tribes. Visited Fort Shahigu – 45.3 miles and alt. 6695 ft., travelled old road across plain for 7.1 miles; new road to work camp (Zhob Militia) for 14.1 miles; went 5.5 miles beyond work camp for lunch. Total miles run – 156. Left Sandeman at 10:00 a.m., returned at 8:00 p.m. Saw Nasra tribe at Shahigu, they are being chased by Suleiman Kheyl but escaped in time. This was mostly Suleiman Kheyl country, but much used by migrants such as Nasras. Water is sometimes scarce in Sandeman, some coming down from mountains and from wells which must be sunk 80 ft. [TRd]

We saw a crowd right off the road near an Indian Army camp. We immediately thought there was trouble but it turned out to be a cremation. A Hindu Army recruit had been killed and his friends had built a bonfire and were disposing of his remains. Hindus are cremated, Muslims buried as is, in shallow graves, Parsis dedicate their dead to the elements and animals, a delectable dish for the vultures … [TL, 9 Dec 43]

DAY 28   SANDEMAN to QUETTA   Thursday, 9 December 1943

Motored to Quetta, leaving at 9:10 a.m., arriving at Residency at 5:05 p.m. Saw large encampments of Karotis outside Hindu Bagh where we had tea in Zhob Militia Post. Karotis part of Gilzais and have special trade of karez digging. Hindu Bah is center of chrome mines (under Pop Wynn), producing 40,000 tons annually for U.S. of 52% ore. Wynn employs some 6,000 camels and donkeys to carry ore from mountains around Hindu Bagh and has one meter-gauge spur track to big working. Has also got some lend-lease trucks. [TRd]

Sheep grazing outside of Quetta, Baluchistan [AN]

DAY 28   SANDEMAN to QUETTA   Thursday, 9 December 1943 (continued)

Shepherd & his sheep near Quetta.
The tail of these "fat-tail" sheep is a great delicacy. Their wool makes the finest carpets

Shepherd & his sheep near Quetta

DAY 28   SANDEMAN to QUETTA   Thursday, 9 December 1943 (continued)

The fat tail sheep – the main breed [AN]

Temporary Government House –
Old G H destroyed in earthquake of 1935. Gov. lives here while new one being built [AN]

DAY 28   SANDEMAN to QUETTA   Thursday, 9 December 1943

Where guests stay when visiting governor [AN]

DAY 29   QUETTA   Friday, 10 December 1943

Saw Major General Money & Col. Bruce-Steer of Baluchistan HQRS, Father Wood, Majors Platt and Alson. [TRd]

DAY 30   QUETTA   Saturday, 11 December 1943

Went for day beyond Hanna Valley for chikhor shoot with the Hays and Benjy. Also present was Major Woods-Ballard, P.A., Quetta [TRd]

Getting out of cars to go on chikhor shoot with governor & his family up the Hanna Valley near Quetta [AN]

DAY 30 QUETTA Saturday, 11 December 1943

We were royally entertained by His Excellency. Had a dinner party for us the night before we left. Took us on a chikhor (partridge) shoot on Sunday with a picnic lunch etc. … [TL, 15 Dec 43]

Unloading lunch (donkey laden with baggage)

Roasting mutton for lunch – Gov's youngest child

DAY 30　　　　　　　QUETTA　　　　　　Saturday, 11 December 1943

Looking up the valley [AN]

A Muslim says his prayers [AN]

DAY 30   QUETTA   Saturday, 11 December 1943

The beaters have their lunch – see letter

Lunch on the chikhor shoot. Enders, Gov. Hays, A. A. Woods, Ballard, Mrs. Hays, Benjie on right. Gov. children in middle

DAY 31   QUETTA to KARACHI   Sunday, 12 December 1943

Leaving Quetta for Karachi. Maj. Alston, Mrs. Wood, Benjie, Tommy, Enders.

DAY 34   KARACHI   Wednesday, 15 December 1943

Zimmermann arrived in Karachi and resumed duties at 1130 (11:30 a.m., local time) 15 December. Reimbursed for expenses on Temporary Additional Duty: first class railroad ticket from Quetta to Karachi (Rs 63/1/0) . . . donkey and lorry hire over Lowari Pass . . . total $40.49.

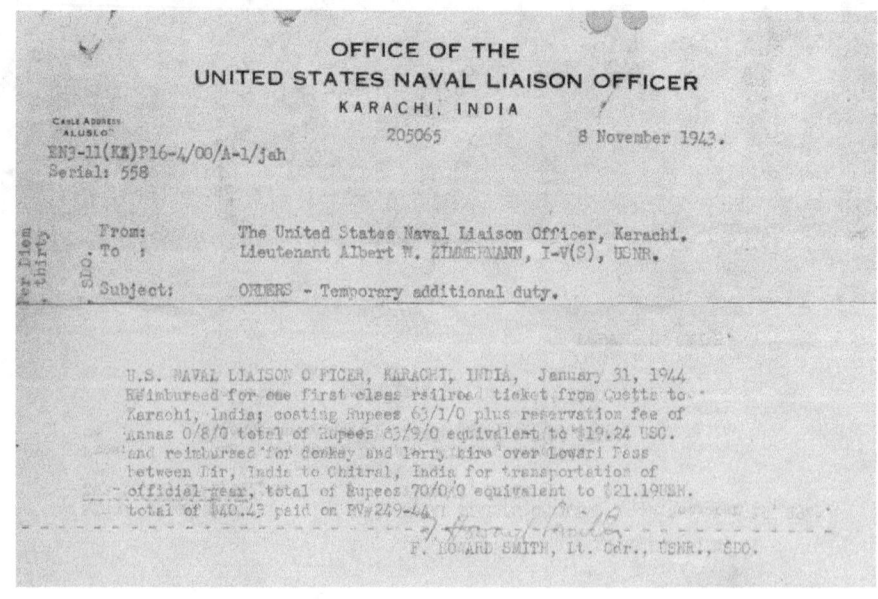

Other Images of Locations Mentioned in Zimmermann's Notes

The Vale of Swat, visited by Zimmermann
November 1943

Images from flidker

Kohat, Waziristan – mentioned in letters of December 1943

Dir and Chitral, North-West Frontier Province

Dir River Valley
Seen by Zimmermann, November 1943

Hunzakut children, said to be descendants of ancient Greeks, who came with Alexander
Similar to those seen by Zimmermann in Chitral, November 1943

Photographs of Lowari Pass – crossing from Dir into Chitral

Zerega Images, from Aspley, 2011

Zerega Images

First ascended in a motor vehicle by Zimmermann's team, 18 November 1943

The Khyber Pass

Illustrated Weekly of India

Illustration (12 November 1944) one year after Zimmermann visited it
Showing a member of the Women's Army Corps (India)
From his Photograph Album

5
AFTERMATH

Albert Zimmermann was relieved of duty in Karachi and returned to the U.S. in May 1945. Soon after he arrived for a long-awaited leave with his family in Pennsylvania, he was found to be suffering from a duodenal ulcer. After hospitalization at St. Albans Navy Hospital on Long Island and convalescent leave at home, he was released from active duty on 30 December 1945. The National Geographic Society bought 42 of his photographs, taken in India, on 14 December 1945, although the Society never published any of them. He retained the rights to use them himself. He continued to serve as a Navy Reserve officer for another two years. He was promoted to lieutenant commander in 1947. He submitted a letter of resignation on 28 April 1947, and he was honorably discharged on 11 September 1947. He resumed his partnership with John Ott, in the wool brokerage firm Ott and Zimmermann, in Bala Cynwyd, Pennsylvania (see photo at right)

Zimmermann returned to Africa, India, and Pakistan in 1950 with his wife and his elder daughter. He renewed friendships that he had made in Karachi during the war and did some business with wool brokers in that area. His photographs and movies taken on that trip complement the many photos and movies that he took in 1943-1945. In later years, he was a generous contributor to his friends who were in need, and to various charities, including Episcopal churches in Pennsylvania and Long Island, the University of Pennsylvania, and the Southern Home for Children, on whose board he served for many years. He died unexpectedly of a cerebral hemorrhage on 24 July 1961, at the age of 59.

Barbara Shoemaker Zimmermann traveled widely with her husband, and after his sudden death in 1961, she traveled with her friend Mrs. Amelie Kane, widow of Commander Jack Kane, USNR. Her last long trip was to Australia and the islands of the Pacific in 1975. Her health was then beginning to fail, and she spent most of the rest of her life at her home in Haverford, eventually being largely confined to bed, where she was attended by nurses around the clock for several years.

Late in life, Barbara was often unaware of much that went on around her, and she was bitterly disappointed by the loss of an eye from infection following cataract surgery. She passed away quietly on the exact anniversary of her husband's death, 24 years earlier. It seemed to her children that she was in fact aware of that date, and that she allowed her life to slip away at that time.

Barbara Zimmermann was buried beside her husband in the churchyard of the Church of the Redeemer in Bryn Mawr, Pa. The two graves were framed by azalea bushes that she had selected. The azaleas overgrew the markers, and were replaced by a flowering dogwood tree, which now shades the gravestones. Barbara (Shoemaker) Zimmermann was an incisive and determined person with a forceful personality and a wry sense of humor. Her autobiography, *Mutterings*, was published privately in 1962. It offers insight into the mid-20th Century world of society and privilege in Philadelphia. She will be remembered vividly by all who knew her, and the stories of "Baba" – as she was known by her grandchildren – will be passed down for generations to come.

Albert Walter Zimmermann

By Bachrach

Albert Zimmermann in about 1950

Barbara Shoemaker Zimmermann

By Bachrach

Barbara Zimmermann in about 1950

Albert and Barbara Zimmermann

At home in Haverford, Pennsylvania

Their children

Their children: Barbara, Albert Jr., Warren, Helene
Philadelphia, Bartram's Gardens, 1983

Before Barbara "Babs" married Mel Johnson, she worked for the C.I.A.
Albert Jr. "Albie" is an ophthalmologist; he married Lenore Lisbinski
Warren was the last Ambassador to Yugoslavia; he married Corinne "Teeny" Chubb
Helene "Lanie" married George Hill; she earned a Ph.D. and was a scientist

Descendants of Albert and Barbara Zimmermann

The Twelve Grandchildren of Al and Barbara Zimmermann

L to R: David Hill, Warren "Tim" Zimmermann, Helena "Lana" Hill, Amanda Zimmermann, Sarah Hill, Barbara Johnson, Alice Johnson (front, center), Susan Zimmermann, Elizabeth "Lily" Zimmermann, Corinne "Quinnie" Zimmermann (at rear), Anne Zimmermann (front), James Warren "Jim" Hill (Thanksgiving, 1983)

Guests at marriage of Alice Johnson and Brian Handwerk, 10 June 2000
The last photograph of Albert and Barbara Zimmermann's children and their spouses
David Hill, **Al & Lenore Zimmermann**, Mrs. Handwerk, Lana Hill, **Warren & Teeny Zimmermann**, **Babs Johnson**, Brian & Alice (Johnson) Handwerk, **Mel Johnson**, **Lanie & George Hill**, Sarah Hill, Amanda & Ray Diffley, Anne (Zimmermann) Crimmins, Barbara (Johnson) and Tom Riley

Albert Walter Zimmermann
Barbara Shoemaker Zimmermann

Their burial sites at Church of the Redeemer, Bryn Mawr, Pa.

6

Zimmermann's Correspondence with Others from Karachi 1943-1945

Including

Gordon Bandy Enders
Sir Benjamin Bromhead
Curtin Winsor
C. Rollins Hanlon

Together with Notes on Their Lives

Gordon Bandy Enders

Major Gordon B. Enders first appears in *Proceed to Peshawar* when he is mentioned in the letter of 26 October 1943 from the anonymous "IB Quetta" to John R. Harris, Esq., Central Intelligence Officer, Karachi. The letter says that Major Enders is to be taken by Major Sir Benjamin Bromhead on a tour of the Afghan-Indian Frontier from Peshawar to Chitral and Quetta, in Baluchistan. IB Quetta asked Harris if Enders and Bromhead could be accompanied by an American officer from the Naval Liaison Office in Karachi. Lieutenant Albert Zimmermann was thus appointed to accompany them. Zimmermann wrote to his wife on 5 September 1944 that the trip had been "instigated" by Enders. Zimmermann's letters to his wife suggest that Enders began his work to accomplish this trip during meetings with officials in Delhi and Peshawar in late 1941. Zimmermann was sent on the trip ostensibly to report to Navy Headquarters on conditions along the Border. However, Zimmermann was also intended to keep an eye on Enders – who had already developed a reputation for hyperbole and less than truthful telling in his reports.[17]

Letters from Gordon B. Enders to His Wife, Betty (Crump) Enders
November & December 1941

The letters from Enders to his wife in 1942 have never previously been published. They show his skill in observing people from other cultures in Asia, and in writing about them sympathetically in the context of their daily lives. When his niece, Dr. Gertrude (Enders) Huntington, learned that I had re-discovered this long-forgotten trip, she kindly provided a copy of the letter from Mrs. Enders to her in May 1942, along with the letters to Mrs. Enders from Gordon in November and December 1941. Maj. Enders wrote to his wife from India, on his way to Kabul, Afghanistan, where he would become the U.S. Military Attaché and the first American diplomat to be stationed in Afghanistan. Enders had recently been recalled to active duty as an intelligence officer, having probably been under deep cover for many years as a secret agent in China and Tibet. In November 1941, he was sent across the Pacific to India with several high-ranking U.S. Army officers, who were assigned to move into positions in North Africa and elsewhere Asia if, as expected, war broke out. Enders crossed the border into Afghanistan, not knowing that the Japanese fleet was at that time secretly approaching Pearl Harbor, and he arrived in the vicinity of Kabul on December 5 – shortly before Pearl Harbor was attacked.

Letter from Mrs. Gordon Enders (Betty) to Trudy [Gertrude] Enders, 9 May 1942[18]

> Mrs. Gordon Enders
> 918 Highland Avenue
> Lafayette, Indiana
> May 9th [1942],

Dear Trudy,
When Gordon was in Iran last summer he visited a city where there was a famous torquoise [sic] mountain – for thousands of years it has been mined, until, now, the supply is very small. But the stones are better than elsewhere, and Gordon selected some unusual one which he bought and wrote me about. Recently he has sent them to me – and a few lapis lazuli ones from Afghanistan – from the Afghan government. I am sending tomorrow three of them to you, in the same box. He told me to do whatever I wanted with them, and I felt we would both like to have you have them. The round one is for "Mother," the oblong for you, and the tiny lapis for Allen. ... Your grandmother wrote me that your father is in Washington. I wish you would tell me doing what, and where? I often think of you all in your big "hostel" home, and the many you are taking / [at least one page is missing here]

... fly at daybreak the next morning "half way across India" and land in Kabul, where the plane was to be presented to the King from our Government – also a bus which was being driven up by his sergeant. Unfortunately, Foreign Devil had a small sale, and we have never yet had any cheque from the royalties! It came out six months after they had agreed to publish it – and too late, I think as all the War books by that time were flooding the country.

I hear you are going to Purdue this fall? Please write and tell me the latest family news, and let me know if the box reaches you safely.

My love to you and the family,
     Affectionately,
       [s] *Betty*
[added in handwriting] *This afternoon saw "Cabin in the Sky" and enjoyed it.*[19]

* * * * * * * *

Excerpts from letters from Major Gordon B. Enders written en route to Kabul, Afghanistan, where he is to be stationed as United States military attaché[20]

Excerpts for W.B.A.A.[21]

En route to New Delhi November 18, 1941

Last night before taking the train I had dinner at the Farbos, an Italian-owned restaurant, which is considered the best in Calcutta. Its windows are covered for blackouts but it is gay and full of cooling fans inside. We saw dining two Indian sisters who are reputed to be the most beautiful in Hindustan. They had been done by famous portrait painters. I took the 9:03 train and had a very comfortable two-bunk compartment to myself, and the whole car is air-conditioned. A bearer brought coffee at 7:30. We breakfasted at 9:00 – the food being wired for at headquarters was brought to the compartment on trays.

I am drinking in the Hindustan that I knew 30 years ago [i.e., in 1911]. The big changes so far have been turkeys. There are large flocks of them in the country tended by half-naked little boys. Of course, there are motor cars everywhere and they are new. The Grand Trunk Road is now paved, and there are Hindu, Mohammedan, and vegetarian restaurants at the big stations. But otherwise I see no marked differences.

Looking at the fields I've seen parrots, pigeons, and the big, blue cranes we call sahnus. The lentils, kaffir corn, sugar cane, and grain, the sisal and mustard are still there. The cattle are on the plains – and the kites and vultures. Little boys still herd the goats and irrigation is still done with Persian wells. There are ponds with water-buffalo and water-chestnuts in them, and the dhobis wash their clothes by beating them on stones, while their sway-backed little donkeys graze with hobbled for legs near-by.

In the villages the trees are still thick, with the cleared off /
[p.2] threashing [sic] floors, the mud huts with the animals tied outside, and the women still making dungcakes for cooking. It is India all right!

New Delhi   November 19, 1941   7:00 P.M.

We stopped to take our lunch at Fatehpur-Haswa,[22] where my father is buried. I saw that the Grand Trunk Road was not paved, and caught a glimpse of the great Peepul Tree in the monkey temple yard where they feed the monkeys in the evening. It's quite remarkable how I seem to feel the texture of the white dust everywhere, and to know the smell of the grasses, and the sound of the trees.

At Etawah, tea was brought in. The same old buck-monkeys were in the station courtyard – with the oxen and bullock carts parked by the road side, with the animals lying along side ruminating. I remembered from childhood hearing the Punjab mail going by our house in the early morning at Etawah, and I remembered where the tracks were from our house. Our bungalow wasn't visable [sic] because it sits back from the road. As the train crossed the Grand Trunk Road I was pleased to find it still unpaved, with nary a motor car in sight, but a line of camels, carts, and horse-drawn ekka, all waiting for us to pass at the same old crossing with its iron gates.[23]

In the fields, during the day, I saw some gorgeous peacocks pecking among the lentils. When evening fell, the cooking smoke streamed out whitely among the mango trees and I could see the little dung fires with dim shapes moving around them. To me it was a kind of reincarnation.

Thursday, November 20, 1941[24]

Have had a full day with calls at headquarters and meeting both British and American officers. My biggest surprise was having my hair cut. I had to go to a ladies' hair dressing place and sat in a cretonne-curtained booth to be shorn. Perhaps on account of the groans and sighs of the ladies everywhere trusted [sic] up in machines or baking under driers my hair cut was particularly bad. However, it is said to be the best place in town. I shudder to think what an Afghan hair-cut will be like! [p.3]

The tonga-drivers careen around on two-wheeled carts – complete with little ponies attached – and take you where you want to go. The drivers sit up in front and when I step in the back seat (which is correct for sahibs) I nearly pull the little beasts off the ground.

This morning I drove out to call on the local Afghan consul general. The house, called "Anad Villa" is new and a longish way out of town. My object was to show him my passport and let him know my plans. Going in, I was seated in a medium-sized room furnished in the modern style by an oriental. There were many doors leading into this room, which was in fact a drawing room, and each door was covered by a dark cloth curtain. These curtains wave back and forth at me. The child is really beautiful – a little pale, but with fine and intelligent features. We were still looking at one another when my guide called me through another dark curtain to the consul general's study. The boy came with me while I talked with his father. The consul general is young, and quite handsome, speaking excellent English. After he had seen my passport and had it sent out to be visaed, we had a long talk. I'm told at headquarters here that Afghanistan wants an American official and I was pleased with the whole atmosphere of our talk.

The ride out here was most interesting. The weather is fine and clear, with a woolen suit very comfortable, and the country is dry but sparkling. We went past the massive tomb of Hanuman which seems to be in fair repair. But everything around it is in ruins – old Moslem burial grounds and shrines, and possibly still older Hindu temples. I think the last time I visited the place was 35 years ago at Christmas time [i.e., 1906]. I recognized it all. As we drove down the road beneath the old, tall trees, the monkeys (some of them with young) sat on the sides and looked at us. Behind Hanuman's tomb the crumbling walls of the old fort stood up, their pink stones a little more weather beaten and the tops of their towers a little more eskew.[25]

From all this we drove out into the huge open spaces of New Delhi which /
[p.4] is planned on truly Imperial dimensions, with its secretarial buildings of the same pink stone and the same Moorish towers, but huge and modern and solid. The plain about is laid out (among ancient ruins) with reflecting pools, fountains, and statues. And yet, remember, an ancient gardiner [sic] twists the tail of an equally time-worn ox, which pulls the modern lawn mower to cut the grass to keep the place looking fine.

Saturday   November 22, 1941   7:25 P.M.

In a few minutes I must dress in uniform to have dinner with General Sir Archibald Wavell whom I met the first day. It's to be kind of an American affair with General Wheeler and his party, some of whom I met in Washington and some in Honolulu.[26]

Sunday November 23, 1941   10:00 A.M.
    I <u>greatly</u> enjoyed the dinner at which were about fifteen – three of us Americans. General Wavell's three pretty daughters were included. The long table stretched in front of a large fire-place, with the commander-in-chief sitting in the middle and Lady Wavell opposite. I was placed at her left. After dinner I talked with General Wavell and his very charming wife.

Wednesday   November 26, 1941
    I'll leave here tomorrow morning at nine thirty and it will take me twenty four hours to reach Peshawar. Yesterday I brought my warm blankets, extra warm clothing, and a steel trunk and went to Chandri Chauk[27] to get them. Before I went I wandered around to watch the people, to smell the cowdung smoke, and to see it wreathe itself among the trees and over the ground. I saw a mother with three children (one a babe in arms) and two goats sitting on the Queens Way, largest shopping street. Her face wasn't covered, and she was taking a spounge bath under her clothing, using a cup and a half of water.
    Another Indian incident was a very well kept cow with a necklace of blue beads, running down the wide walks and roads. She was obviously lost and stopped occasionally to moo anxiously and finally disappeared down an alley. As I waited on a corner an uncovered woman with a little switch in her hand /
[p.5] came striding along, stopping people and asking questions. She wore a bright yellow chuddar with bold red figures on it. Her chemise was pale blue and her skirt a russet brown. She had a golden stud in her right nostril, bracelets, and ring. There were two rings on her bare feet. She wasn't pretty but her glance was direct and self-respecting. She strode like a man, her very full skirt swaying wide and rhythmically – she was looking for her cow. About ten minutes later I came upon her again. She had found her cow which had heard her call and was coming back at the trot. She looked at me, for I had stopped to watch, and we both smiled.
    Going to Chandri Chauk brought us there at dusk and I saw all the sights and smelled all the smells. We went and came in a tonga and the din and confusion of lamp-lighting-time was indescribable. But it was Indian and therefore familiar to me and I could talk to them in their own language and was understood. The camel carts were making up for their night trips, the sacred bulls were settling down to their cud chewing usually in the most inconvenient places for traffic, and the homing throngs shouted and joked under tonga wheels, in front of tram cars and between bicycles. In the midst of all this my thoughts go back to Lafayette [Indiana] and Purdue.

Saturday, Nov. 29, 1941
Dean's Hotel, Peshawar
    I'm all set to cross the border into Afghanistan on Wednesday or Thursday of next week. The trip from Delhi up took its allotted 24 hours and the journey was very dusty. I had to use my new bedding-roll . . . There were altogether three Indian officers, two British officers (one knew Shanghai) and a civilian with an American wife. Arrived here (and it was quite chilly) I was met by the local magistrate who is to entertain me tonight.

Monday, Dec. 1st, 1941.
    After tiffen [usually spelled tiffin, a light mid-day meal] we took a car and drove to the Kabul River which is in tribal territory and about 14 miles from here. The country was a bit dusty but the air was crisp. We went through several villages with their water buffaloes and cattle, and saw many camels along the road. Many of the fields were bare, but I saw a good deal of standing sugar cane. The tribal

(Pathan) women are picturesque, dressing almost uniformly in black and scarlet, and carrying baskets and bundles on their heads. We saw several tribesmen with their rifles (a great many of them have firearms). The Kabul River was wide and swift, coming off the high, bare hills which rim the flat plains of Peshawar. We were always in sight of the big peak, Tatarar, behind which lies my road to Afghanistan and the Khyber Pass. Although expected momentarily, there was no snow on the hills yesterday.[28]

My today's schedule is tiffen with the R.A.F. and dinner with the governor, Sir George Cunningham, to whom I carry Colonel Benson's letter of introduction.[29] Tiffen and dinner tomorrow are also booked. This is tremendous[30] [p.2] hospitality, particularly considering that these people work very hard. They look very gay at the club, but they do work hard to keep order and quiet on this always uneasy frontier and to be ready for contingencies. Except in the evenings, soldiers wear uniforms, so I have kept mine on most of the time.

I was told at Delhi that I wouldn't hear Hindustan spoken here, but I do. I've walked the nearby bazaar and find everyone speaks it. They speak Pushto also, but not exclusively apparently. Although it is cold in the mornings and evenings, and fireplaces aren't really adequate, I think you would like Peshawar in November and December. I think you would like the bazaar here also. The place swarms with rug shops and fur shops, all very reasonable. This morning I saw a gentleman driving a protesting lady water buffalo through the crowds (you'd have liked that!) and the other evening another gentleman wrapped a long-haired, black and white goat around his neck, mounted his bicycle and rode through, the bleating goat acting as bell. A little urchin in a fez, the other evening, saluted me with "God Damn!"

Tuesday, Dec. 2nd, 1941

I have had my dinner with the Governor and am now set to go through the Khyber Pass. By a stroke of luck, I've found the civilian Englishman, named Robinson, who has his own Dodge and has driven the passes more than 30 times. He has invited me to accompany him and we'll be going either tomorrow or Thursday. Whichever day it's going to be, we will stop and lunch (leaving here at about 11:00 A.M.) with the officers of the Frontier Force [these are local soldiers, not British] whom I've met here and have found very helpful and friendly. Most of them are youngsters with prodigious complaints of their way of living, but very fit and ready.

About yesterday, I lunched with the local R.A.F. acting chief, his wife and some friends. We talked a lot of shop (my host and I) up to tiffen, and, after eating, we went out to see his farm. He keeps ducks, chickens, pigeons, rabbits and dogs and has a largish garden. His prize duck / [p.3] is a big one which he got by frightening the hawk which was carrying it off. It was very young and badly hurt but now is huge and very tame. My host chased it and the others into the irrigation ditch for a swim. At this point a distinguished Pathan landowner from Kohat [a city in North Waziristan] drove up (by previous arrangement) and two carloads of us went into the native city of Peshawar. The landowner was a "Rai Bahadur,"[31] the holder of a title and has two sons in the Frontier army. Both of them are Oxford. The old gentleman is bearded, handsome in his way, and speaks good English. He took us first to an old church built by the English nearly a hundred years ago. It was really fine, with its heavy, carved doors and lacey wooden altars, both of which are originals and made in Peshawar. There is some old stained glass in some of the windows. The church isn't large, and its exterior is very Moorish with balconies, niches, and slender pillars.

After seeing the church, we went atop the Police Station tower to get a bird's eye view of the city which is purely native and is separated from the cantonments (where we live) by the railroad. It was a Moorish city (the Mohammedan influence) amazing with homes jammed close together and shooting up three and four stories, each story with its carven overhanging balconies. On the roof-tops were mud walls without roofs, the places where the ladies of the harem take the air. In these "fairy towers" we saw the ladies relaxing without their heavy burnouses which cover their heads completely and have small cotton screens for looking through. From poles stretched across the tops of the walls, hung round, iron basket-things, the local refrigerators, and also cradles of wood, full of infants. The whole place (looking from above) was an indescribable jumble of steps, terraces, and towers.

Later, in the bazaar, my host ran out of gas, so we looked around while Rai Bahdur's chauffer went for more. It was straight from the "Thief of Bagdad," colorful, confused, dusty, crowded, noisy, and opposite of fragrant. We watched a potter run his wheel and there was a new twist to it. / [p.4] The turning platform with which he worked was at the floor level, but over a narrow pit. In the pit was another turntable. The potter hung his right leg into the pit and with it gave the lower wheel an occasional kick. This turned the revolving table on which he formed his pots. We also saw them stretching rugs, making Afghan hats and coloring fan handles. We also got dozens of urchins (the usual onlookers) to push our car and turn it around in the narrow and packed streets. Much fun and laughing.

We went back to tea with our party because we were all late, and then back here to dress for the Governor's dinner and the dinner was most satisfactory. There were Lady and Sir George Cunningham, a Mr. and Mrs. Joyce of the Civil Service, two A.D.C.'s and myself. The Cunninghams are delightful people and my after dinner talk with him, most informative and helpful.

Wednesday, Dec. 3rd, 1941

The drive to Kabul takes two days. The trip is only 160 miles, but very rough going after leaving the good Indian road. We spent the night in Jelalabad [sic] and expect to reach Kabul by 3:00 P.M. on Friday.

I've got some cigarettes and arranged for some Afghan money, which I'll get tomorrow. At the tailor's I've ordered a pair of the heaviest riding breeches imaginable . . . I'm hoping to hear my car has reached Karachi so that I can get going within the month . . . In Delhi I sat in a conference with our General Wheeler[32] and British G.H.Q. and have received what are practically orders to drive up to Russia and into Iran and down to the Indian border again. It's about 2000 miles. / [p.5] Peshawar has become much warmer, and the violent shivers of the first two days are gone. The sun is gorgeous, with nippy, misty, mornings and evenings.

Thursday, Dec. 4th, 1941

Coming home, last night, we drove around the Peshawar perimeter, which is to say the barbed wire and steel gates which enclose the cantonments and are barred at night. There was a bright moon so that, even with the mist of smoke over the ground, we could see the outline of the mountains which ring Peshawar. Occasionally we came across the Indian night patrols, the men marching in pairs and looking picturesque in their high turbans and carrying their rifles across their backs.

The Khyber Road has a steel gate across it – in fact a pass is required for anyone to stray out of the cantonments after (I think) 9:00 P.M. The barbed wire is not so much a military defense as a means of stopping tribesmen from stealing at night. In a country where everyone carries a rifle, and where said rifle is worth more than the blood-fine for a man's life, thieves go to some pains to steal army weapons. There have been times, of course, when there was a good deal of shooting to enliven the cantonments, but everything is quiet now. The tribesmen have to surrender their guns before coming in.

During these busy days here, I've taken a tonga only once. The remainder of the time I walk. Perhaps I average six to eight miles a day. This morning I was trying to take in some of the details to pass on to you. There is a profusion of flowering shrubs – poinsettias, bougainvilleas, and a low bush with bright henna flowers which are trumpet like and grow in clusters. The trees are magnificent – huge peepuls with smooth whitish bark and light green and glossy leaves; pines, tamarisks and sheesham. I believe I even saw a eucalyptus tree. Then, too, there's a pepper tree which looks much like a willow. Out on the dusty plain there are real autumn tints, although no trees seem to be losing their leaves. The flowers, too, are interesting. There are roses everywhere in bloom, and considered almost a weed. They come in all colors. There are many deep red lilies in bloom in / [p.6] a park I pass. The big thing, however, is chrysanthemum. They are everywhere in purples, russets and yellows. The birds are very noisy, especially the grey-necked Indian crow … The big hawk, called chiel, has a note of his own, and he is very bold, like the crow swooping right down on the sidewalks for tidbits. There are sparrows, of course, and minas (which you have seen in Honolulu and Hongkong). This morning I saw 6 brilliant green parrots (very noisy), with long pointed tails and (I suspect) grey-green heads.

On the roads one sees tongas, bycycles [sic] galore not so many cars, but plenty of bullock and buffalo carts with heavy wooden wheels. Then there are the donkeys … Sometimes you'll see a six-foot Pathan sitting on one and holding his foot up so they / [p.7] won't drag. For the most part, these donkeys go in groups of four or five, carrying wood, coal, sugar cane or anything else and in charge of an urchin who is no bigger than they. Men and boys alike, however, sit on the very end of the donkey – directly over its hind legs. I suppose this is to prevent breaking its back in case of a huge man. I have a donkey-boy friend who is very dirty and loves to stare at me out of his good eye as he passes the hotel gate. One also sees sheep and goats being led singly through the bazaars on leashes like dogs.

This morning a boy had a pet (and huge) ram which he was sicking onto his boy friends, much to their terror and delight. The timid ones ran behind a grey-bearded barber who shaved a customer squatting in the dust all unconscious of what transpired. Last evening, after tea and writing, I walked through the crowded bazaar to see the remarkably fine fruit and vegetable stands, and to watch good Moslems drinking tea out of large and lovely brass samavars [sic]. There was what sounded like a riot down the street, but when I went over I found it was only a lot of school boys holding a silver athletic trophy aloft and celebrating at the top of their lungs.

But you should see the carrots and cabbages, the grapes and pomegranites [sic] they have here. I should think they are unsurpassed anywhere. This (and Afghanistan) is real fruit and vegetable country. The other shops were interesting, too; but rather a hodge-podge of shoddy European stuff mixed in with the native brass, cloth, etc. Plenty of tailors seem to live in the bazaar, and I saw a shop where the big Singer Sewing machine had been sunk into the floor so that the pedal / [p.8] was out of sight, its operator appeared to be legless. Coming out (and passing the old blind beggar who asks Allah to bless you) I heard some tentative drum beating. Ahead, there was a bright light as for a celebration and presently I saw a band of bearded Pathans of the most remarkable aspect. On their heads were bright yellow hats meant to look like tam o'shanters, but managing to look more like chef's caps. They had abbreviated khaki coats on, and their baggy trousers (believe it or not) were fashioned from Scottish plaid. There were four drummers and about six pipers with bagpipes under their arms. They stood in a circle facing inward, and blocking the sidewalk. When he saw me, the No. 1 drummer exercised his English. He said, "bon, thoo, tree!" With that, the whole crowd burst into allegedly musical action. The old blind beggar came hobbling over, calling on Allah and hoping to find a generous crowd, and I escaped.

Still Thursday, 6:15 P.M.

Have just finished my evening walk through the bazaar, and because this is a look-see letter, I'll try to note some of the sights. First, there were a tremendous dust being raised over by the railroad – it was really thick. There was a lot of sweepers (men and women) sweeping the dust of the road and side paths. I don't know why they did it, for all they got was a lot of sugar cane pulp which the population spits out all day long.

Leaving the sweepers to their artificial dust storm, I went to the "business district." This time I went where they sold grain and food and sweetmeats, where the beggars got handfuls of flour put in their tin cans, where gentleman sat in the dust eating pilau and curry with big chapattis; where / [p.9] children bought jalobi, or ludoo, or parra or two. I saw a woman in a bright red bournouse, holding a baby and looking out through her screen like a night in armor. I saw a goatskin bellows in a blacksmith shop – the kind that is lifted from the floor with the left hand and holding the end open, and is then pressed down, the end having been squeezed shut. I'd forgotten about this bellows, and got the usual "reincarnation" feeling. I saw them grinding spices for curry, using a stone roller on a flat stone slab; and I watched a Mohammedan cook skewering mutton on iron spits about twice as big as knitting needles and putting them over a charcoal fire. I saw any number of native doctors and in a back alley, an "Xray and Electrotherapy Institute"!

I saw a Gurkha in a Boy Scout hat and shorts, carrying his big kukri (knife) hanging by a rainbow sash looped over his shoulder and falling to his side. I saw full fifty bearded and blanket-wrapped tribesmen tumble out of an incoming Khyber bus and make off down the street with their big staves. They were very silent, and looked me over carefully. The tribesmen seem to be bold, proud people who don't easily give you the pathway. Their lives are harsh (what with feuds and vendettas) few of them expect to die in bed. The youngsters aspire to beards – very feebly sometimes – while some of their elders are content with thick mustaches. There's a good sprinkling of reddened beards in Peshawar, each marking a man who has made his pilgrimage to Holy Mecca.

As I came home, the evening prayer was on, and I came on a crowd of worshippers who had spread their rugs in a small brick enclosure on the sidewalk. They faced Mecca, which this evening was a bright pink sunset cloud standing up over a / [p.10] distant blue mountain peak. In front was a leader (rather disreputable looking in dirty, loose white garments and possessed of a red beard) who intoned and called the time for genuflections, etc. Behind him was a straight row of eight, behind that a line of four, and last of all a pair – all men of course. Nothing stopped in the bazaar during the prayer. A gent squatted in front, blowing a fire; at the side there was a bare native bed with a little white cock (tied by the log with a string) roosting on it alone; and the kids played everywhere. As the worship progressed, a latecomer dashed through the confusion. He ran to a hydrant to do the necessary ablutions before prayer, which consisted of wetting his right hand and swuzzling his mouth out twice. He then took his place in the second row, nearly knocking over a brother in his haste. The last I saw of him he had his head back, and his up-flung, open hands were silhouetted against the bright pink sunset.

Tonight I came back to the hotel by the round about way, seeing my familiar after tea haunts, and I found out about the prayers I saw yesterday. On our main shopping street just off the very crowded bazaar, there's a very wide sidewalk which gives access to provision shops (a'la China), rug shops, etc. On this sidewalk the pious have erected brick enclosures facing Mecca. These places have walls about two feet high. Tonight, being Friday and therefore the Mohammaden Sabbath, I discovered little niches or shrines within these enclosures. There were (today only) little oil burning lights in the dusk at these open air altars. In one, a worshipper was still praying long after the evening call to prayer. He kept his red fez and candles on and was down on his knees. He laid his open hands / [p.11] palms upwards, close together on the ground, and put his forehead down on the bricks. He had no prayer rug, and he held this attitude a long time.

Gordon Enders' correspondence with Albert Zimmermann – 22 December 1943

> Kabul, December 22, 1943
>
> Dear Al:
>
> Herewith the length of Chitrali putthu which I promised you. Luckily for me, Thayer is able to bring it down to you by hand due to his sudden transfer to London.
>
> Thayer's transfer was not the only surprise I got on my return. I'm leaving day-after-tomorrow for New Delhi where I'll probably spend a week or two.
>
> Tommy and I got out of Quetta on schedule and made our trip back as planned without hitch. We were pretty tired after doing the run from Kandahar to Kabul in one day, but we got in before dinner. Fortunately we encountered no snow on the road, nor rain, nor anything else disagreeable.
>
> Please remember me to Commander Smith and the boys, and if you have time, drop me a line. I'll be glad to hear how the movies turned out and whether or not I will be able to get prints. Thayer expects to hit Karachi about midnight of the 26th and I'm asking him to call on all of you. He's out again for London on the 28th.
>
> With best regards,
>
> Sincerely,
>
> *Gordon*

This letter was carried to Karachi by Charles Thayer, who delivered it to Zimmermann on 26 December 1943 (see AWZ letter to BSZ, 28 December 1943). I have not been able to translate the word "putthu" in the "gift of Chitrali putthu," but it is probably a length of cloth.

Gordon Enders – His Life

Gordon Bandy Enders was the elder son of Emmanuel Allen and Frances Marie (Seibert) Enders. He was born in Essex, Page County, Iowa, on 7 May 1897. His father came from a long line of hard-working, pious Germans in America, so-called "Pennsylvania Dutch. Emmanuel was different from the rest of his family, in that he was precocious and independent. Emmanuel married the daughter of Swiss immigrants, whose middle name, Marie, suggests that she may have been a Roman Catholic. She spoke a different dialect of German, and it was considered scandalous for them to marry. Emmanuel was raised as a Lutheran, but he became a Presbyterian minister, and Gordon was born in his first parish, in rural Iowa. Rev. and Mrs. Enders had previously adopted a daughter, who they named Miriam, presumably derived from the middle name of her adoptive mother, Frances Marie Enders. Miriam was a foundling, who was abandoned by her own mother at birth. Emmanuel and Frances later had a son, Robert, known as "Bob." Emmanuel applied to become a missionary, and his application was accepted. The Enders family of five sailed from Philadelphia to India via Liverpool in December 1903. Mrs. Enders was a teacher and managed their many household servants. The Enders family lived initially at Etawah, on the Grand Trunk Road, but they soon moved to Almorah, at the foot of the great sacred mountain, Nanda Devi, on the Indian-Tibetan border. From Jowar Singh, their "number one servant," a warrior caste Hindu, and one of their servants, Masih Ulla, a Muslim, Gordon learned to speak the local languages – Urdu and Hindi – and from Jowar's father-in-law, Chanti, a Tibetan Buddhist, he learned some Tibetan, and the customs and religions of the people of that area – Hindus, Muslims, and Buddhists. While still a child in India, these men introduced him to Wu Ming-fu, a Chinese trader from Chengdu, and thus he learned to speak a bit of the trader's Chinese dialect. He would later learn many other languages, including speaking, reading and writing in Mandarin, and many other Chinese dialects, as well as Persian, Arabic, Pathan (also called Pashtun), in addition to retaining his boyhood German, French in World War I, Japanese in World War II, and finally Spanish.

Gordon was sent back to America to finish high school in 1910 at about the time his father died and was buried in India. His mother and siblings lived in India for several more years, while his mother continued her work as a teacher. Gordon lived in Wooster, Ohio, with the children of other Presbyterian missionaries, where they attended a preparatory school associated with The College of Wooster. One of his school mates was William Eddy. Within a few years, Bill Eddy became famous as a highly-decorated and badly wounded young Marine Corps officer in World War I, and he was also a Marine Corps colonel in the O.S.S. and a diplomat in World War II. Gordon entered The College of Wooster in the fall of 1913. He became interested in flying as a result of seeing airplanes at a fair in the summer of 1914, soon after World War I began.[33] He was in his third year at Wooster in 1916 when he left college to join the fight. He joined the Norton-Hayes ambulance unit of the Red Cross, and he "drove for six months in Picardy and Verdun in 1916." He then returned to the U.S. and enrolled as a junior at the College of the University of Pittsburgh. Once again, he didn't finish his junior year. He returned to the war in Europe, intending to sail from New York on 7 April 1917, which by coincidence was the day after the U.S. declared war on Germany.[34]

Enders joined the French Foreign Legion in April 1917, and he soon soloed as a pilot.[35] His U.S. Army service record shows that four months after he arrived in France, he joined the U.S. Army. He was appointed as a corporal in the Signal Corps, and as a Squadron Leader, A.E.F., on 10 August 1917 and he served in that rank until 25 April 1918. He was commissioned the following day, 26 April 1918, as a Bomber Pilot with the 223d Bomb Squadron. The *Air Service Journal* for June 1918 says he accepted the rank of 1st lieutenant in May 1918.[36] He continued in service after the Armistice on 11 November 1918 until 9 October 1919, at which time his commission was terminated. In a formal photograph in May 1919, he is wearing the badge of the U.S. 2d Army, which had by then been disbanded.

He survived many close calls and crash landings, including one over Brittany in which his plane crumpled and he fell without a parachute from an altitude of 3,000 feet. He was unconscious and was thought to be dead. However, he was soon found to be alive and was taken to a nearby U.S. Army hospital, probably Base Hospital No. 8 or Base Hospital No. 9 in Saveney. Base Hospital No. 9 was later relocated to Chateauroux, about 400 km to the east. Patients were transferred between these two hospitals, both of which were staffed by American medical school faculty members. The faculty and staff of Cornell Medical School/New York Hospital organized Base Hospital No. 9, and that hospital was supported by many upscale charitable organizations in New York City. He was nursed back to health by an American Red Cross worker, Elizabeth "Betty" Crump, daughter of Samuel Crump, of Montclair, N.J. Her father was a wealthy industrialist who held several patents for inventions in printing. His company, Crump Label Co., had a large plant in Montclair, and it had offices in six cities across the country. Gordon and Betty were married in La Rochelle, France, on 22 April 1919, with one of his fellow aviators as best man. Photographs of Gordon and Betty taken at about that time show her in a Red Cross uniform with three chevrons on her sleeve, each denoting six months of wartime service, and a black arm band, in memory of her brother, Lt. Samuel Crump, U.S.A., who had been killed in France on 29 September 1918, six weeks before the Armistice. Four gold chevrons were on Gordon's lower left sleeve, indicating two years of service in the war. On his left breast, he wore the U.S. Aviator's badge. Above his pocket, inboard, was the solid color ribbon of a Chevalier of the Legion of Honor; and outboard, the Victory Ribbon with four bronze battle stars.

Gordon Enders' service record shows that he was credited with two years and two months of service in World War I.[37] In 1923, Betty dedicated her first book, *Swinging Lanterns*, to her husband, whom she called "Pierre." The nickname may refer what was then a well-known ribald joke that was told by WWI veterans. The story was about the amorous adventures of "Lucky Pierre, the famous French fighter pilot."

Gordon and Betty Enders returned to the U.S. in 1919. After visiting her home in New Jersey, he studied for, and passed, the examinations needed for employment by the U.S. Department of Commerce. He was sent to China in 1920 as an assistant commercial attaché for the Commerce Department, to be based in Beijing (then called Peking). Enders made many contacts with wealthy people, both Chinese and foreigners. Enders says that he became acquainted with Prince Dilowa, the third ranking Lama in the Tibetan pantheon; Liang Shih-yi, the "richest man in China"; Ku Hu-ming, the "Bernard Shaw of Asia"; Hsu Chih-chuan, president of the Republic of China; Prince Kung, uncle of the boy Emperor, Pu-yi; and Tsu Hai-san, who later became foreign minister of Tibet. From Shen Shung, his first tutor in China, and a "Mr. Bamboo," he learned to speak Mandarin and write in Chinese characters. He later added dialects spoken in other parts of China. Enders ostensibly left government service in 1923 to become an agent in China for several U.S. corporations. There is good reason to believe that he also became, under cover, a U.S. government intelligence officer. And that he continued his status as an intelligence officer after he returned to the U.S. from China in 1937, initially under cover, and then openly with the U.S. Army in 1941. He was discharged from the Army in 1945, but he soon returned to service and continued to work as an intelligence officer with U.S Army Intelligence (G-2) until he retired as a colonel at age 65 in 1961. Gordon and Betty made several trips back to the U.S. between 1921 and 1937, and they also went their separate ways in China. Betty wrote two books about her experiences in China, and she became a widely-traveled lecturer in the U.S. Her father died in Shanghai in June 1925 on a trip to visit his daughter and son-in-law, and he was buried there. In 1935, Gordon published his first book, *Nowhere Else in the World*, which he wrote in collaboration with a well-known author and editor, Edward Anthony.

For his commercial business as a salesman, Gordon equipped Chiang Kai-shek's forces with a fleet of twenty Corsair bombers, and he trained Chiang's officers to use them. He also served as a personal pilot for Generalissimo and Mde. Chiang. At the country home of the Chiangs, in the hills near Shanghai, Gordon was introduced to the spiritual ruler of Tibet, the 9th Panchan Lama. The Panchan believed that the future of Tibet rested on having a good relationship with its historic protector – China – and that this would require Tibet to pay tribute to China. With Chiang's blessing, the Panchan appointed Gordon to be a high-ranking Tibetan official. He received a "Golden Passport," and flew Tibetan gold dust to banks in Shanghai, where it could be used by the Chinese government.[38] This relationship collapsed when the Panchan died while returning to Tibet on 1 December 1937. Tibetan governance was then divided between the followers of the late Panchan and those who controlled the little boy (born in 1935, and still alive in 2018) who was identified as the 14th Dalai Lama. At about the same time, in 1937, the Japanese army reached Shanghai. After a brief military engagement, during which Gordon earned the British Shanghai Emergency Medal, he returned to the U.S.[39] From 1938 until late in the summer of 1941, Gordon's principal occupation was as a professor of history at Purdue University. He and Betty traveled together, and independently, as popular lecturers. They spoke about recent events in China, and about the future of Asia. They often visited each of their families – the Enders and Crumps – and their home was near one of Betty's relatives in Lafayette, Indiana. The children of Gordon's brother Bob especially appreciated the exotic tales that their uncle Gordon and aunt Elizabeth told them. They recalled that Elizabeth, who had grown up in a grand home with servants, was a stickler for decorum.

Everything changed for Gordon and Elizabeth in 1941. Japan had already put a curtain of secrecy around the territories that it had acquired from Germany after World War I as a result of the mandate by the League of Nations. Japan was advancing steadily in China, and there seemed little doubt that the colonies of Britain, France, and the Netherlands would soon be attacked. The U.S. military had prepared contingency Plan Orange, which presumed that Japan would strike at the Philippines and Guam. General Douglas MacArthur was dispatched to organize the defense of the Philippines. He was expected to hold those islands until reinforcements would arrive to push back the Japanese away. In anticipation of the forthcoming conflict, the U.S. Army planned to send officers to Asia and the Middle East, to become acquainted with the local leaders. They would work with their counterparts in Egypt, Syria, Iran, Afghanistan, and India, and prepare for action if and when the U.S. entered the war. We can presume that Gordon was identified early in 1941 as one of those who should be one of the leaders for the U.S. in Asia in the war that was expected to come. Gordon Enders' qualifications were unique, with his language skills, his record of bold and effective action in combat, and his knowledge of central and eastern Asia. Enders' experience was not unlike that of his school mate, Bill Eddy, who was born in Syria and spoke Arabic fluently, and who had probably been undercover as a spy since World War I. Eddy would be recalled for duty as a Marine Corps colonel, O.S.S. officer, and diplomat. Gordon Enders received the call he was expecting, and he immediately proceeded to Washington, where he was briefed at the War Department, and given equipment and orders. He was sworn into service as a major, effective 15 September 1941. As a military attaché, he would report both to a senior officer in the Army's Military Intelligence Service (MIS), known as G-2, and also to a reporting senior in the State Department. He would be the first U.S. diplomat to be stationed in Afghanistan.[40] He was to find a suitable location for the U.S. mission and prepare the facility for the arrival of other diplomats; and he was instructed to travel widely across that wild and mysterious country. It is said that no man can have two masters, but Gordon Enders would often utilize the dual reporting requirement to both State and Army, and act as if he had no master. While in Washington, he had the foresight (or was instructed) to visit the British Embassy and introduce himself to Lieutenant Colonel Rex Benson, and to receive a card of introduction from Benson to the Governor of the North-West Frontier Province of India.

Then, in late October 1941, with a very high priority, Major Enders boarded a flight to the west with several other Army officers. They departed from Washington, and after stopping in California, Hawaii, various Pacific islands, and Manila, they proceeded over the Himalayas to Calcutta, India, where they arrived in mid-November. Gordon was the only one in the group who had any personal knowledge of India, and he was the only one who spoke the languages of that country. Although he may have been the most junior officer in the group, he was the center of attention when they dined with Field Marshal Sir Archibald Wavell and his wife and children. He and Wavell were both combat veterans of World War I, in which both were injured and decorated. Enders dressed in uniform, showing his Pilot's wings, the French Legion of Honor and the WWI Victory ribbon with four battle stars. He was seated beside Lady Wavell. He then proceeded by train across India to Peshawar, on the North-West Frontier, savoring a return to the land of his childhood. In Peshawar, he presented the card of introduction from Col. Benson to Sir George Cunningham. The doors to Peshawar society were flung open to him, and he enjoyed every minute there. He located a car and proceeded over the Khyber Pass into Afghanistan in the first week of December 1941. He passed through Jalalabad along the historic route to Kabul, on the southern branch of the ancient Silk Road. He arrived in Kabul on 5 December 1941. Three days later, on 8 December 1941 in Asia, he learned that America was now at war. Enders soon received a new American invention which would facilitate his travels in Afghanistan – the ubiquitous four-wheel drive vehicle known affectionately as the "jeep." He would call his personal jeep "Ma Kabul." It came to Kabul with a driver, Sergeant Tommy Nicholson. But Major Enders enjoyed driving it himself, with his aviator's hat, top crushed and jauntily tipped to one side, an unlighted cigar clamped in his mouth, and a holstered .45 cal. Army pistol at his right hip.

Enders' second book, which went to the publisher in the summer of 1941, was published in 1942, soon after the war broke out. It was *Foreign Devil: The Adventures of an American 'Kim' in Modern Asia*. The title implied that Enders' goal was to emulate the fictional boy-spy who was the hero of Rudyard Kipling's Nobel prize-winning book, *Kim*. To do this, Enders would need to travel along what was known to Kipling and "Kim" as The Border, the Durand Line, that the British drew in 1896 to separate Afghanistan from India. The British disregarded Afghanistan's objections, and established the border a bit north of the highest point of the Khyber Pass. Enders would eventually accomplish this Trip two years later, traveling from Chitral to Quetta, in November and December 1943. Like the fictional "Kim," Enders would play the Great Game. It became his mission in life. Kipling wrote, and Enders believed, that Russia was the Enemy. In the original Great Game, the goal of the Russian Empire, and then of the Soviet Union, was to control the highlands of Central Asia. Enders correctly foresaw that the conflict between the U.S.S.R. and Great Britain would end with the independence of India. However, as Kipling prophesized, the Great Game would continue. Enders was one of the earliest to see that a new Great Game would begin, in which the U.S. would replace Britain as the adversary of Russia.

Enders had an additional appointment as Assistant Military Attaché in Iran, where the Chief of Mission, Louis Goethe Dreyfus, also had responsibility as non-resident Minister in Afghanistan. Enders traveled to Tehran soon after he found a building in which to locate the American mission. He returned to Kabul with another intrepid American – Charles Thayer – who would be the chargé d' affaires and secretary of the mission. Thayer was a West Point graduate, and after leaving Kabul, he was a O.S.S. officer in Yugoslavia. He had previously been stationed in Moscow, and after the war, he served in senior positions in the U.S. Foreign Service. His career suddenly collapsed as the result of false accusations from Senator Joseph McCarthy. Enders and Thayer had a good relationship, although Enders did not get along well with Cornelius Engert, the U.S. minister to Afghanistan. Neither Enders or Engert were known to be truth-tellers, but both of them had friends in high places. Both of their wives were on good terms with Eleanor Roosevelt, so both of them could play the FDR card. Enders' reports to the Army have been lost, but his reports to the State Department show that he irritated many diplomats in Washington. Enders drove his jeep all over Afghanistan, and he had some hair-raising adventures. His reports to the State Department show many minor corrections marked in the margins by unnamed readers, none of whom had personal knowledge of the country, but who showed their irritation with Enders.

Meanwhile, anonymous antagonists were waiting in the wings, ready to pounce on the U.S. minister to Afghanistan, Cornelius Engert. After FDR died in May 1945, Engert was soon relieved of his duties in Afghanistan and was forced to retire. He lived a long life after that and he had a successful career in Britain (receiving the O.B.E.) and in non-profit organizations, but his ambition to become an ambassador was never realized. Gordon Enders, on the other hand, competed the journey of his dreams along the border of India and Afghanistan on 12 December 1943, and returned to Kabul. He was promptly ordered to proceed to Delhi to be the senior Army intelligence officer in the China-Burma-India Theatre, with additional duty as senior Army intelligence officer for the South East Asia Command in Kandy, Ceylon. His title was Military Observer. He returned twice to Kabul; once to accompany General Patrick Hurley on his visit to Kabul in January 1944 as the personal representative of FDR, and then to accompany his replacement as military attaché in Kabul, Major Ernest F. Fox. Enders' duty as Military Attaché in Afghanistan officially ended on 1 April 1944, and he was assigned on that date to be a Military Observer in Delhi. Enders contracted malaria in India, and on 7 May 1944, he was sent back to the U.S.

to recover. He probably never returned to India, Afghanistan, or China during World War II. However, he came across the Pacific again in 1945 to Hawaii, Okinawa, Japan, and Korea. He returned to Korea during the Korean War, and as an intelligence officer until 1961, he could have gone almost anywhere.

When he arrived back from India to be treated for malaria, Enders was assigned to temporary duty with the Military Intelligence Service in Washington, D.C., for two months. The Army then sent him to Military Government School at Charlottesville, Va., from 1 August until 31 December 1944, and to Civil Affairs Training School at Northwestern University in Chicago until 11 February 1945. He did not complete the course at Northwestern, however; his heavily redacted official service record cryptically says "no" to the question of whether the course was successfully completed. He was given two weeks of leave, probably to go home, put his affairs in order, and prepare to enter the war again. He returned to the Pacific Theatre, assigned to Headquarters, 10th Army in Oahu, Hawaii, on 23 February 1945. The 10th Army planned to invade Okinawa on 1 April 1945. Enders was an Assistant Military Government Officer while the division was in Oahu, and he was promoted to Military Government Officer in March, at the time that 10th Army Headquarters was completing its plans for the Okinawa campaign. The battle for Okinawa was one of the fiercest in the war in the Pacific, with Japanese suicide bombers (kamakazi pilots) claiming many U.S. casualties, and American flame throwers working their way slowly across the island. Enders became Executive Officer of the 10th Army's Military Government Special Team #1 on 15 April. The 10th Army's 27th Division had been held in reserve, and three days later, on L [for landing day] +18, the 27th Division joined the battle. The 27th Division had a storied history, as the New York National Guard Division, with an impressive battle history in World War I, and again in World War II.

Enders was transferred to the 27th Division on 25 May 1945 as a Military Government Officer, and he officially remained in that post until 10 September 1945. The Japanese army in Okinawa had continued to fight fiercely for nearly a month after the war had officially ended. The first atomic bomb had been dropped on Hiroshima on 6 August, and the second bomb was dropped on Nagasaki on 9 August. In the meantime, the Soviet Union had declared war on Japan on 8 August, as promised three months earlier in the Potsdam Agreement, and on 9 August the U.S.S.R. had invaded Manchuria and North Korea. Emperor Hirohito famously broadcast on the radio, for the first time, on 15 August [Japanese time], announcing that Japan would cease fighting. Allied forces began to enter Japan peacefully on 28 August. President Truman announced the surrender in a radio broadcast at 10 p.m. on 1 September, calling for a day of prayer and thanks on 2 September. The surrender of Japan was formally recognized on the deck of the U.S.S. *Missouri* on 2 September, and that date is commonly called V-J Day. Nevertheless, the Japanese Army in Korea had not yet surrendered. And recalcitrant enemy troops continued fighting on Okinawa. After conducting "cleanup operations," the 27th Division finally left Okinawa for the main island of Japan on 7 September. At about the same time, Gordon Enders, now a lieutenant colonel, was sent under deep cover, and probably with only verbal orders, to negotiate the surrender of the Japanese forces in South Korea. How he accomplished this is unknown, but it is said that he delivered the document of Japanese surrender in South Korea to General MacArthur in Tokyo.[41]

From 11 September 1945 until 14 May 1946, Enders was assigned to the Foreign Affairs Section, Military Government, U.S. Army in Korea. He was initially appointed as Secretary in the Foreign Affairs Section, and on 15 November 1945, he was made Chief of Section. On 1 January 1946, he became Director of the Office of Foreign Affairs. His experiences during this period of time would serve him well when he returned to Korea in 1951. His next appointment was typical of the post-war period for many Army officers: from 15 May until 23 September 1946, he was assigned to duty as a Liaison Intelligence Officer with the Civil Affairs Division of the U.S. Army at the Pentagon. We can presume that he was given some well-earned leave to return to his wife, and he was also briefed on what his future jobs might be in the world of intelligence. He was released from active duty on 23 September 1946, and he was recalled on 17 May 1948 for what would be the rest of his career.[42] He spent much of the two years between September 1946 until May 1948 on what is called "inactive" duty with the Army; his service record shows 180 days on "Active Duty for Training" as a Reservist. He probably used this so-called "duty for training" for cover as an intelligence officer, perhaps travelling in civilian clothing, to places that are undisclosed in the record. He was recalled to active duty on 1 September 1948 as Officer-in-Charge of a unit that was called, cryptically: "ORG Tng (TID, G2)." G-2 is the intelligence section of a general staff; i.e., at the level of a brigade or higher; in this case, the 2d Army. His post was at Fort George Meade, Maryland, with Headquarters, 2d Army – which had been his command in World War I – and he continued in this position until 11 July 1949. Betty Enders probably moved to Fort Meade at that time. Gordon was assigned to duty in many places in the United States over the next thirteen years, and Betty made her home with him in these places.

Gordon Enders was assigned to travel for shorter periods, outside of the continental U.S., and Betty may have accompanied him on some of those trips. There is no official record of when he went, or what he did, but in about 1961, he typed the list of additional countries that he had visited as: Britain, Belgium, Germany, and Italy (all probably before the end of World War I); Okinawa, Japan, and the Philippines (during World War II); and North Africa (UAR, Tunisia, Morocco), Suez Canal (four transits), Singapore, Thailand, Burma, Haiti, Cuba, and the Panama Canal. And we also know that earlier in his life he had been to India, Afghanistan, Iran, Tibet, China, Manchuria, Mongolia, and Korea.

Enders complained while he was in India in World War II that he was one of the oldest majors in the Army, but sometime in the period between 1945 and 1952, he was promoted to lieutenant colonel. The dates of his promotions are redacted in his service record. He was in various positions with Headquarters, 2d Army, from 1 September 1948 until November 1951. His titles include: Acting Director, Tactical Intelligence Division, G-2 Section (from 11 July 1949 until 1 November 1949), then Assistant G-2, in the same Division until 10 February 1950, when he was given TDY (temporary duty) for three months at Strategic Intelligence School, Washington, DC. This duty ended on 6 May 1950, and he returned to his previous position as Assistant G-2 in the office of the Director, Tactical Intelligence Division, HQ, 2d Army. When North Korea suddenly invaded South Korea on 25 June 1950, Korea must have immediately become the principal focus of what was the Army's most important intelligence base – Fort Meade, Maryland. There is no record of where Gordon traveled over the next year, but he was officially transferred to duty in Korea on 14 November 1951 as a member of the U.S. Military Advisory Group (USMAG/K). From 17 November 1951 until 10 December 1952, he was senior advisor to the Assistant Chief, G-2, Army of the Republic of Korea. For his service in this position, in developing the intelligence branch of the Korean Army, he received the Legion of Merit on 26 January 1953. He received the award during his next assignment, as Executive Officer, G-2 Section, Headquarters [HQ], XVI Corps. He continued in this position until 18 November 1954, with an interruption to serve as G-2 Air for HQ, XVI Corps, from 9 February until 15 October 1953.

He was assigned to duties away from Fort Meade for the next four years. Betty probably followed him to his assignments at Fort Riley, Kansas, (27 January 1955, Director of a division at Army General Staff School), Fort Benning, Ga. (1 May 1955, Assistant Director of a department), and again at Fort Riley (19 August 1955). On 30 September 1955 he was "Reld AD" (cryptically, this should mean "released from active duty"). However, the next day he was assigned to HQ Detachment 5021 SU at Fort Riley, but two weeks later, he was en route to Norfolk, Va., to serve from 7 November 1955-7 December 1956 as an Intelligence Analyst (MOS 962.60) with 8696 DU, Amphibious Group 2, Naval Amphibious Base-Little Creek, Norfolk, Va.

In 1957, he returned to Fort Meade and spent the rest of his career there: with HQ, 2d U.S. Army (until 31 January 1957); with 319th MI Bn (until 14 May 1957); again with HQ, 2d U.S. Army (TDY until 24 July 1957); 525th MI (TDY until 8 October 1958); 177th MilCensDet (until 14 May 1959); and finally "from 15 May 59 to [date illegible] April 61 with 2d USA (Enl) (2000) FGGM, Md." The last item was his final assignment, but the abbreviation "Enl" is a puzzle. Does it mean "Enlisted"?

In April 1961, the month that Enders retired, Cuba was invaded by more than a thousand Cubans who hoped to spark a full-scale rebellion against Fidel Castro and his Communist government. The invasion was planned and executed with the help of the C.I.A., and after it collapsed, some C.I.A. officers called it a "perfect failure." It began with air strikes on 15 April by Cuba rebels flying in American B-26 bombers, repainted to look as if they had been stolen from Cuba. On 17 April, the rebels landed at the Bay of Pigs. Castro's forces killed 114 and captured 1,100 of the invading force. The plan was hatched during the Eisenhower administration, but it was carried out when President Kennedy was in office, and

Kennedy was thereafter forever edgy about the C.I.A. It is unknown if Enders, working in Army intelligence, had foreknowledge of the plan, or participated in the planning, but it was a disastrous event that occurred just before he retired. He wrote in 1961 that he had visited Cuba at some point in time.

One of his neighbors, a boy at that time, recalled that Gordon was known as "Sergeant Enders." The last months at Fort Meade must have been difficult for him. His wife, Betty, died of "emaciation" at Fort Meade Army Hospital on 13 February 1961. They had been married for 41 years. Various dates are given for the year of her birth, but if she was born on 16 May 1879, she would have been 81 years old.[43]

Gordon Enders remarried in 1962, to a widow, Elizabeth (Schwalbe) "Liz" Garrahan, with three daughters of her own. They moved to Albuquerque, N.M. Gordon was an excellent photographer, and he had a large collection of cameras, although none of his photos have been discovered. He especially enjoyed outdoor photography, and it said that he spent the last full day of his life on Sandia Mountain, near Albuquerque. He was buried in the Santa Fe National Cemetery. His second wife, Elizabeth "Liz" Enders, died 13 June 1987 in Springfield, Pa., while visiting relatives, and she was buried in Pennsylvania. The urn containing Betty's ashes was kept by Gordon, and it then passed to one of his step-grandchildren. A grand-niece of Betty's located the urn, and it was interred beside him in the summer of 2017. A new stone was erected there, to memorialize both of them.

Gordon Bandy Enders
(7 May1897 - 2 Sep1978)
Buried 1978
Santa Fe National Cemetery

COL
U.S. ARMY
WORLD WAR II

[original tombstone]

Elizabeth Crump Enders
(16 May 1879 - 13 Feb 1961)
Ashes
Interred 30 June 2017
Santa Fe National Cemetery

A.R.C.
DEVOTED WIFE

Illustrations for Gordon and Elizabeth (Crump) Enders

Enders Family in Iowa, ca. 1900, with three small children.
Gordon is on step, right, above his brother Bob. Miriam is on step, left.
(from Gertrude [Enders] Huntington, Ph.D.)

Enders Family in India, ca. 1905. Gordon is 2d from right, next to his sister Miriam
(from Gertrude [Enders] Huntington, Ph.D.)

Gordon Bandy Enders

Gordon Enders as a boy
(courtesy of F. T. Treichler)

Undated photograph
(courtesy of Gertrude Enders Huntington)

Lieutenant Gordon Enders in 1919, at the time he married Elizabeth Crump.

His U.S. Army dress uniform shows that he had been in the 2d Army; an upward-pointing red chevron, known as the "honorable discharge stripe," on his left sleeve indicates that he has been honorably discharged; four gold chevrons on his lower left sleeve, each of which denotes six months of wartime service in Europe; a winged propeller on his collar, indicating assignment to the Air Service; and on his left chest, the badge of a Pilot, and two ribbons: inboard, in solid color, the French Legion of Honor, and outboard, the ribbon of the Victory Medal, with several battle stars. He is wearing a Sam Browne belt, which was used to support a pistol at his left hip, although the pistol is not shown. His right lower sleeve is not visible, so it is unknown whether or not he has the "wound chevron."

More Images of Gordon Enders

Photos upper L and center are from Findagrave.com.
Upper R, from Zimmermann's photo on the North-West Frontier Trip, 1943

Below: L, from *Nowhere Else in the World*. Center, with Panchan Lama. R, Enders' "Golden Passport"

GORDON B. ENDERS – CITATION FOR THE LEGION OF MERIT
Awarded 10 February 1953, for exceptionally meritorious conduct from
17 November 1951 to 10 December 1952

HEADQUARTERS
UNITED STATES ARMY FORCES, FAR EAST

CITATION FOR THE LEGION OF MERIT

Lieutenant Colonel GORDON B. ENDERS, 0426933, Infantry, United States Army, distinguished himself by exceptionally meritorious conduct in the performance of outstanding service as a member of the United States Military Advisory Group to the Republic of Korea, in Korea, from 17 November 1951 to 10 December 1952. As senior advisor to the Assistant Chief of Staff, G-2, Republic of Korea Army, Colonel ENDERS materially assisted in developing the intelligence branch of the Korean Army into a well organized, competently staffed section capable of coordinating all functions pertaining to intelligence and counterintelligence. Proffering wise counsel and timely recommendations, he evinced a comprehensive knowledge of oriental politics and customs in his tactful and diplomatic approach to projects dealing with highly classified data, winning the confidence and respect of the Republic of Korea General Staff. Colonel ENDERS' notable achievements contributed significantly to the United Nations' campaign for international peace, reflecting great credit upon himself and the military service.

Curriculum Vitae
Prepared by Gordon Enders, ca. 1961

ENDERS, GORDON B. (Colonel, AUS Retired)
539 Solano Drive
Albuquerque, New Mexico

BIOGRAPHICAL

India & Pakistan	--	Childhood (son of missionaries) 1903 - 1910; Military Observer, 1944
France	--	French Foreign Legion, 1916-17; U. S. Army (Pilot), 1917-19.
China (including the Manchurias & Mongolia)	--	U. S. Department of Commerce, 1921 - 1923; business, 1923-1930; Aviation Advisor to Chiang Kai-shek, 1930-31; Foreign Advisor to Panchan Lama of Tibet, 1930-1937.
Iran	--	Asst Military Attache, U. S. Legation
AFGHANISTAN	--	Military Attache, Kabul, 1941-44
Korea	--	Director, Office of Foreign Affairs for U. S. Army of Occupation, 1945-46 Senior Intelligence Advisor to the Republic of Korea Army, 1951-52.
Japan	--	Deputy G2 (Intelligence) XVI Corps, Sendai -- 1953-54.

Countries visited (less than 6 months):

- Britain
- Belgium
- Germany
- Italy
- N. Africa (UAR, Tunisia, Morocco)
- Suez Canal -- four transits
- Aden (Protectorate)
- Various Pacific Islands
- Okinawa
- Philippines
- Singapore
- Thailand
- Burma
- Puerto Rico
- Haiti
- Cuba
- Panama Canal -- one transit.

Author of two books: Nowhere Else in the World (with Edward Anthony)
Foreign Devil

Professional Lecturer -- W Colston Leigh Bureau, 521 Fifth Ave, NYC -- 7 years, some 500 lectures.

Professor -- Purdue University, 1937-41 on Far Eastern History and U. S. Foreign Relations.

Broadcast Time -- Approximately 235 hours, including Purdue classroom lectures and miscellaneous news commentaries.

NOTE: Due to early education in British Schools, accent in speaking is British (modified), similar to Standard English as used by Hollywood.

Gordon B. Enders
Passport Application
April 5, 1917

Gordon B. Enders
AUS, ORC, USAR, Service Number O 426988
Service Record from National Archives, from FOIA

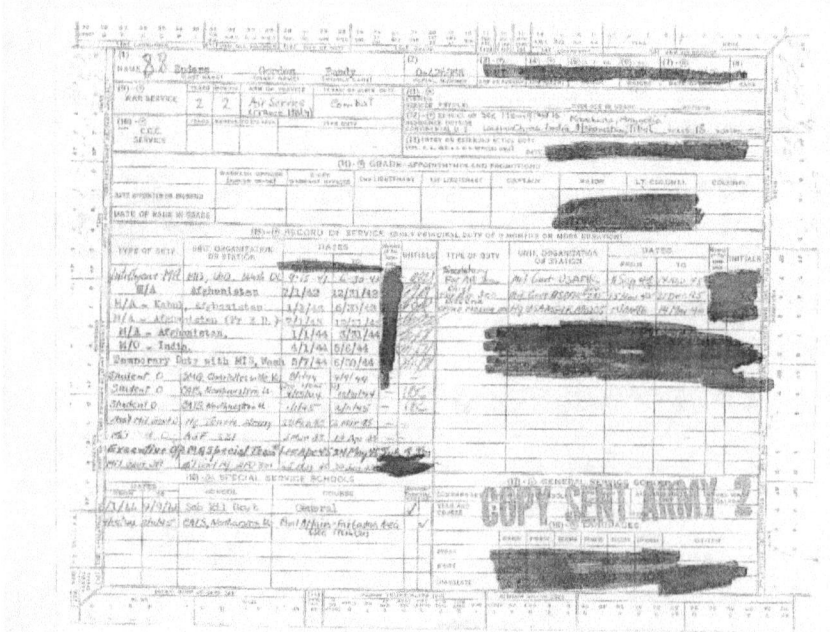

Gordon Bandy Enders

Service Record from National Personnel Records Center (Redacted)

Page 1

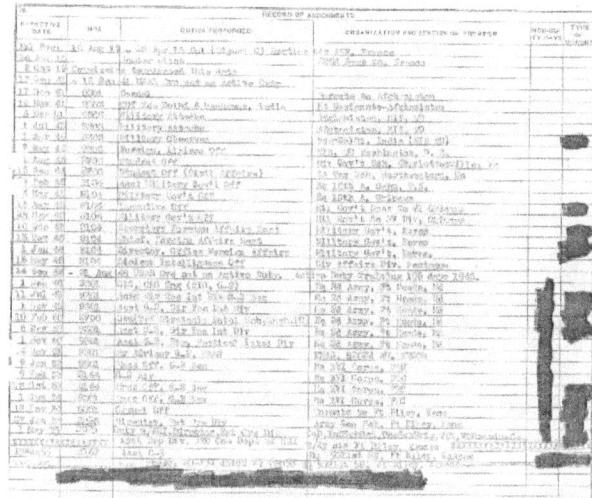
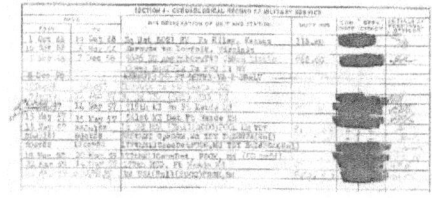

Gordon Bandy Enders - Service Record from National Personnel Records Center
Pages 2-4

Col. Gordon B. Enders, U.S.A.F. (Ret.), died in Albuquerque, New Mexico, September 2, 1978. He spent four years in Wooster Prep before entering the college with the class of '18. At the end of his sophomore year he joined an ambulance unit in France. Later he transferred to aviation and was inducted into the United States Air Force when it reached France. After combat duty he served with the Army of Occupation. It was 1920 before he returned to the United States. He soon moved to China with the Department of Commerce. Between 1921 and 1937 he studied Mandarin and traveled all over the Far East. During the later part of this period he helped train the Chinese air force, acted as a foreign advisor to the Panchen Lama, and served in the forces defending Shanghai against the Japanese. He returned home in 1938, taught history at Purdue, and served as a commentator for their radio station. During this period he visited Wooster, lecturing on the Far East. Apparently two years of academic life was enough: he went to Afganistan as the United States Military Attaché. By this time he could read and speak Arabic, Persian, Urdu, and Hindi, as well as French, German, and Mandarin; yet here he began the study of Japanese. He returned to the United States in time to join the troops who went ashore at Okinawa and was underground in Korea when Japan surrendered; he carried the surrender of the commander in Korea to MacArthur. A period of comparative peace followed. At the outbreak of the Korean War he served there with United States Army Intelligence. When he reached retirement age, he was asked to continue in service, but his evaluation of conditions in Indochina led him to choose retirement. He published two books and some short stories. Only an old Inky hand could have survived such a varied life.

Obituary, probably written by his brother, Robert Enders, from the Alumni files at the College of Wooster

Elizabeth Crump in May 1919, in the
month after her marriage to Gordon Enders

Elizabeth Crump Enders
publicity file photo as Chautauqua Lecturer
(from archives of University of Iowa)

Mentioned in Enders' Letters

Lt.-Col. Reginald Lindsay "Rex" Benson,
after he was knighted, Sir Rex

Major General Patrick Hurley

Major Ernest F. Fox

General and Mrs. Wavell, from Connell, p.289

Field Marshal Wavell, Lewin p.134

Selected Bibliography for Gordon and Elizabeth (Crump) Enders

Elizabeth Crump Enders, *Swinging Lanterns* (New York, N.Y.: D. Appleton and Company, 1923)
_____, *Temple Bells and Silver Sails* (New York, N.Y.: D. Appleton and Company, 1925)
_____, "Bits of Old China," *Good Housekeeping* (date and page number unknown)
_____, [title unknown] *Asia* (1934)

Gordon B. Enders, "Danzy," *Asia*, v.20 (October 1920): 871-6.
_____, "Prohibition in Old India" (with decoration by Herb Roth), *Asia: The American Magazine on the Orient* v.20 (No. 10, November 1920): 1003-4.
_____, "New Role for Panchen Lama," *Asia*, v.34 (August 1934), 462-7.
_____ and Edward Anthony, *Nowhere Else in the World* (New York: Farrar & Rinehart, 1935)
_____, "The Last Theocrat: The Panchan Lama of Tibet," *World Youth* (30 November 1935), 7.
_____, "Night Raiders in China," *Liberty* (3, 17, 31 July 1937).
_____, *Foreign Devil: An American Kim in Modern Asia* (New York: Simon and Schuster, 1942)
_____, "The Nomad Woman," *Collier's Magazine* (7 October 1950), 16, 62-5.
_____, "Open Up Files," *Albuquerque Journal* (28 March 1968), A5 (letter to editor)

_____. Letter to Department of State, 30 March 1943, re Dodge "carry all" for the King of Afghanistan, State Department Archives, Microfilm Publication #1219, Afghanistan, 1940-45, Box 4.
_____. "Afghanistan's Strategic Geography." 8 pages (4 double-sided) RG 165, Entry 7 Afghanistan 5000-6400 Box 4 NND 745008, Geography (5900), 5930- Report 107-A, mimeographed, 6 September 1943.
_____. RG 165, Entry 7. 128 Kabul. Memorandum for the Minister, Report No. 103, 4 September 1943; 8 November 1943, addendum to 22 Oct and 4 Nov, ref M.A. Report # 103 of 1 Sept 1943; Memorandum for the Minister, 4 Nov 1943.
_____. Memorandum, "Afghanistan in General" to General Patrick Hurley, 13 January 1944, Kabul (5 pp., typed), from Hurley Collection, University of Oklahoma, Box 83, Folder 7.
_____. "Report of Military Attaché, Kabul, Afghanistan, on General Hurley's Visit to Afghanistan," 1/18/44 (5 pp., typed), in Hurley Collection, Box 493C, Folder 4.

References for Gordon and Elizabeth (Crump) Enders

• Elizabeth Enders to General Patrick Hurley, letter with her signature, 22 April 1944, in Hurley archives, University of Oklahoma, Box 84, folder 4
• [Enders family] Patty Hoenigman, "Striking Genealogical Gold: Tapping into Descendants of Second Marriages," *The Chisholm Trail*, vol. 37 (no. 4, Fall 2017), 23-26, with cover photo of the family of Samuel Crump – parents & five children, with Elizabeth at age 13.
• Gordon Enders – typed obituary, 2 pp. Wooster College records. Probably prepared by his brother Robert. (obituary obtained by Patty Hoenigman).
• Taylor, Mark. "What happened to the Sandia bighorn herd? Gordon Enders Focused on Sheep" probably *Albuquerque Journal* (22 December 1986). In Hurley Collection, University of Oklahoma Libraries, Box 493C, Folder 4.
• William Hoy, "A Yankee Adventurer and the 'Living Buddha'" *Chinese Digest* (28 February 1936), 11, 14. Says Gordon B. Enders was "a youthful adventurer and soldier of fortune" was "commissioned by the Panchen Lama, spiritual ruler of Tibet, to convey the gold dust of the region into currency . . . for the modernization of that province."

- Notes from NARA II summarizing interactions between Gordon Enders, Cornelius Engert, and Charles Thayer: RG 165 (Records of the War Department General and Special Staffs); RG 165, Entry 7 Afghanistan 5000-6400 Box 4 NND 745008. Also:
  - Geography (5900), 5930-Afghanistan's Strategic Geography, Report 107-A, eight pages (4 double-sided), mimeographed, of 6 September 1943.
  - 128 Kabul 8 November 1943, addendum to 22 Oct and 4 Nov, ref M.A. Report # 103 of 1 Sept. Memorandum for the Minister on 4 Nov 1943
  - Report No. 103, 4 September 1943 is a 3-page report with appendices / State Department Archives / Microfilm Publication #1219, Afghanistan / Boxes 4-7, Afghanistan 1940-45 / Box 4 / GBE letter of 30 March 43 requested a Dodge "carry all" for the King, and Enders requested it be the U.S. Army Air Force's Dodge, diverted to be used by the King.
  - Letter from the State Department to Thayer at Am Embassy, London, 23 December 1943.
  - Engert letter 15 July 1944, says that the rebel leader Abdurrahman, known as "Pak," was next in importance to the Faqir of Ipi.
  - Engert letter 13 Sep 1944, says Mazrak Khan was a rebel in the South, in addition to "Pak" and the Faqir of Ipi. He was known as "Mazrak.

- In 1944, as the new Military Attaché, Ernest Fox requested military training in the U.S. for Afghan Army officers, including medical officers. This request was made after he met with MG Daoud Shah, a senior Afghan military officer. On 26 March 1944 Fox was introduced by Gordon Enders to MG Shah, Commander of the Central Army forces in Kabul, when Enders was saying farewell to General Shah.
- Social Security Death Index. Gordon Enders / SSN 305-42-3155 / born 7 May 1897 / died Sep 1978
- Gordon Enders Obit. "Former Adviser to Chiang Dies," *Albuquerque Journal* (n.d., September 1978).
- "Mead Should Represent People, Voter [Col. Gordon B. Enders] Asserts," *Albuquerque Tribune* (2 February 1963, p. A2.
- Edmundo Carrillo, "'Adventurous spirits' reunited – international tale has a happy ending in New Mexico," *Albuquerque Journal* (9 June 2017). Based on an interview with Patty Hoenigman, a great-niece of Elizabeth (Crump) Enders.
- Patrick J. Hurley Collection, Western History Collection, Oklahoma University Libraries, Box 493C, Folder 4. Gordon Enders with photo of the Panchan Lama.
- Anon., "Flying Gold Out of Tibet: Planes Invade Land of the Lamas," *Modern Mechanix Hobbies Inventions* (November 1936), 76-7, 120 (accessed via Ancestry.com, on a search for Gordon Bandy Enders, 8 August 2011).
- [Enders] *Purdue Exponent* issues in which Gordon Enders is mentioned (pages copied and sent from David M. Hovde, Purdue University hovde@purdue.edu, subsequent to inquiry in e-correspondence by George Hill, 19 May 2009)
- Fox, Jean. "Gordon Enders, Confidant of Chinese Leaders, Visits Shore: Eminent Author Is Foreign Counsellor to Grand Lama of Tibet" Winnetka [Ill.] Talk (21 January 1937), 27. In Hurley Collection, University of Oklahoma Libraries, Box 493C, Folder 4.

Sir Benjamin Gonville Bromhead, 5th Baronet Bromhead of Thurlby Hall

His Ancestral Line[44]

Edward Bromhead, of Thurlby Hall, Auborn, Lincolnshire, England, married Anne Eyre, daughter of Anthony Eyre. They had a son, **Benjamin Bromhead**.

Benjamin Bromhead lived in Lincolnshire; died 21 January 1782. He married Margaret Bordman, daughter of James Bordman. They had seven children. Their eldest son was **Bordman Bromhead.**

Col. Boardman Bromhead was baptized 26 September 1728 at St. Margaret, Lynn, Norfolk, England; died 7 December 1804. He was the third son and fourth child of Benjamin Bromhead and Margaret Boardman. He married, 18 May 1756, at St. Mary Magdalene's, Lincoln, Frances Gonville, daughter of William Gonville. They had the following children:
1. Edward Bromhead (b. 1757).
2. **Lt. Gen. Sir Gonville Bromhead,** 1st Bt. (1758-1822).
3. Ann Bromhead (b. 1760).
4. Frances Bromhead (b. 1762).
5. Elizabeth Bromhead (b. and d. 1764).

Lt.-Gen. Sir Gonville Bromhead, 1st Baronet, was born 20 September 1758; died 18 May 1822. He was the son of Col. Boardman Bromhead and Frances Gonville. He was baptized 30 September 1758 at St. Margaret in the Close, Lincoln, co. Lincoln. He married, 18 July 1787, Jane ffrench, daughter of Sir Charles ffrench, 1st Bt., and Rose Dillon, Baroness ffrench of Castle ffrench. He fought at the Battle of Saratoga, N.Y., in 1777, where he was wounded and captured. He was created 1st Baronet of Thurlby Hall, co. Lincoln, on 19 February 1806. Sir Gonville Bromhead and Jane ffrench had the following children:
1. **Sir Edward ffrench Bromhead,** F.R.S., 2nd Bt. (1789-1855).
2. **Maj. Sir Edmund Gonville Bromhead,** 3rd Bt. (1791-1870).
3. Rev. Charles ffrench Bromhead (1795-1855).

Sir Edward ffrench Bromhead, 2nd Baronet, was born 26 March 1789; died 14 March 1855, unmarried. He succeeded to the title on 18 May 1822. He was educated in 1806-8 at the University of Glasgow and he graduated B.A. from Gonville and Caius College, Cambridge. He was admitted to the Inner Temple to practice as a barrister. He was made a Fellow of the Royal Society (F.R.S.) in 1817 and was High Steward of Lincoln. On his death, being childless, his title passed to his brother.

Major Sir Edmund Gonville Bromhead, 3rd Baronet, was born 22 January 1791; died 25 October 1870. He was the son of Lt. Gen. Sir Gonville Bromhead, 1st Bt., and Jane ffrench. He married Judith Christine Wood, daughter of James Wood, 15 September 1823. He fought at the Battle of Waterloo. He succeeded to the title of 3rd Baronet on 14 March 1855. Edward Gonville Bromhead, 3rd Bt., and Judith Christine Wood had the following children:
1. Frances Judith Bromhead (d. 1917).
2. Victoria Gonville Bromhead (d. 1909).
3. Elizabeth Frances Bromhead (d. 1921).

4. Capt. Edward Bromhead (1832-1869).
5. **Col. Sir Benjamin Parnell Bromhead,** 4th Bt., (1828-1935).
6. Col. Charles James Bromhead (1840-1922).
7. Lt.-Col. Gonville Bromhead, V.C. (1845-1891). As a lieutenant, he was awarded the V.C. for his actions at the Battle of Rourke's Drift. He is buried at Allahabad, India. He was portrayed by Michael Caine in the movie, *Zulu*.

Col. Sir Benjamin Parnell Bromhead, 4th Baronet, was born 22 October 1838; died 31 July 1935. He succeeded to the title on 25 October 1870. He married Hannah Smith, who died in 1902. They had a son, Edward Gonville Bromhead.

Major Edward Gonville Bromhead was born 2 September 1869; died 18 December 1910. He was the eldest son and fourth child of Col. Sir Benjamin Parnell Bromhead, 4th Bt., and Hannah Smith. He married Emily May Hosking, daughter of Edward Hosking, on 8 November 1897. He achieved the rank of major in the 2nd KEO Gurkhas, Indian Army.

Edward Bromhead and Emily May Hosking had the following children:
1. Dorothea Janetta Gonville Bromhead, b. 1898.
2. **Lt.-Col. Sir Benjamin Denis Gonville Bromhead,** 5th Bt. (1900-1981).
3. Lt.-Col. Edmund de Gonville Hosking Bromhead (1903-1976).
4. Anne Marie Gonville Bromhead, b. 1905.

Lt.-Col. Sir Benjamin Denis Gonville Bromhead, 5th Baronet, O.B.E., was born on 7 May 1900, son of Maj. Edward Gonville Bromhead and Emily May Hosking; he died in 1981. He was educated at Wellington College, Berkshire; and at Sandhurst. He fought in the Iraq Campaign in 1920, and in the Waziristan Campaign between 1922 and 1924, in which he was wounded. He was mentioned in despatches in fighting on the North-West Frontier in 1930. He succeeded to the title of 5th Baronet Bromhead, of Thurlby Hall, co. Lincoln, on 31 July 1935; the title was created in 1806. He fought in the Waziristan Campaign in 1937, where he mentioned in despatches. He was Commandant of the Zhob Militia in Baluchistan from 1940 until 1943, and he was invested as an Officer of the Order of the British Empire (O.B.E.) in 1943. He was Assistant Political Agent in the office of the Governor of the North-West Frontier Province from 1943-1945, and he was Political Agent for North Waziristan from February 1945 until the Independence and Partition of India in April 1947. "Benjie" Bromhead married Nancy Mary Lough, daughter of Thomas Seon Lough of Buenos Aires, Argentina, on 6 August 1938. They had three children: Diana Jane Gonville Bromhead, born 20 January 1940; Anne Kathleen Gonville Bromhead, born 1 September 1942; and John Desmond Gonville Bromhead, born 21 December 1943.

"Benjie" Bromhead, Nov 1943
Photo by A.W. Zimmermann

Sir Benjamin and Nancy Mary Lough had the following children:
1. Diana Jane Gonville Bromhead, b. 20 January 1940.
2. Anne Kathleen Gonville Bromhead, b. 1 September 1942.
3. **John Desmond Gonville Bromhead**, b. 21 December 1943.

John Desmond Gonville Bromhead, son of Lt.-Col Sir Benjamin Denis Gonville Bromhead, 5th Baronet, and Nancy Mary Lough, was born 21 December 1943 in Peshawar, North-West Frontier Province, India (now Pakistan). He was educated at Wellington College in Berkshire, England. He succeeded to the title of 6th Baronet in 1981 but does not use his title. In 2003, he lived in Thurlby Hall, Auborn, Lincolnshire, England.

BROMHEAD GENEALOGY
From thePeerage.com

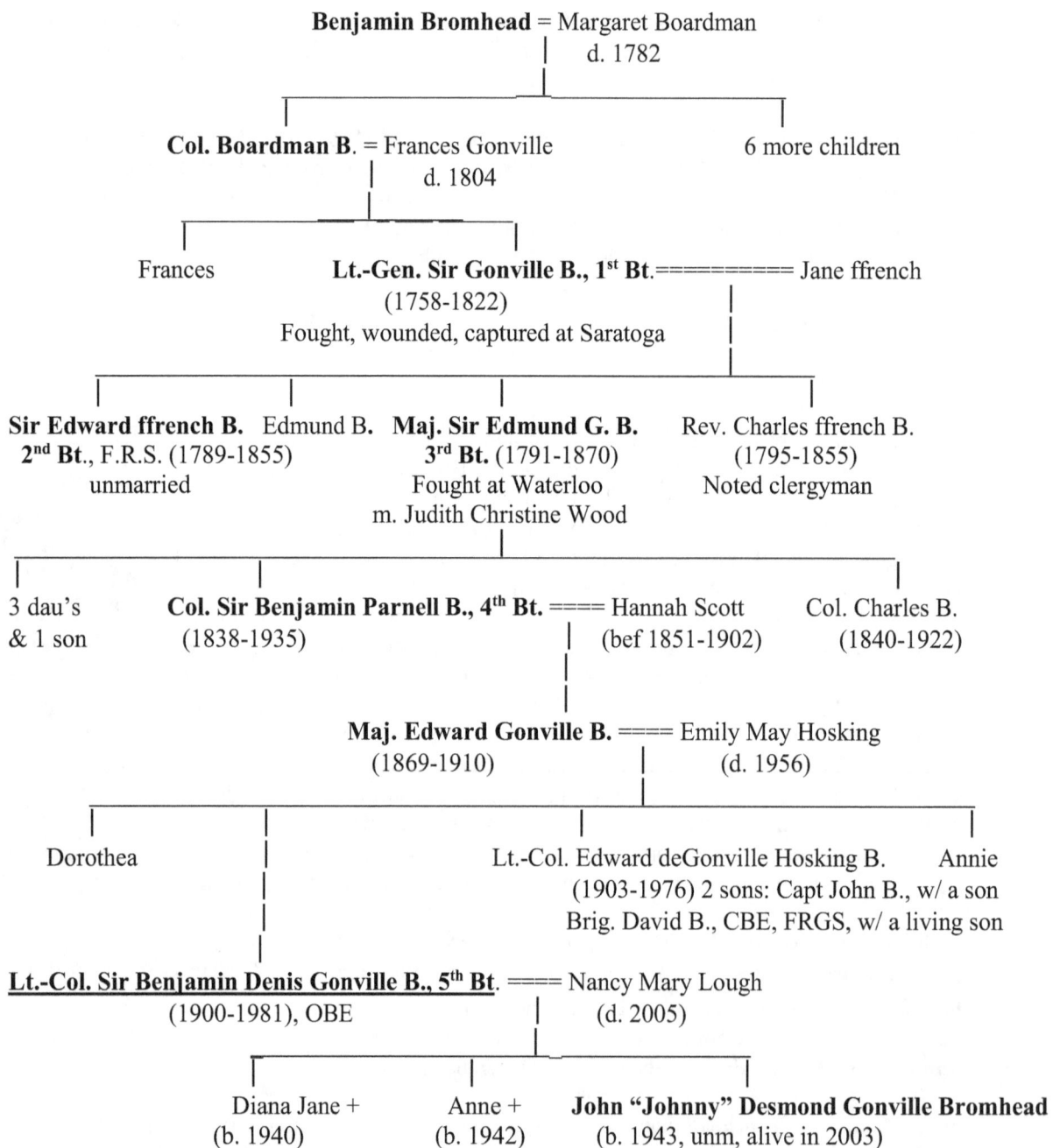

Bromhead Letter File[45]

<u>Transcriptions</u>
BB wrote on AZ's business card (undated):
Major Sir B Bromhead
c/o Grindlay & Co., Ltd / Peshawar
c/o Grindlay & Co., Ltd / 54 Parliament St., London[46]

[BB to AZ, ALS-I]

> The Services Hotel.
> Peshawar.
> January 15. 1944.

Dear Al

Thank you so much for your letter which I should have answered sooner. It was good of you to write when you were suffering from that cold – It seemed a real bad one when I left you in the train, and I was glad to hear that you were getting over it. I found it rather hard to do all I should do in recent weeks, as Nancy suddenly produced her baby, a son, in the early hours of the 21st Dec: and on account of various complications has only just left hospital – so all my spare time has been spent in visiting her. She was terribly distressed at first poor girl as the boy was born with a cleft palette and hare lip – it was a sad anti climax for her – but she has now recovered, as she realizes that the matter can be put right by a good surgeon. The trouble is to find out where the operation can best be done under present circumstances – it's beginning to look as if we shall have to send the child to England, as after three weeks of writing and writing we have had no news of any plastic surgeon out here – and as our nurse wants to go to England in any case this year to see her parents who are old, she will be able to take the boy with her. Nancy will probably go to the Argentine with the two other children [their daughters], as her parents are out there, and there seems no sense in risking the whole family on the journey home. We had a rather quiet Christmas and New /p.2/ Year owing to Nancy being in hospital – but she herself is now alright and the boy seems to be doing well.

I enjoyed our tour together and it was nice of you to be so appreciative – I hope that if ever you want to come north any time again you'll let me know – or anyone else from your Karachi Mess. I've not heard or seen anything of Gordon [Maj. Enders] since Quetta, but I believe he's been down to Delhi recently, as I met Tommy [Enders' sergeant] the other day. I've not had the films from Bombay yet, I hope they've not lost them. Incidentally, I went into Holmes, the photographers, the other day, and they told me that they had been unable to print your pictures that you and Gordon had ordered as they had run out of printing paper – I said that perhaps your office could let them have enough to cover your requirements?

I saw HH [His Highness], the Mehtar of Chitral here the other day and he told me that your H.Q. out here had promised him a jeep – he was delighted, and is going to start improving the Lowari [Pass, from Dir into Chitral]. Well Al, I hope you've quite recovered from the cold and are fit – if you ever want to come up to these parts again please let me know – I'd like to see you again, tho' I don't see much hope of Karachi myself for the moment. I'm off to Delhi on the 3rd or 4th February and then to Sili[?] and Quetta for a few days. Nancy joins me in sending our best wishes and good luck for 1944.

Yours aye, Benjie.

[BB to AZ, ALS-I]

                                                      The Services Hotel.
                                                                  Peshawar.
                                                              January 27th '44.

Dear Al,

    Here's a Mess Bill from the South Waziristan Scouts. I have paid it by M.O. and enclose the receipt, so that you can pay me back whenever you have time. It is the only bill that I have received for any of you, so I imagine that this is all there is. I hope that everything goes well with you. Karachi must be pretty good at this time of the year. We have been having real cold weather, with a good deal of rain, but today it is a clear sky and grand. Do you remember 'Doc' Hassett at Miranshah? He is here on leave, and we are dining with him tonight. He said your visit was very welcome to the lads in the Scouts Mess – they like to meet people from outside – and he hopes you be able to go back one day. 'Father' Wood[?] has also been here, he had drinks with us last night – he has been transferred from Quetta, and is District Officer Frontier Constabulary at Hanger – perhaps you remember a rather small place we passed through between Kohat and Thale. Nancy joins me in sending our best wishes.

Yours sincerely Benjy, P.T.O.

/p.2/ Sorry, my babu has sent yours and Gordon's and Tommy's bills on one M.O. (Money Order no 0007) and your share came to Rs12/5/- (Rupees twelve annas five only). I've sent the M.O. receipt to Gordon as his and Tommy's share came to most.

[NB to AZ, ALS-I]
[Envelope]

[Front] Lt. W. A. Zimmerman, / U.S., Naval Mess, / 254, Ingle Rd. / Karachi
[Back] From. Lady Bromhead / Sevices Hotel / Peshawar / N.W.F.P.
[Letter, on small note paper]

                                                                Services Hotel
                                                                    Peshawar
                                                                 Feb 23rd 1944

Dear Al,
    Thank you ever so much for the lovely shawl, it is very much appreciated & more than useful. [I believe this was the "length of Chitrali putthu" that was mentioned by Enders in his letter to AWZ, that was carried by Thayer to AWZ in December 1943]
    I expect Ben back tomorrow evening, he has been away more than three weeks. I do not suppose that he will be here long.
    We have no plans for the summer yet. Everyone says I'm quite crazy! But how can I plan until we settle up what we are doing about the boy?
    I hope you have good news from home, & that your purchases up this end were a success.
    With renewed thanks,
        Yours sincerely,
            Nancy.

[NB to AZ, ALS-I]

                                                               Services Hotel
                                                            Peshawar. N.W.F.P.
                                                              April 12th 1944

Dear Al,

    I enclose a cutting sent to me by my people. I suppose this was an uncle of yours? I am sorry.

    It was John C. Zimmerman who was my grandmother's friend.

    You will be glad to know that John Desmond is now out of hospital; I wish I could see him before he goes home. [Their son has obviously gotten to England with their nurse and had plastic surgery for his cleft palate and hare lip. The trip was probably made by going to Karachi, and then across the Indian Ocean to the Suez Canal, across the Mediterranean, and through sub-infested waters of the Atlantic Ocean to London.]

    I am busy with the other two, & weary! Anne has just had sand fly fever.

    Ben has been to Delhi lately, then was here about 3 days & is now in D.I.K.[47] until Monday.

    I shall have to wait for him to get your address.

    I hope you have good news from home.

        Yours sincerely

        Nancy Bromhead

[a clipping is enclosed from an unknown paper, probably from Argentina, 2 June 1943, telling of the death on 31 May of John Edward Zimmerman of 25 Summit Street, Chestnut Hill, Pas., eldest son of "John C. Zimmerman of this city" and of the late Anna Cecelia Mackinlay de Zimmermann." They are, however, not relatives of AZ.]

[BB to AZ, ALS-I]

24/4 [4 April 1944]

Dear Al

    I have been meaning to write to thank you for your letter and the cheque – I will do so more fully from Quetta, but in the meantime please accept my apologies.

    If you see Mrs. Dwane Thomas whom you know, would you give her a message to the effect that her 'Ghilzai'[48] dress has not been forgotten and I hope to pick it up in D.I.K. next week and send to her from Quetta.

    Have been abnormally busy – poor old Robbie died the other day.

    Hope you are very fit.

        Yours sincerely,

        Benjy

Telegram from BB to AZ on 22 September 1944, giving him permission to publish the photos on the N.W.F.P. trip; original is in the Scrapbook:

Indian Posts and Telegraphs Department
O PE   PESHAWAR   25 STE 31 LT ALBERT
ZIMMERMANN   254 INGLE ROAD KARACHI =
MANY THANKS YOUR LETTER THAT ALL RIGHT GO AHEAD
HOPE VISIT KARACHI SELF NEXT MONTH WRITING FROM
KASHMIR BEST WISHES = MAJOR BROMHEAD

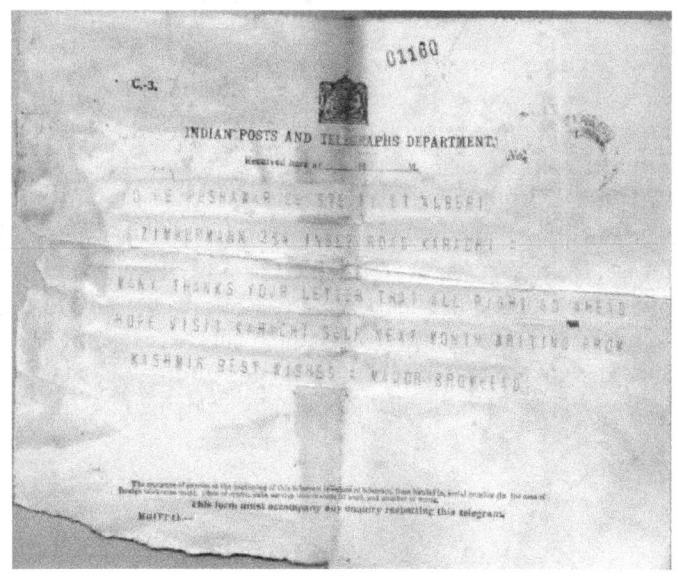

Additional Notes, based on AWZ's letters to his wife

24 November 1943 – The Bromheads had dinner in Peshawar with AWZ and others with the deputy commissioner of the NWFP, Khan Sahib, and his wife, at the Services Hotel. Bromhead (BB) was probably in a khaki uniform with 5 ribbons on his blouse: The baronet ribbon, OBE, and three campaign medals, with clasps. Two of them also have oak leaf clusters. His wife is very pregnant; she was delivered of a child on 21 December. Enders is not there, and it is unknown where he is.

25 November 1943 – The Bromheads had dinner with AWZ and other at Khan Sahib's house. Dress is probably the same for BB, but his wife, Lady Bromhead, would be wearing a different frock than the previous night. Enders is not there, and it is unknown where he is.

26 November 1943 – The Bromheads and others have dinner as guests of AWZ at Dean's Hotel. Bromhead would be in civilian clothes, and his wife would also have on a different frock. Enders is not there, and it is unknown where he is.

27 November 1943 (Saturday). After a day in which BB took AWZ to the Khyber Pass (AWZ in his khaki uniform, same as on the trip to Chitral; BB in civvies), they return to Peshawar, change clothes, and go to Government House for the garden party with the Viceroy. BB is probably in mess dress, with red jacket, miniature or full-sized medals, neck ribbon for the Baronet emblem. Lady Bromhead wears perfect afternoon cocktail attaire. Enders is wearing a khaki uniform with blouse (coat) and tie, with the U.S. Army's Aviator's Badge. He has four inverted chevrons on his left sleeve for 48 months of overseas service in WWI, and at least three bars on his lower sleeve for 36 months of overseas service in WWII. He may also be wearing a wound chevron on his right lower sleeve. His other ribbons include the WWI Victory ribbon with four stars, WWI Occupation ribbon, American Defense Ribbon, Asia-Pacific Theatre Ribbon, European-Middle East Theatre Ribbon, and French Legion of Honor.

The Bromheads go to the Peshawar Club with AWZ to a formal dinner dance – also very fancy – although AWZ wears the same Navy blue uniform (white shirt, black tie) that he wore to the garden party. He has no service ribbons at this time, although he was entitled to wear two: American Theatre, and Asia-Pacific Theatre. Enders is not there.

24 September 1944 – BB gives permission by telegram to AWZ to publish his article about the Trip. The telegram was saved, and photos were sold to the *National Geographic*, but the article was never completed or published.

A Later Report about the Bromhead Family

A correction to *Proceed to Peshawar* (page 111) was kindly sent to me in a letter from Miss Romilly Greer Leeper, now of London, on 29 June 2014. Miss Leeper is the daughter of the "Mrs. Liepur" (which I transcribed incorrectly) who was the host for Sir Benjamin and Lady Nancy Bromhead and LT Albert Zimmermann on two occasions when he was in Peshawar in November 1943. Mrs. Leeper of Pesahwar was Mrs. Dorothea Evelyn Letitia Leeper (née Lloyd), who was at that time married to Lt. Col. William Cluff Leeper of the Indian Political Service. Miss Romilly Leeper says that her birth certificate shows him as Home Secretary North West Frontier Province. After the Independence and Partition of India in 1947, Mrs. Leeper and her daughter Romilly returned to England. Mrs. Leeper worked as a secretary in London, and later remarried. In the meantime, Romilly lived until the age of nine at Thurlby Hall, and shared a governess with the Bromhead children: Diana-Jane, Anne and Johnny. She added that "I lost touch with the Bromheads many many years ago but am everlastingly grateful to them for taking me in."

MEDALS AND DECORATIONS OF SIR BENJAMIN BROMHEAD, 5th BARONET

The medal of a baronet is worn on a neck ribbon,
depicting the Red Hand of Ulster,
for English, Irish, GB and UK baronets.

Officer of the Order of the British Empire
(awarded to Benjamin Bromhead, 5th Bt., in 1943)

General Service Medal (1918)
With clasp for Iraq Campaign 1920

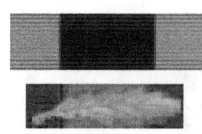

India General Service Medal
(1908-1935)
With clasp for Waziristan Campaign 1921-24, Wounded
With Clasp for North-West Frontier 1930
Mentioned in despatches, so entitled to oak leaf cluster

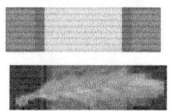

India General Service Medal (1936)
With clasp for Waziristan Campaign 1937
Mentioned in despatches, so entitled to oak leaf cluster

IMAGES ASSOCIATED WITH BROMHEAD AND GONVILLE FAMILIES

Thurlby Hall, seat of the Bromheads

Google Images

THURLBY HALL

Lawn at Thurlby Hall

(From Google Images -> Thurlby Hall -> Drakes Family)

Thurlby Hall is no longer the residence of the Bromheads

"Admiral [sic: Maj. Sir Edward Gonville] Bromhead [3rd Bt.], the father of Gonville Bromhead, VC, whose part in the film *Zulu* was played by Michael Caine. The Bromhead family later moved to their London home and I understand that they leased Thurlby Hall to Frank's father, John Thomas Drakes . . . it is possible that, as a tenant farmer, John Thomas Drakes and his family actually lived in the adjoining cottage … Thurlby Hall was the second home of John Thomas Drakes (1851-1908) and his wife Alice Martha Wildsmith (1852-1920). It was where their three younger children were born: Alice Margaret Drakes (1883-1926), Sidney (1885-1957), and Frank (1889-1957)."

It appears that Thurlby Hall is now for sale by Savills, which shows interior views:

(From Google Images -> Thurlby Hall -> Savill)

Sir Benjamin Bromhead's Schools in England

Wellington College – One of England's oldest and most prestigious secondary schools

Sandhurst – the Royal Military Academy

Google Images

Correspondence with Curtin Winsor and others at ONI – 1943-1945

Lieutenant Curtin Winsor, USNR, was a Desk Officer in the Office of Naval Intelligence (ONI) – Far Eastern Division. He was the point-of-contact for Lieutenant Albert Zimmermann, USNR, and other the Naval Liaison Officers in India in 1943-1945. The ONI was in the Office of the Chief of Naval Operations, known as OPNAV. It was OP-16. The Far Eastern Division (aka Far Eastern "desk") was referred to as "FE," so Winsor's division or "desk" was OP-16-FE. The Naval Liaison Officers were referred to as ALUSLOs, or as Aluslos.

<u>Abbreviations and Notes</u>

AL = Autograph letter, unsigned (a handwritten letter or document, unsigned)
ALS-I = Autograph letter in ink, signed (i.e., handwritten, in ink)
ALS-P = Autograph letter in pencil, signed
ALUSLO = U.S. Naval Liaison Officer. The Senior ALUSLO for India-Burma Theatre
    was based at New Delhi, India.
AZ = LT Albert W. Zimmermann, USNR
BSZ = Barbara Shoemaker Zimmermann (Mrs. Albert W. Zimmermann)
CW = LT Curtin Winsor, USNR[49]
JANIS = JANIS (Joint Army Navy Intelligence Studies)
JICA = Joint Intelligence Collection Agency[50]
NLO = Naval Liaison Office or Naval Liaison Officer
ONI = Office of Naval Intelligence[51]

SACSEA = Supreme Allied Command Southeast Asia. Mountbatten's HQ, originally in New Delhi, relocated to Kandy, Ceylon, in October-November 1943.
SEAC = Southeast Asia Command (equivalent to SACSEA), but derisorily referred to by some Americans as meaning "Save England's Asia Colonies"
SEAT = Southeast Asia Theatre (also India-Burma Theatre [IBT], or India-Burma-Ceylon [IBC])

Contents of this file

15 Sep 43	CW to AZ [all but one of CW's letters were written from ONI]
25 Sep 43	Col. Off to AZ [Off is at ONI]
15 Oct 43	CW to AZ
9 Dec 43	CW to AZ, written when CW was at the office of JICA, New Delhi
21 Feb 44	CW to AZ
21 Feb 44	CW to AZ [a copy of original; receipt greatly delayed; see notes]
6 Apr 44	CW to AZ
21 Apr 44	CW to AZ
23 May 44	CW to AZ
28 July 44	CW to AZ
22 Aug 44	Commodore Milton Miles (probably) to Director of Naval Intelligence [DNI] (probably) [copy w/o name of originator or of recipient; comments about CW's letter of 21 Feb 44]
12 Sep 44	CW to AZ [permission to publish in *Life*]
16 Oct 44	CW to AZ
20 Nov 44	CW to AZ
19 Jan 45	CW to AZ
23 Jan 45	CAPT Habecker at SACSEA to AZ and Voorhees
9 Mar 45	CW to AZ [written at ALUSLO, Delhi]
9 Mar 45	CW to Col. Bales at ONI (unsigned blind copy to AZ) [written at Delhi]

Winsor's usage of capitals and lower case letters is very irregular. I have attempted to preserve his usage as well as possible, but it is also a bit difficult to decipher. I have lightly edited the punctuation and spelling when I think he intended something else but simply failed to do it.

[CW to AZ   ALS-I]

2712 35th Pl. NW.   15 September [1943]

dear Al:

It's about time I wrote you a few lines. I have enjoyed reading your first report and compliment you on them.[52] I hope you like your job and "ain't let the climate git you" yet.

We have just finished the hottest dryest summer Washington ever had – 53 days with the thermometer in the 90's. but since Labor Day it has been almost cold and everyone has colds as a result.

I now have my children with me and my wife here and we are holding down a small house out "Mass." Ave. way (near "Wis.").[53] Morris Duane[54] lives with his family right behind us. Henry Pemberton is coming as a boarder when my children leave.[55]

Jim is at an Air Transportation Training School at Fort Slocum [Long Island], while sick and tired of training, he feels that he will have a real job when he gets out in a field which might offer much after the War. Sal goes up to see him on week-ends.[56]

/p.2/ ONI staggers along. I could write much about JICA, JANIS and other alphabetical improvements but I guess it's not my cue to do so. I haven't been home for two months and know nothing of your family. I did phone Barbara as soon as I heard of your arrival.[57]

I still toy with the idea of coming out, basicly to Delhi, but with a new little wife and the uncertain position of my kids etc., I jist don't think I should.

I lunched with the Duke last week. He likes his Ordnance job. Briggs now heads a Wave Training School.[58]

In connection with issuing India Port Summaries I'd like to compose an appropriate limerick as a dedication to each summary. Bombay and Madras would be easy and Rangoon has several ditties built around it, but have you any rhyme scheme for Karachi? And Vizafatan[?] throws me for a loss.[59]

Drop me a line and if there's any thin I can do at this end, let me know.

Curt

[TLS]

NAVY DEPARTMENT
OFFICE OF THE CHIEF OF NAVAL OPERATIONS
WASHINGTON

25 September 1943

My dear Lieutenant:

I received your letter dated July 29 some time ago and have been unable to answer it due to the press of circumstances. It is a pleasure to know that you reached your destination safely and that the situation is as favorable as I assume it to be. We are sending out at a very early date Lieut. (jg) Howard Voorhees[60] who will be the relief for Lieut. (jg) Burns. I trust that by this time Lieut. Browning has been detached without relief although we will send a replacement at an early date.

Regarding any future assignment, I believe it would be best to wait and see what develops. You will be there for at least 18 months unless you become ill or are transferred for any other reason. Naturally, it is impossible to even estimate where the need for you will arise after that time. Please give my very best to everyone on post.

Sincerely yours,
[s] Clarence Off[61]

Lieut. Albert W. Zimmerman, [sic]
Office of the U.S. Naval Liaison Officer,
Karachi, India.

[ALS-I]

**NAVY DEPARTMENT
OFFICE OF THE CHIEF OF NAVAL OPERATIONS
WASHINGTON**

15 Oct [1943]

dear Al:

Lt. Voorhees is taking this with him. Thanks for your nice letter. I haven't seen Barbara but Sally tells me she is well although lonesome.

Jim is very happy with his Air Transport job in N.Y. City. I expect to be making a trip as courier next month and will linger at the major ports to get some first hand impressions. I should be due your way before too long perhaps. As I get horribly air sick I'll probably have a puking good time.

We're holding up a revision of our Port Summary till your new stuff arrives.

There's little news since I last wrote. The papers announce that B29's with 1/3 increased range and twice the size of our Liberators are in Mass Production. Penn has a damn good football team and has cleaned up on Yale and Princeton. Bill Bullitt is not expected to beat Barney Samuels in the Phila mayoralty campaign.[62]

Best –   Curt

[ALS-P]

[no header]

JICA   New Delhi   9 Dec [1943]

Al:

Sorry to have missed you coming in. Barbara and your kids are fine and send their love. You're going to kill me when you read that she gave me some letters for you and I forgot them! However, I've sent a dispatch to Washington asking that they be forwarded to you be the next pouch. I have a personal liquid token of my esteem for you.

I should hit Karachi <u>Friday, Jan 7th</u> – for a couple of days before leaving for home and want to talk to you then about many things concerning you and others.[63] Try to reserve some time for me these days so I can see you alone. Perhaps, you can arrange to show me the port facilities in person.

I'm off on the 12th on a general tour of the area. If you want to write me do so c/o JICA and they will forward. Brother James is studying for CuMgot[?] at Northwestern University and may be sent overseas the end of January.   Merry Xmas,

Curt

[ALS-I]

**NAVY DEPARTMENT
OFFICE OF THE CHIEF OF NAVAL OPERATIONS
WASHINGTON**

2/21 [1944]

dear Al:

You will be #1 as things stand now and it looks pretty definite. I didn't recommend this in so many words for fear I might be accused of personal favoritism. Sadler, however, did so recommend. Naturally I'm delighted. I don't yet know whether you'll get a spot promotion although this I did recommend when it had apparently been decided to give you the job. I should say the chances were good.[229]

---

[229] AZ was recommended to become Commanding Officer of Naval Liaison Office Karachi, and this appointment eventually came through. He was not, however, promoted to LCDR at this time, although he achieved that rank as a drilling Reservist after the war was over.

Cdr. S. will be recalled without an accounting at Karachi but his vouchers here have been carefully checked and he has been directed (1) to submit a certified inventory property account. He's also been directed to (2) eliminate counter intelligence activities and discharge Sheikh. (This means discharge Mme. Dubash too but they didn't want to sully the lady's name by mentioning her.) (3) To cut down the style of operation of the mess. All this by wire.[64]

You will not be appointed till after he's left and it may take time to get his orders thru.

Sadler said you wondered whether you are confirmed to RAF for courier service to Colombo. This is correct. If you want to use Tata or Quantas [sic] write or wire setting forth need, and that these lines are safe and we'll authorize. Back it up with facts and figures.

Tell Voorhees they wanted to transfer him to Aden but I talked them out of it. This was a compliment because Aden is busier.

Best o' luck — Curt

[ALS-I][65]

N D
O o CNO

21 Feb. [1944]

dear Al

By now the bombshell from S and O. should have burst over Karachi and if so I have no doubt you are all concerned about the docking of pay. I was merely a cog in the workings of the machine that produced the bomb. The explosion would have come regardless of me or Sadler.

<u>It has been suggested that all of you junior officers probably boarded at the mess under oral or written orders.</u>

<u>I believe Cdr. Off is writing Voorhees to the same effect.</u> [underlining in original]

I gather your couriers will go to Bombay. The change [to be Commanding Officer at Karachi] topside mentioned in my last note is definite. Congrats and good luck. Probably no spot [promotion] at least for some time. /p.2/ Jim is believed to have gone over.

All well here and our new CO is fine. I now handle the affairs of the lands of kangaroos and of 18th rainbow trout in addition to my other areas and two assistants have been given to me, Castle being one of them. I am busier than ever.

I hope to have a skit I wrote on "They Joys of Wartime Air Travel" published. If so will send copy. Sorry that I can't make this more explicit for obvious reasons but read it carefully.

C W

NAVY DEPARTMENT
  WASHINGTON DC
   ~~OFFICIAL BUSINESS~~
     <u>Personal</u>

                          Lt. Albert Zimmermann
                             Aluslo Karachi   (Censor mark)

Deliver <u>personally</u> to Lt   (& initials)
Zimmermann or Lt (jg) Voorhees <u>only</u>
[underlining in original]

[ALS-I]

NAVY DEPARTMENT
OFFICE OF THE CHIEF OF NAVAL OPERATIONS
WASHINGTON

6 April [1944]

dear Al:

I was hoping to hear from you via Leavitt but guess you had your reasons. A letter is en route (or should be shortly) designed to clarify thing by directing:
(1) Continuance of mess in present quarters but at reduced scale
(2) Making of new separate leases for mess + office space.
(3) Use of furniture by mess free.
(4) Discontinuance of orders at airport directing all Navy personnel to come to mess.
(5) Check aged officers less credit for rent paid for quarters + furniture. They don't want the C.O. to assume this because officers signed statements that govt quarters not available in asking for rental allowances.

    This is all the doing of A. branch. If I'd had my way the C.O. would have been quietly transferred weeks ago without all this delay and fuss which must be far from pleasant to endure for you and Howard [Voorhees]. However, orders recalling him should be on their way unless Bu Pers [Bureau of Naval Personnel (BUPERS)] has gone to sleep entirely.

    All well here. Castle helps me and believe I need his aid. A revised summary on Karachi is in the works and with the good reports you and Howard sent in should be one of our best.

    I understand full Lts. of 1½ yrs. standing become eligible for promotion, but this is just scuttle.
/p.2/
    Jim writes cheerfully from England. He is apparently near Oxford. Sal bears up fairly well.

    The scuttle is that once a month an officer in the India area will be sent back as courier as was Thomas, so you may get a trip back some time in the next 12 months. I imagine Markey[66] decides who goes. The colonel has hinted that I might go out again in the fall on another junket.

    Best of luck. Dewey has the nomination now but I doubt the election. Show this to Howard if you deem it advisable and give him my best.

                               Curt

P.S. Col. Bales is working hard to try to eliminate <u>any</u> pay checkage which was Bu Per's idea anyhow, not ours. I'll let you know if there are any changes.

[ALS-I]

NAVY DEPARTMENT
OFFICE OF THE CHIEF OF NAVAL OPERATIONS
WASHINGTON

21 April [1944]

    dear Al:
Sid. Sweet[67] the bearer of this is a close friend of mine in Fe. You can talk to him about <u>anything</u> that you would discuss with me if I were in his shoes and I hope you will spill everything that's on your mind. The A. branch has bungled thing – I can't help the delay in recall etc. but if there's anything I <u>can</u> do let Sweet know. Brother James is in S.W. England attached to the British Army now. Sal is pretty down.
        Best,
                               Curt
P.S. If because of the [unnamed] disaster at Bombay you need additional officers to handle the increase in activity at Karachi I would suggest working it out with Curren and sending dispatch to us. Curren is about to get Potter, and extra officer, from Washington.
[AZ wrote on this in pencil: "Ans May 25"]

[ALS-I]

NAVY DEPARTMENT
OFFICE OF THE CHIEF OF NAVAL OPERATIONS
WASHINGTON

23 May [1944]

dear Al:

Thank God you're rid of him at last. I hope now you get the responsibility and opportunities that you desire. Off tells me he will put in a strong letter for you soon on a regular not a spot promotion to counteract any thing he may have put in your fitness report. You can count on me to back it up with a good recommendation from the Colonel.[68]

If you want a trip home as a courier I suggest you contact Markey.[69] There is supposed to be one trip a month open to the boys in the area and while others who have been out longer will no doubt have precedence you ought to get in on it eventually.

Jim likes England but says he ain't done nothing yet and doesn't ever expect a raise. He thinks his work will be dangerous and hectic after the big push.

Give my best to Howard and let me know if there's anything I can do. We would like you as a long range project to send us material on 604, 100 to 600 of the Index Guide for your area. You may have to be careful about getting this dope for fear of misunderstandings, but it might prove useful if the area were some day to fall into enemy hands.[70]

Curt

[TLS]

NAVY DEPARTMENT
OFFICE OF THE CHIEF OF NAVAL OPERATIONS
WASHINGTON
28 July 1944

Dear Al:

This will introduce Lt. (j.g.) Marshall Green[71] of Op-16-FE Jap shipping desk, a good friend of mine. Anything you can do to make him comfortable will be appreciated. Tell him anything you want to pass on to me or to others on this side.

I suggest that you initiate (since you are your own C.O.) a request for consideration for your own promotion, citing relevant dates, length of service in grade, nature of work, etc. Babson did this recently. From what Off tells me, promotions are slow and not too likely because ONI is top heavy with rank, but without such a request your chances may be nil.

Capt. Markey is not returning to India and his successor has not been named so I don't know who will determine courier trips home in his absence. However, Capt. Baltazzi tells me arrangements "have been made" so that Aluslos can return for leave on these trips. I suggest you write a letter to JICA-CBI asking to go on one.[72]

Jim writes cheerily from Normandy. Best to Howard.

[s] Curt
Curt Winsor

The next letter was probably written by Commodore Milton Miles, USN (see Note on Winsor's letter to DNI, RADM Schuirmann, which Winsor enclosed to Col. Bales on 9 Mar 1945). The addressee is not named, and there is no signature on it, so I can only speculate that this letter was from Miles to Schuirmann. I believe the letter is not the original document; it appears to be in the handwriting of AZ (see my comments in fn to CW's letter to AZ on 21 February 1944, and my comments on CW's letter to Col. W. L. Bales, 9 March 1945).

The writer says he has "met Z & V," but he does not say when or where. It was probably when he was the "commodore" who AZ wrote about to BSC, saying he passed through NLO Karachi on 31 Jan-1 Feb 44.

[AL-unsigned]
22 Aug. 1944

If I am right, I think you may have to [fire?] somebody. If you will read the photoed [letter] a few times and take into account what has been going on, I think you will agree that there's skull-duggery about.

I don't know Winsor. I have met Z & V [Zimmermann & Voorhees]. My personal opinion of Z is that he is a very nice fellow personally but he does not have much on the ball either from an administrative or naval standpoint. However I don't know much about him and withhold judgement on him because I might do him an injustice.

The original letter was delivered to a naval officer at the airport at Karachi in the latter part of Feb. It was delivered by a USA Airforce pilot with a request that it be put into the hands of Z or V only and that it not be opened by any other officer because of the private nature of the matter included therein. If you really wanted to know who opened it I could tell you, but I prefer not to. It was entirely legal.

Z + V were particular good friends and buddies of Winsors. How many others are involved I don't know. How many other letters went back and forth between W, V + Z by the hands of pilots + passengers is not known but they are attempts to /p.2/ defeat official mail & censorship.

Apart from the first W may have abused the trust placed in his official position as Col Barnes representative I personally believe he must have distorted the facts of the Karachi situation to suit his own ends. It is apparent from his part that he was instrumental in getting Z assigned as CO. He apologizes for not getting Z a spot promotion. In any case that sort of clandestine correspondence between a desk man in ONI and the junior officers of a LO is highly irregular and is subversive to discipline and the proper relations which should exist in any command. The whole thing is rather fantastic and I cannot believe it happened right under our own eyes in the U.S.N.

I have nothing more to say about this except that I would recommend that you put your espionage to work on this particular little job. It is not really my business and I don't want to do anybody an injustice but I do think Smith has had an injustice done to him. That is why I have spread myself at considerable length in my forwarding endorsement to his unsatisfactory fitness report and in this note to you.

[ALS-I]

NAVY DEPARTMENT
OFFICE OF THE CHIEF OF NAVAL OPERATIONS
WASHINGTON

12 Sept. [1944]

dear Al:

I see no objection to your sending in your photos + an article to Life. I'll be glad to clear it with the Public Relations Office here. Col. Bales concurs.

We've written a stiff letter to JICA-CBI directing them to let you fellows come home. Cdr. Ladd who probably had just reported at Delhi as acting C.O. will probably take more active steps in this line than O'Connor so keep your oar in.

Jim still was doing nothing in Normandy as of Aug. 25 but he may be now. Cdr. Covington spoke well of you when he returned. Sorry to hear Howard [Voorhees] has been ill. Capt. Collins who will take Capt. Markey's place is an older man formerly on duty in Australia.

Best,

Curt

[ALS-I]

NAVY DEPARTMENT
OFFICE OF THE CHIEF OF NAVAL OPERATIONS
WASHINGTON

16 Oct. [1944]

dear Al:

I saw Barbara briefly yesterday and talked to her on the phone. Barbara is disappointed, of course, that they won't let you back now but she took it bravely. I did everything I could, Al, even to the extent of sticking my neck out and getting my ears pinned back, to get you home for Xmas. The front office, however, will not hear of leaving only one officer on a post.

I'm going to try to persuade Capt. Balthazzi [sic: it is typed Baltazzi in the next letter] to let you and Arthur Babson arrange for temporary duty at your posts (while you and he come over) by officers from Bombay or JICA. You'd better not do anything about this till I let you know. Meantime – I'll try to talk to Balthazzi at a favorable moment.

I wouldn't count on your return till mid winter, however. As for your coming back permanently, I should imagine they'd pull you, if you request it, before midsummer. I'd write Bowen[73] off, if I were you. I could probably get you a job here in Washington with FE if you wanted it, & by then or before. Let me know and I'll advise you of openings.

Bob Faye is replacing Castle as my assistant. He's O.K. Delaplaine is also coming into Fe – probably N.E.I. desk.

The whole set up out your way may be shaken at the top. A certain Commodore is reaching out his tentacles I believe, but we hope some Captain will be sent out from here to replace Gene[74] /p.2/ Your beloved former CO is scrapping with Benk. in C. [Ceylon] and this doormat of an organization we slave for won't stand up for its officers.[75]

Jim is still doing nothing constructive overseas.

If you see anything wrong with our K. summary advise us.

Best to Howard. How is he?

            Curt

[TLS]

NAVY DEPARTMENT
OFFICE OF THE CHIEF OF NAVAL OPERATIONS
WASHINGTON

20 November 1944

Dear Al:

I have your letter of 12 October and we have likewise received your request for change of duty. As soon as I have an officer at had whom I can recommend to take over in that area, we will start him on his way. I hope that it can be arranged that you may be back in this country by Christmas but frankly I cannot hold out much hope along those lines. Captain Baltazzi feels that the officer at Karachi as at the other posts should be the rank of lieutenant commander and at the present moment we have no one at hand for that particular duty.

As you probably know, a regular Navy captain has been appointed Senior Naval Officer, India-Burma-Ceylon and has reported in today in preparation for assuming his duties. He will have full authority over all naval groups in India. This should clear up many situations in that part of the world.

It was unfortunate that you did not get your courier trip in but I believe things will work out better in this way as we would have relieved you at the end of eighteen months anyway. There is nobody that feels you have been moping around or bemoaning your fate. Everybody appreciates to the fullest the situation of officers on foreign station and are only too happy to do all in their power to bring about the best possible conditions. This courier trip business will be straightened when Captain Davis reaches there inasmuch as he will have authority to shift officers from Jica and between the posts in order to better the situation in any way he believes to be in the best interests of the naval service.

Please give my very best to Howard and warmest personal regards to yourself.

            Sincerely,
            [s] Bowen

Lieut. A. W. Zimmerman [sic],
Naval Liaison Officer,
Karachi, India.

AL-P [in Winsor's hand]

Office Memorandum – UNITED STATES GOVERNMENT

TO: LT. ZIMMERMANN
U S Naval Liaison Officer   KARACHI, India
FROM: Lt Winsor - FE
SUBJECT:

Tell Zimmermann he can expect to be relieved on or about 1 March by Lt Cdr. Henry Groman, formerly with JICA Chungking. His orders call for his getting 25 days leave upon his return and then to go on duty at Port Director's School in N.y. City – a good school. Course lasts about 9 weeks. Can live where he wants. Thereafter foreign duty again, probably. Lt. Winsor has notified Mrs. Zimmermann.

[ALS-I]

NAVY DEPARTMENT
OFFICE OF THE CHIEF OF NAVAL OPERATIONS
WASHINGTON

19 Jan [1945]

dear Al:

I talked with Barbara on the phone today and also with Henry Groman. the latter now plans to leave here about 10 Feb. Barbara says you don't want to make any decisions till you return. She asked if you could split your leave and save some of it till the Summer and I said I though this could be arranged.[76]

If you want to avoid going out again on foreign service I advise you strongly to line up something definite elsewhere i.e. with the D.I.O. 3ᵈ N.D. if you want to stay in N.y. or with FE if you want to stay in Washington.

Both Capt. Davis + Col. Bales want me to go overseas as a courier leaving about 1 or 15 March for about two months. I need a replacement to work with Bob Rye while I am gone. (Incidentally, he is a hell of a nice guy). If you would be interested in having this two month temporary duty in Washington on my job while you line up a permanent berth for yourself let me know promptly because Eliot Norquist may wish to fill my shoes if it can be arranged.

Jim at last has his promotion.

Best of luck,

Curt

[no header]

[ALS-I]

Hdq. Sacsea, A.P.O. #432
Jan 23ʳᵈ [1945]

Dear Zimmermann and Voorhees:

My attempt of last Thursday was successful and at the risk of creating an international incident, I must admit that the "York" is the closest approach that I've had to pre-war flying – and that is based on almost twenty five thousand miles of air travel. Even lunch was served, although the sandwiches from your kitchen far surpassed those offered on the plane. The seven and a half hour trip was the least boring of any.

You were all most kind in providing arrival and departure transportation at all hours, to say nothing of various shopping trips and help in preparing for the wilds of Ceylon. I am also most grateful for the hospitality of your mess during my lay-over. That is probably my last memory of real luxury.

I am quartered at the Queen's Hotel,[77] in a room with private plumbing, which is a compromise to make up for the lousy food. Luckily, there is an America Army mess where I can take an occasional meal, but I anticipate no trouble in keeping my figure.

Yesterday I meat Peachie Durand and delivered the message, which was duly appreciated. She is one of many attractive girls in this immediate theatre.

If any of you have any occasion to make a trip this way, I hope you will let me know ahead of time. We aren't as palatial as the place to which you have become accustomed but you can always count on me doing my best to make you comfortable.

Again, many thanks for your courtesy and kindness and my best wishes to you both and to Halla.
    Sincerely,
    Habecker[78]

P.S. Captain Linaweaver[79] will probably arrive Karachi via R.A.R. plane, either Tues (30th) or Wed. (31st). Huntingdon will confirm by wire.

[no header, on ruled tablet paper]   [ALS-I]

A.P.O. 885.   9 Mar. [1945]

    dear Al:

I enclose a copy of a letter I wrote to Col. Bales on the advise [sic] of Capt. Davis. The Captain said he had talked to you at length about the mess etc. but that he could only get you to answer direct questions. While I can well understand your reticence, I do think that the Captain is the person to whom one should tell the whole story since it is his specific duty and he has orders from CNO to get to the bottom of it. Perhaps you feel that you have told him everything but he seemed to think you had not.

I'm very sorry on your account to hear that Groman isn't coming.[80] He was certainly all lined up to come when I left. I hear that the Captain has requested orders for Lt Cdr O'Connor now in Calcutta to replace you. This shouldn't take too long. I hope to drop in at Karachi about the 5th or 7th of April for a day or so but will advise you later. If you wish to reach me I'll be at Colombo from the 16th to the 24th, then Madras, then Delhi. I had an interesting week in Calcutta and got down to Akyab. I didn't see Smith. He's laid up in the Hospital with a bad case of Dengue and pneumonia.

    Best regards to you and Howard.
    Curt

This letter is an unsigned blind carbon copy of Winsor's letter to his supervisor, Col. W. L. Bales, at ONI. The letter is marked in CW's hand in pencil: "For Zim." Winsor passed it on to AZ as an unmentioned enclosure with his ALS-I of 9 March 1945 (supra). In this letter, CW attempts to rebut the statements that were made in an unsigned AL of 22 August 1944 (cf), which CW says was written by Commodore Milton Miles to the DNI, RADM Roscoe E. Shuirmann, in support of LCDR F. Howard Smith, who was the previous CO at NLO Karachi. The matter of LCDR Smith and the Officers Mess in Karachi may have even come to the attention of the CNO, for Winsor wrote to AZ on 9 March 1945 to say that CAPT Davis, the new Senior Naval Officer in IBT "has orders from CNO to get to the bottom of it."

In this letter of 9 March to Col. Bales, CW tries hard to cover his tracks, and to show that he acted entirely in accordance with standard Navy policies. He was probably dissembling, and the senior officers who read this letter would likely have doubted some of his protestations. Commodore Miles thought Winsor should be fired for favoritism to his friends Zimmermann and Voorhees, defaming Smith, and communicating outside of Navy channels. Whether or not this letter to Bales was accepted as being complete and accurate is unknown, but he apparently survived this crisis and was eventually promoted to LCDR. He had a distinguished career in post-war civilian life. AZ wrote his last letter home from Karachi on 31 March 1945, one day after his relief arrived and began to get oriented. AZ returned to America in April 1945.

[TL-unsigned, probably typed by a yeoman; by context, it can be seen that it was typed in Delhi]

9 March 1945

Colonel W. L. Bales
Room 4625
Navy Department
Washington, D. C.

Dear Colonel:

I have been to Karachi, Delhi, Calcutta, and Akyab so for [sic] and believe I've been able to help clarify the picture of what and of what we do and do not want in the way of reports. I've also been of some value, I think, to Captain Davis who certainly has taken hold of all the loose strings and pulling them together into a coordinated, unified outfit. Both he and Castle are now well. I saw quite a bit of Eddie O'Connor[81] at Calcutta and hope he gets the Karachi job, but pursuant to your instructions I have steered my course away from all matters of administration and personnel.

There is one matter which I believe I should call to your attention pertaining to the Smith affair at Karachi. When I made my trip last year, Colonel Boone wrote for me a letter of authority in which he designated me as his "personal representative," in just those words, to take up all matters affecting FE with the several posts out there. Colonel Boone, A4B, 16F, and Captain Green (ALUSLO Eastern Fleet at the time), directed me to inquire into the mess situation at Karachi concerning which reports had been coming in from officers returning from duty in the theaters and couriers. You saw my report in which Lt. Comdr. Sadler "heartily concurred" and for which Captain Schrader praised me in writing.

The junior officers in Karachi were in a dither about themes for which they expected to be checked on the pay and they weren't getting along with their C.O. So after my return I wrote them a latter [sic] explaining that a change would be made top side, that Zimmermann would probably take over and that their pay would be checked unless they could show that they had been ordered by Comdr. Smith to live and eat at the mess. To write such a personal letter may have been indiscreet but Comdr Off and Lt. Higgins and Comdr. Sadler thought someone should advise them and I volunteered. This letter was given to an Army courier officer in late February with specific instructions to deliver it personally to either Lts. Zimmermann or Voorhees. It was marked on the outside of the envelope "Personal, to be delivered personally to Lt. Zimmermann or Voohees only." <u>This letter was never received by either of these officers.</u> It was intercepted and opened by Comdr. Smith, photostated by him and given to Commodore Miles who wrote a letter about it /p.2/ to DNI saying that the Director might want to fire me, and that Smith was being persecuted whereas he should be promoted, etc. The Commodore added that the letter had been obtained legitimately. I don't see how this could be. I knew nothing about this until Captain Davis showed me the photstat of my own letter and told me about Commodore Miles who may now be in Washington, and for all I know, "gunning for" me. [underlining in original]

One thing, Colonel, I wish to make clear: Neither Zimmermann nor Voorhees are particular friends of mine. I knew neither of them before I entered the Service. I never recommended nor "worked for" Zimmermann to succeed Smith. When asked I said I thought he could handle the job. Furthermore, I had nothing against Comdr. Smith personally nor did I ever accuse him of any wrong misconduct [sic]. Throughout this whole affair I was concerned only with obeying my instructions. I could not help but see and know that things were not well handled nor in good order at Karachi and I said so. I have never gone into this with you because I had not thought my conduct demanded justification. I now give you the whole story. Plans now call for my leaving for Bombay, Colombo, and Madras; returning to leave from here for Washington on 15 April so that I should be back 19 April. This is later than I desire or had expected by Comdr. Sadd says the courier situation cannot permit of any earlier date.

Captain Davis, who approved of my writing this letter, sends warm regards in which Comdr. Sadd, Comdr. Egan, and Lt. Castle join. Looking forward to getting back and working under you again, I am, sir

Very sincerely,
[unsigned, but by CW]

[Editor's comment, by GJH: CW refers here to his letter of 21 Feb 1944 to AZ, which began, "dear Al." See above for a transcript of that letter. CW's statements in this paragraph suggest that the letter that AZ has filed with his papers may not be the original letter from CW, but instead a copy of it. Furthermore, although the text is typical of the language that CW used in his correspondence, the handwriting is not CW's; instead, it looks like AZ's handwriting. In addition, the paper on which it is written is coarse, heavy, unwatermarked paper, such as some of AZ's letters were written on; page 2 is written with a top flip, as AZ always did it, not a side flip, as CW did it; the header is abbreviated, showing just initials, and the envelope has been replaced by a drawing of an envelope, in handwriting and printing that is undoubtedly Zimmermann's, not Winsor's. Also, I believe the unsigned letter of 22 August 1944 from Commodore Miles to the DNI, which CW mentions to in this letter to Bales, which was preserved by AZ (and is transcribed above), is also in AZ's hand; it is not Miles' original letter. CW must have provided these two letters (CW's of 21 Feb and Miles' of 22 Aug) to AZ, who copied them and returned them to CW. If my hunch is correct, CW showed copies of these two letters to AZ on 25 March, when he passed through Karachi on his way to cities in eastern India. I believe AZ copied them and gave the "original" photocopies back to CW, who took them when he continued his trip. It is clear that CW and AZ had much more personal communication than CW acknowledged in his letter to Bales.]

United States Naval Training School, Dartmouth College, Hanover, New Hampshire, November 24, 1942 – January 28, 1943 (Hanover, N.H.: U.S. Naval Training School, 1943). 4to, 11 x 8 ½ in., cardstock covers, heavy glossy paper. This book consists of photos and names of the students at the USNTS Dartmouth, and the faculty and staff, grouped by Company and Platoon, with group photos of the students and of the staff. Individual photos of the Commanding Officer (CAPT H. M. Briggs, USN, a submariner), Executive Officer (LCDR Daniel Stubbs, USNR), and the President of Dartmouth (Ernest Martin Hopkins).

AZ is in Fourth Company (p.45ff), Platoon 1 (p. 48). The group photo for this company follows page 45, and AZ is in Platoon 1, second row from the bottom, third from right. Each student is shown alphabetically within his platoon, with rank, a home address, college and year, and occupation shown for each. AZ: "Lieutenant / Haverford, Pa. / University of Pennsylvania '23, B.S. / Wool Dealer." LT(j.g.) Curtin Winsor, Ardmore, Pa., Princeton, AB '27; University of Penna, LL.B. '30, Attorney and writer, is in Fourth Company, Platoon 6 (p. 54).

[A copy of this photo is in the AWZ Wartime Scrapbook]

Google Images
Curtin Winsor

Ancestry.com
Elizabeth Browning Donner
(1911-1980)
She married (1) Elliott Roosevelt,
and (2) Curtin Winsor

Google Images

2712 35th Place, NW, Washington, DC
Curt Winsor's home, 15 September 43

Lieutenant Albert W. Zimmermann's invitations

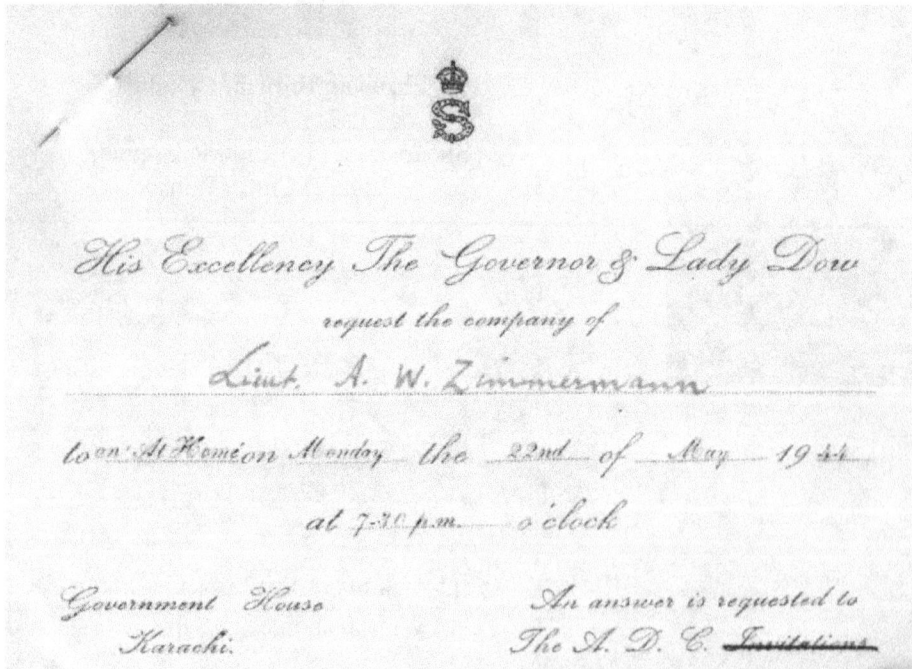

Invitations, from Upper L, clockwise Acceptance by Governor and Lady Dow, for 27 October 1944
To the Iranian Consulate, for the "Birthday of the Shah," 26 October 1944
From Governor and Lady Dow, "at home," for 22 May 1944.

12 October 1944.

Captain Burrows Sloan Jr. accepts with pleasure the kind invitation of the United States Naval Liaison Officers for Friday October 27th at 1900 hours.

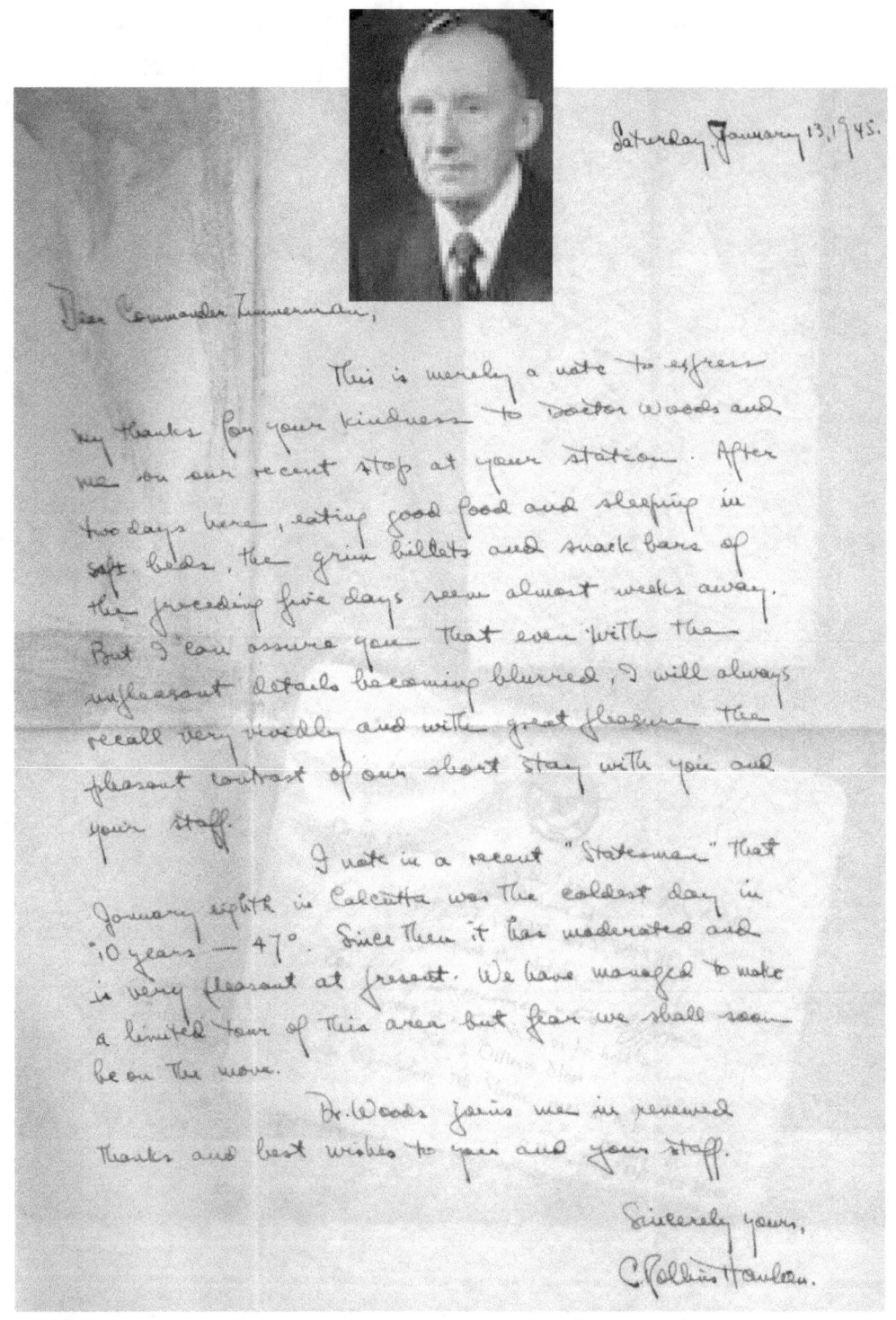

Saturday, January 13, 1945.

Dear Commander Zimmermann,

This is merely a note to express my thanks for your kindness to Doctor Woods and me on our recent stop at your station. After two days here, eating good food and sleeping in soft beds, the grim billets and snack bars of the preceding five days seem almost weeks away. But I can assure you that even with the unpleasant details becoming blurred, I will always recall very vividly and with great pleasure the pleasant contrast of our short stay with you and your staff.

I note in a recent "Statesman" that January eighth in Calcutta was the coldest day in 10 years — 47°. Since then it has moderated and is very pleasant at present. We have managed to make a limited tour of this area but fear we shall soon be on the move.

Dr. Woods joins me in renewed thanks and best wishes to you and your staff.

Sincerely yours,
C. Rollins Hanlon.

C. Rollins Hanlon, M.D., F.A.C.S., was on a secret medical mission to the Pacific Theatre when he stayed with Zimmermann in Karachi. He was already famous as a surgeon and scientist, and he had a distinguished career after the war as a surgical educator and chief executive officer of the American College of Surgeons

End Notes for Chapters 1-6

[1] Additional information about the parents, siblings, and ancestors of Albert Zimmermann and Barbara Shoemaker can be found in George J. Hill, *American Dreams: Ancestors and Descendants of John Zimmerman and Eva Katherine Kellenbenz, Who Were Married in Philadelphia in 1885* (Heritage Books, 2016); and Hill, *Quakers and Puritans: The Shoemaker, Warren and Allied Families. Ancestors and Descendants of William Toy Shoemaker and Mabel Warren, Who Were Married in Philadelphia in 1895* (Heritage Books, 2015). See Appendix B for a summary of the biographical sketch that Zimmermann submitted to the Navy with his application for a commission.

[2] One of the mutual friends of AZ and Curtin Winsor, his desk officer in the Far Eastern Division of O.N.I., was Henry Rawle Pemberton. He is one of those named by Edward Digby Baltzell, *Philadelphia Gentlemen: The Making of a National Upper Class* (New Brunswick, N.J.: Transaction Publishers, 2009 [Free Press, 1958]), 298. Pemberton, Zimmermann, Curtin, and others like them are described in the book, "The quantitative backbone of the book is based on the 770 Philadelphians of various class and ethnic backgrounds listed in *Who's Who in America in 1940*, an index of the elite leadership structure."

[3] The photos were later returned to AZ with a letter dated 21 June 1949 from L. V. Boardman, Special Agent, F.B.I., enclosing two photos taken by AZ of a wedding at St. Paul's Church, Philadelphia, before 1937, showing German Bund members in uniform.

[4] On 13 June 1942, FDR "suddenly dissolved the COI and established in its place the Office of Strategic Services (OSS), transferring control of it from the White House to the JCS," and although Donovan "lost control of the Foreign Information Service," which became the Office of War Information (OWI)," he was named director of the OSS (from Anthony Cave Brown, *Last Great Hero*, 237).

Freeman Lincoln = Joseph Freeman Lincoln, prolific author, was a Major, later a Lt. Col. in the OSS. He was in London at this time, and he later went to Kandy, Ceylon; he received the Bronze Star. Freeman Lincoln was the son of Joseph C. Lincoln, known as "Daddy Joe" to the Zimmermann girls, a prolific author of books about Cape Cod. Freeman and his wife Ginny's daughters, Anne and Crosby, were good friends of theirs. Anne Lincoln would be the "glamour girl of the year" in 1945 (*Life Magazine*, 5 March 1945, 118-121); Crosby nearly died of polio in the summer of 1942. Lincoln also was a neighbor of "Bitz" Durand, whose daughter Virginia ("Peachy") Durand went to Kandy as a civilian OSS secretary.

Clarence Lewis = in the U.S. Marine Corps as an OSS officer. His daughter Susan was the best friend of Zimmermann's daughter Helene. AZ learned on about 28 October 1943 that Freeman had been promoted, and on 4 March 1944, he wrote that Freeman would be going "somewhere else" in his theatre, meaning Kandy, Ceylon, where OSS had its headquarters.

[5] Dorwart, *Conflict of Duty*, Chapter 15 (162-171). For background on "the Room," see: Christopher Andrew, *For the President's Eyes Only: Secret Intelligence and the American Presidency from Washington to Bush* (HarperPerennial, 1996); H. Montgomery Hyde, *Room 3603: The Incredible True Story of Secret Intelligence Operations During World War II* (New York City, N.Y.: The Lyons Press, [1962] 2001) Foreword by Ian Fleming; Joseph Persico, *Roosevelt's Secret War: FDR and World War II Espionage* (New York City, N.Y.: Random House, 2002); Richard Harris Smith, *OSS: The Secret History of America's First Central Intelligence Agency* (Guilford, Conn.: The Lyons Press, an imprint of The Globe Pequot Press, [1972] 2005); William Stevenson, *A Man Called Intrepid: The Secret War* (Guilford, Conn.: The Lyons Press / Globe Pequot Press, [1976] 2000); and Jeffery M. Dorwart, "The Roosevelt-Astor Espionage Ring." *Quarterly Journal of the New York State Historical Association* 62 (No. 3, July 1981): 307-22 (at http://cryptome.info/fdr-astor.htm, 7 pp.; accessed 4/27/2009).

[6] **Rear Admiral Harold Cecil Train** (1887-1968)'s Navy career included command of USS *Arizona* (BB-39) until 5 February 1941, when he was relieved by Captain Van Valkenberg, who was killed by a direct hit on the bridge on 7 December 1941. Captain Valkenberg's Naval Academy ring was all of him that was recovered; he was awarded the Medal of Honor. On 7 December 1941, Captain Train was Chief of Staff to Vice Admiral William S. Pye, Commander, Battle Force, Pacific Fleet. In the movie "Tora! Tora! Tora!" Train was portrayed by Karl Lucas. In that movie, Captain Train was said (incorrectly) to have commanded USS *California*, which was sunk at her moorings. The photo of RADM Train is from the FDR Library. Harold Train was promoted to RADM and served as Director of Naval Intelligence from July 1942 until September 1943, when he was relieved by RADM Roscoe E. Schuirmann (Sept 43-Oct 44), who in turn was relieved by RADM Leo H. Thebaud (Oct 44-Sept 45). Zimmermann served under all three men when they were DNIs.

RADM Train was Commander, Panama Sea Frontier and Southwest Pacific Area, with additional duty as Commander, Fifteenth Naval District from October 43-June 44. He received the Commander-in-Chief's Award from the Military Order of the World Wars in 1962. Rear Admiral Train is the father of Admiral Harold Depue Train II, USN (ret), who served as the North Atlantic Treaty Organization's Supreme Allied Commander, Atlantic; and Commander-in-Chief, U.S. Atlantic Fleet. ADM Harold D. Train II wrote a favorable comment about *Proceed to Peshawar* which appears on the back cover of the book. He was known as "Harry" Train. He is the father of RADM Elizabeth L Train, who was at one time the Deputy Director of Naval Intelligence.

See Jeffery M. Dorwart, *Conflict of Duty: The U.S. Navy's Intelligence Dilemma, 1919-1945* (Annapolis, Md.: Naval Institute Press, 1983), Chapter 18 (194-206). For Zacharias' perspective, see: **Ellis M. Zacharias**, *Secret Missions: The Story of an Intelligence Officer* (Annapolis, Md.: Naval Institute Press, [1946] 2003), where he says derisorily that his chief [RADM Train] "never spent a day in intelligence work"; and Zacharias, *Behind Closed Doors: The Secret History of the Cold War* (New York City, N.Y.: G. P. Putnam's Sons, 1950). An authorized biography which restates much of the same in the third person was written by Maria Wilhelm: *The Man Who Watched the Rising Sun: The Story of Admiral Ellis M. Zacharias* (New York City, N.Y.: Franklin Watts, Inc., 1967). A more balanced view of the subject, but still tilted towards him, is in David A. Pfeiffer, "Sage Prophet or Loose Cannon? Skilled Intelligence Officer in World War II Foresaw Japan's Plans, but Annoyed Navy Brass" *Prologue Magazine* [National Archives] 40 (No. 2, Summer 2008), 11 pp. The photo of Zacharias was taken when he was a CAPT; it is on the cover of Wilhelm's book.

[7] **Snow = Edgar P. Snow**, correspondent of the *Saturday Evening Post*, who became famous for his close relationship to the leaders of Communist China. He encouraged the visit of President Nixon but died of cancer just before Nixon arrived. His ashes were divided between America and China.

Trainer = Tom Treanor, a reporter for NBC and the *Los Angeles Times* (his name was misspelled by AZ), died in France on 21 August 1944. His book about humor on the front lines, *One Damn Thing After Another* (Garden City, N.Y.: Doubleday, Doran & Co., 1944) was dramatized on NBC on 14 November 1944 on the program "Words at War," which honored 24 war correspondents killed in World War II. His survivors included a wife and children, to whom his book was dedicated. Quotations from *One Damn Thing*: "If you want ..." (p. 83); "Despite protestations" (p.140).

AZ, Intelligence Report to JICA/CBI and DNI, 5 July 1944, in Appendix 5 (*infra*).

[8] **Thornton = Thomas A. Thornton** was CDR and later CAPT, USNR. He was Deputy Chief, JICA-ME in many documents in RG 38, Serials JICA 1943-44, Box 19; and in Box 20, he was O-I-C, Naval Section. He later became Chairman, Joint Intelligence Collection Agency, Middle East, and under that title a folder of his papers is filed in the FDR Presidential Library, Hyde Park, N.Y.

Markey = Eugne Willford "Gene" Markey. See Markey in Chapter 2 (*infra*) for his profile up to November 1943. Briefly, he was at this time a Commander, USNR. He was en route to become the Senior Naval Liaison Officer in the India-Burma Theatre, based in New Delhi. He had additional duty with JICA/CBI, which was then located in New Delhi. I believe JICA for the theater moved to Colombo, Ceylon, and/or Calcutta, as JICA/IBT after China was detached from the South East Asia Command. His name and the title of O-in-C, Naval Section, JICA/CBI, New Delhi, is on several reports, including one immediately after the end of AZ's trip to the NEFP (21 December 1943), in File series 277-329, Boxes 24-25. AZ's trip to the NWFP is not mentioned in any of these reports. Markey was promoted to CAPT, USNR, while he was in India. He was later advanced to Commodore and after World War II, he became a Rear Admiral. He had previously been married to the actress Hedy Lamarr (fn, AZ to BSZ, 7 June 1944).

Quentin Reynolds = journalist and World War II war correspondent.

General Strong = Probably Maj. Gen. George Veazey Strong, Chief of U.S. Army Intelligence (G-2).

Henry Hotchkiss = "Henry Hotchkiss" appears in the list of Foreign Service appointments for 1943 as Assistant Naval Attaché and Assistant Naval Aviation Attaché in Cairo. He does not appear in any list thereafter, and I cannot determine what became of him. Because of his connection with aviation, I suspect that he may well have been Henry G. Hotchkiss, a lawyer from Washington, D.C., whose father was born in New Haven, Conn. He applied for passports in 1920 and 1930, the latter naming his wife Gladys. Henry G. Hotchkiss was the author of two books which have been frequently cited: *United States Aviation Reports* (1942), with Arnold W. Knauth (first author) and Emory H. Niles (third author); and *A Treatise on Aviation Law* (1938).

[9] Mrs. Dubash's invitation to AZ for this dinner on 20 July 1943 is in the AWZ Scrapbook, Page 12(V). She appears in several subsequent letters from AZ. Her given name, Nadia, first appears on 24 October 1943 when AZ refers to her as the "girl friend" of his Commanding Officer, LCDR F. Howard Smith. Her maiden name, presumably Russian, is unknown. She disappears from the record after February 1944, about the time that Smith was relieved and transferred. A letter from LT Curt Winsor at ONI, 21 February 1944 (in Winsor folder) says: "Cdr.

S. will be recalled without an accounting at Karachi but his vouchers here have been carefully checked and he has been directed (1) to submit a certified inventory property account. He's also been directed to (2) eliminate counter intelligence activities and discharge Sheikh. (This means discharge Mme Dubash too but they didn't want to sully the lady's name by mentioning her.) (3) To cut down the style of operation of the mess. All this by wire." Sheikh also disappears from the record at this point, but his business card, with the NLO address stricken out, is in the AWZ Wartime Scrapbook, showing Sheikh's persistence. A Google search for Nadia Dubash was fruitless.

[10] Charles H. Harris III and Louis R. Sadler, *The Archaeologist was a Spy: Sylvanus G. Morley and the Office of Naval Intelligence* (Albuquerque, N.M.: University of New Mexico Press, 2003), Appendix 5 "General Instructions," 381-2, e.g.: "In employing natives or residents to obtain information the greatest care must be used." Jeffrey M. Dorwart, *Conflict of Duty The U.S. Navy's Intelligence Dilemma, 1919-1945* (Annapolis, Md.: Naval Institute Press, 1983), 211: In December 1944, "the department ordered that funds for the collection and classification of information could not be used for dinners, luncheons, and entertainment." AZ wrote to BSZ, Letter #55 on 25 March 1944, "it seems we have been living in sort of a fool's paradise, the fool being the one that set it up and has paid us full per diem during our stay. Evidently we've been living on much too high a scale and the day of reckoning has come."

[11] **Brigadier (later Maj.-Gen.) Neville Godfray Hind**, CSI, MC. His wife was Marguerite "Poppie" Hind. AZ first mentions going to dinner at the residence of Brig Hind on Saturday, 31 July 1942, with Maj. D. Montgomery Smyth, Military Secretary to the Gov. of Sind, and his wife Joan Smyth.

[12] **Senator Richard Russell** = Richard Brevard Russell, Jr., a Democratic Party politician who was Governor and a long-time United States Senator from the state of Georgia.
Ambassador Gauss = Clarence Edward Gauss was a career Foreign Service Officer. He was ambassador to the Republic of China from 1941-44. He resigned the post in November 1944, and was replaced by **Patrick Hurley**.
General Haddon = Brig. Gen. Julian B. Haddon. The XIV Air Force Service Command came into existence in May under Gen. Haddon. On 20 August 1943, HQ, China-Burma-India Training Unit (Prov) was activated at Karachi, India, with Gen. Haddon as Commander.

[13] Details in Albert W. Zimmermann's Wartime Papers (AWZ WP). He will be referred to as AZ or AWZ.

[14] Jeffrey M. Dorwart, *Conflict of Duty: The U.S. Navy's Intelligence Dilemma, 1919-1945* (Annapolis, Md.: Naval Institute Press, 1983), 207-9: At the height of the war, early in 1943, "the ONI directed twenty-nine attaché posts manned by 156 officers (mostly reservists), twenty-two observer posts with seventy-nine officers, forty-three liaison offices with eighty-four officers, and thirty-five men attached as shipping advisors, assistant consular officers, petroleum observers, or some such cover designation. No matter what the title, each officer received from ONI a little kit including a Hoey Position Plotter, various stationery and supplies, a .38 revolver, belt and holster, a *World Almanac*, fifty rounds of ammunition, and sometimes a gas mask and steel helmet.... The designation 'naval liaison officer' arose from war experience, probably at the behest of the British as part of a ship routing agreement." Three of the offices were in India: Bombay, Calcutta, and Karachi.

[15] His RAF flight departed Karachi at 0715 on 29 October; he arrived in Colombo at 1830 on 30 October; he departed Colombo at 0700 on 1 November, and arrived in Karachi at 1500 on 3 November.

[16] AZ to BSZ, Letter #86, 6 Sept 1944.

Notes for Gordon B. Enders

[17] *Proceed to Peshawar*, 112, provides references to negative comments made about Gordon Enders by his contemporaries: Alghan Lucey, a O.S.S. officer who knew him in China, called him a "bag of wind," and Roderick Engert, who knew him personally, told me that he was a "blowhard." Roderick's father, Cornelius Engert, was the U.S. Minister to Afghanistan. Enders and the Engert family lived near each other in Kabul. As a young man in WWII, Roderick became a translator for the O.S.S. in India. Other anonymous critics were in the State Department. Zimmermann wrote on 6 December 1943 that Enders "did the talking [to the Viceroy] as he has a flare for it and modesty isn't one of his virtues." My reading of Enders' two books shows there are many inconsistencies in them, and several rearrangements of events which are inconsistent and counterfactual. Enders was nevertheless a gifted writer, and it is easy to imagine that he was a spell-binding lecturer, too.

[18] Copies of these letters were sent to George Hill by Mrs. Gertrude Enders Huntington (nee Trudy Enders) on 11 January 2010, from: G. E. Huntington, 129 Kendal Drive, Kennett Square, PA 19348.

[19] This movie was released in April 1942, which makes it possible to date the letter as 9 May 1942.
Editor's comment: Betty Enders uses the European spelling of "cheque," and her handwriting is distinctly different from the usual Palmer method used by Americans at that time. Note, too, that at this time in 1942, the following

members of the Enders family were still alive: Trudy's father Robert Enders was in the O.S.S. in Washington, D.C., and her grandmother Enders was with him there.

[20] WBAA is now Purdue Public Radio. These excerpts were probably read on the air by Mrs. Enders. Elizabeth Crump Enders also wrote two books about her experiences with Gordon in China. She and Gordon were lecturers on the same Chautauqua Circuit as First Lady Eleanor Roosevelt. Some of the dramatic incidents described in her books are probably embellished, but her audiences must have appreciated them. She could tell the true story that before they were married, as a Red Cross worker in France in World War I, she nursed Gordon back to health after he fell out of an airplane, without a parachute, when it was 3,000 feet above the earth.

[21] From the website for WBAA, FM 101.3, AM 920, Public Radio from Purdue, http://www.purdue.edu/WBAA/inside_wbaa/history.shtm (accessed 25 January 2010): **Our History** President Harding was in the White House, Sinclair Lewis's novel, "Babbit," was newly published, theater patrons were frightened by F. W. Murnau's Nosferatu and amused by Harold Lloyd in Grandma's Boy. The year was 1922 — the year that WBAA received its license to broadcast.... **New Home** WBAA was back on the air on January 25, 1930. The next big move for WBAA came in 1941 when the studios moved to their present home in Elliott Hall of Music. Just as WBAA studios moved from place to place, WBAA's location on the AM radio dial also shifted over the years. Originally licensed at 834 kilocycles, it changed to 1100 kilocycles in 1925, to 1400 kilocycles in 1928, to 890 kilocycles in 1934, and finally to its present location at 920 kilocycles in 1941. **Strong Foundation** WBAA has a rich history as the longest continually operating radio station in Indiana. The people and programs that listeners have heard over the years on WBAA are too numerous to mention. From news, educational programs, sports coverage, and music broadcasts, WBAA has enriched the lives of countless listeners since 1922.

[22] Fatehpur, Haswa, is in the state of Uttar Pradesh, in north central India, about 60 miles NW of Allahabad.

[23] Etawah, in Uttar Pradesh, is about 110 miles NW of Fatehpur; it is about 50 miles E of Agra and the Taj Mahal; and approximately due south of Nanda Devi.

[24] Enders is now in New Delhi, adjacent to [Old] Delhi. It is about 150 miles NW of Etawah, through which he passed at tea-time on 19 November.

[25] The tomb of Hanuman is a temple that is dedicated to a Hindu god. It is said to have been originally by Maharaja Man Singh I of Amber (1540-1614) during Emperor Akbar's (1542-1605) reign.

[26] **Archibald Percival Wavell** was then Commander-in-Chief India; he later became Viceroy and was ennobled as Lord Wavell, which was his title and position in November 1943 when Enders met him again in Peshawar. **Brigadier General (later Lieut. Gen.) Raymond Albert Wheeler** and his staff were en route from Washington to Teheran via Karachi, where he was to establish an Iranian Mission for the U.S. Wheeler returned to Delhi after Pearl Harbor was attacked and became the senior U.S. military officer in India.

[27] Chandri Chauk, on Connaught Place, is the largest outdoor bazaar in Delhi.

[28] The scenes described by Enders would be similar to those seen by AWZ in late November 1943 when he went from Peshawar to the Khyber Pass. Enders' notes could therefore be quoted and cited to amplify what AWZ wrote about his trip.

[29] Editor's note: "Colonel Benson" was **Lt.-Col. Sir Reginald Lindsay "Rex" Benson**, who was the military attaché of Great Britain in Washington from 1941-1944.

[30] Lt. Col. Rex Benson was of the merchant banking family of the ex-Quaker, Robert Benson, whose bank later became Kleinwort, Benson. The Benson Bank had corporate relationships with J. P. Morgan and Lehman Brothers. Sometime after 1932, Rex Benson married **Leslie Foster Nast**. She was the second wife of Condé Nast, who married her in 1928 and from whom she was divorced in 1932. By Nast, she had one child, a daughter, Leslie, who married firstly, Peter George Grenfell, 2nd Baron St. Just, and secondly, Lord Bonham Carter. From www.thePeerage.com (accessed 1/24/10): Lt.-Col. Sir Reginald Lindsey Benson married Leslie Foster, daughter of Albert Volney Foster and Grace Leslie, after 1935. Lt.-Col. Sir Reginald Lindsey Benson lived at Chichester, Sussex, England. He also went by the nick-name of Rex. Leslie **Foster** was born on 19 June 1907 at Winnetka, Illinois, U.S.A.² She died on 25 May 1981 at age 73 at Sussex, England. Child of Leslie Foster and Condé Montrose **Nast** : Leslie **Nast**+. Child of Lt.-Col. Sir Reginald Lindsey Benson: David Holford **Benson**+

[31] Rai Bhadur Tirath Singh was awarded the Indian Service Order by King George V. He died in February 1946. His sons were Raghunath Singh and Sardar Ranbir Singh. Rai Bahadur Tirath Singh was the second son of S Hoshiar Singh and Charian Devi. He joined the United Province Police Force and became Deputy Inspector General of Police. He was awarded the ISO for his role in the capture of Sultana, a fearsome bandit. He was honored with the title of Rai Bahadur by George V. He died in February 1946 and was buried at Dholpur.

[32] BG (later LTG) Raymond Albert "Spec" Wheeler was en route to Tehran to establish an Iran Mission for the United States. After Pearl Harbor was attacked, he was reassigned to Delhi where he became the senior U.S. military officer in the headquarters of SEAC as Deputy Supreme Commander under Lord Mountbatten; in June

1945 he became Commanding General, India-Burma Theatre. His chief of staff, with whom Enders met in Honolulu, was **Lt. Col. (later Brigadier General) Don G. Shingler**, also an engineer, who was "sent to Iran to direct the establishment of a supply route into the Soviet Union." See Arlington Cemetery website for details about Wheeler and Shingler.

[33] He had just finished his 1st year at college and was working at Chautauqua Assembly, at Lake Chautauqua. He saw his first airplane flight "in the month that war broke out." (*Nowhere Else in the World*, 145).

[34] On April 5, 1917, he applied for a passport in New York City, to sail to Europe on the *Chicago* on 7 April. He had planned well in advance for this, having obtained depositions from Clarence F. Eddy of Cornell Medical College on 30 March, and from his sister Miriam, in Wooster, O., on 3 April; she said their mother was still living in India. He "solemnly swore" that he intended to be an American ambulance driver, and that he intended to return within six months; instead, he joined the French Foreign Legion as a pilot, and did not return to the U.S. for more than two years. The original application and supporting documents are in the National Archives, accessed via Ancestry.com, on Passport Applications, January 2, 1906-March 31, 1925; 1917; Roll 0355 – Certificates: 49501-0300, 28 Mar 1917-07 Apr 1917.

Clarence Ford Eddy was the older brother of William "Bill" Eddy. Both men were housemates of Enders in Wooster, Ohio. Clarence graduated from the College of Wooster in 1917 and Bill graduated in the same year from Princeton. How and when Enders actually traveled to France is unknown. The S.S. *Chicago*, of the French CGT line, left NYC for Bordeaux on 23 July 1917 with 329 listed passengers on board, but the name of Gordon Enders is not shown on the list. A British passenger ship, S.S. *Chicago*, was torpedoed off of Flamborough Head on 7 August 1918, losing 3 of the 123 on board, but it is not known to been in New York in April. And yet another S.S. *Chicago* served as a U.S. Army troopship to return the 348th Infantry Regiment to New York on 7 March 1919 after six months of service in Europe.

[35] Enders tells of his wartime service in WWI in *Nowhere Else in the World*, 147-164. On p. 147, he says he "drove for six months in Picardy and Verdun" with the "Norton-Hayes ambulance unit." The name of unit that Enders refers to is ambiguous, and unintentionally confusing. Richard Norton, a Harvard archaeologist, is said in many accounts of American ambulance services in WWI to have been associated with two similar-sounding Red Cross Units: the Norton-Hayes Red Cross unit, and the Norton-Harjes unit. A few citations state that Norton-Hayes was in association with an American financier, Herman H. Hayes, of Morgan, Hayes & Co., of Paris. Many sources state that in the Norton-Harjes Unit, Richard Norton was associated with a French millionaire banker, Henry Herman Harjes, son of John H. Harjes, was senior partner of Morgan, Harjes & Co. Hayes-Norton recruited volunteer drivers in January 1916, to leave as a group, immediately: (http://www.thecrimson.com/article/1916/1/13/more-ambulance-drivers-needed-pwe-invite-all/ (accessed 12/2/17). The Norton-Harjes Unit was established in 1914, and it persisted until 1917. Norton-Harjes drivers included many who were later famous, including E. E. Cummings and John Dos Passos. Other histories show them to have been in Norton-Hayes. The fact that Herman is the first name of both Hayes and Harjes complicates the story. The answer to this conundrum is surely that Herman H. Hayes and Henry H. Harjes are one and the same; and that there is but one unit, usually called Norton-Harjes, which became a part of the ambulance service of the American Red Cross in the fall of 1917. The history of Norton-Harjes is the true history of this unit. For their services, both Richard Norton and Henry Harjes were awarded the Croix de Guerre and Mde. Harjes was made a Chevalier of the Legion of Honor.
See: Henry Herman Harjes (http://wc.rootsweb.ancestry.com/cgi-bin/igm.cgi?op=GET&db=phco&id=I11657) and Richard Norton (https://net.lib.byu.edu/estu/wwi/comment/AmerVolunteers/Morse4.htm) (accessed 12/5/17)

Enders drove an ambulance with this service for about six months, and he then returned to the U.S., to live in Pittsburgh. He enrolled in the College at the University of Pittsburgh in 1916-1917 as a junior at in the Class of 1918: "The present school year of 1916-1917 will soon be a thing of the past" and "College juniors [include] Gordon Enders - Pittsburgh, Pa. We would give half of our inheritance to attempt a pungent pun on Enders, but we're so fearfully certain that the thing's been done before that we can't muster the nerve" (from *The Owl* [University of Pittsburgh, 1918], 14, 180).

He left Pittsburgh before the end of the school year and obtained a passport on 5 April 1917, to depart for France from New York City on 4/7/17. His passport application was endorsed by his sister, Miriam, who was then living in Wooster, O.; their mother was still in India. The application was also endorsed by Clarence Ford Eddy, of Cornell Medical School. He was the older brother of William "Bill" Eddy; both men had been at the College of Wooster with Enders. The original application and supporting documents are in the National Archives, accessed via Ancestry.com, on Passport Applications, January 2, 1906-March 31, 1925; 1917; Roll 0355 – Certificates: 49501-0300, 28 Mar 1917-07 Apr 1917. Enders enlisted in the French Foreign Legion to take flying lessons with a squad of 15. He soloed in three weeks and became a French Bomber Pilot. Four months later, he became a corporal in the U.S. Army Signal Corps. He may have been listed as a U.S. Army corporal, while continuing to fly a French bomber, under French command. His first period of service as a commissioned officer in the U.S. Army can be dated precisely from 26 April 1918 until 9 October 1919.

[36] His service record says that he when he was commissioned on 26 April 1918, he was a "Bomber Pilot" in "223d Bomb Sq. France." I have not been able to locate the record of either a U.S. or French Bomb Squadron 223. Maynard Creel found Enders' record of acceptance of commission as a 1st Lt in May 1918 in the *Air Service Journal* (June 1918), third column, third block down. All of his appointments and promotions are redacted from his service record, and FOIA says only that he was a Colonel, retired.

[37] Gordon Enders' records were transcribed in 2016 by NARA for FOIA. The first record begins with Dates of Service 09/10/1917 to 10/09/1919. From WWI, his Decorations and Awards were: "French Order of Merit" and "WWI Victory Medal w/4 Bronze Service Stars." At that time, there was no such decoration as the "French Order of Merit," so the correct translation should be Order of the Legion of Honor, of which the lowest rank was Chevalier. His original service record at National Personnel Records Center, shows: "War Service / 2 years / 2 months / Air Service (France Italy) / Combat." The service in Italy is a puzzle, because he does not mention it. Many other items are redacted. His Record of Assignments shows: "Enl Svc: / 10 Aug 17 - 25 Apr 18 Cpt (Signal C) Section Ldr AEF, France / 26 Apr 18 / Bomber pilot / 223d Bomb Sq., France / 9 Oct 19 / Commission terminated this date"

[38] During this period, Enders was also engaged in a protracted lawsuit against United Aircraft Exports, regarding his claim that the company owed him $48,594 as commission for the sale of twenty aircraft to the Chinese government in 1930. He filed the claim in 1933, and initially won, but the company appealed, and on 13 February 1936, the decision was reversed, in favor of United Aircraft.

The testimony of T. V. Soong, Finance Minister of the Chinese Nationalist Government, was said by the *New York Times* to have been responsible for the setting aside of the initial verdict. Enders appealed the decision, but the judgment was unanimously affirmed, "with costs." The final decision is in New York (State) Supreme Court. Appellate Division, *Reports of Cases Heard and Determined in the Appellate Division of the Supreme Court of New York* vol. 250 (Banks & Bros., 1937), 709.

It was a messy business, and it has been suggested by Anthony R. Carroza that Enders may have been double-dipping: "Widespread payment of 'squeeze,' or bribes, to secure government contracts had led to upheavals in the Chiang Aviation Commission . . . Soong attempted to solve the problem by dealing directly with American **aircraft** manufacturers, but he quickly became involved in a controversy with **Gordon B**. **Enders**, who had offered his services to deal with US. companies without receiving a commission. Enders secured a purchase agreement with **United Aircraft** Exports for **planes**, but he then asked for a fee from the company. Realizing that his direct-contact method had failed, Soong demanded that **United Aircraft** reimburse the commission fees that had been paid. Soong quickly learned that eliminating graft in aircraft deals had no easy solution." In Anthony R. Carroza, *William D. Pawley: The Extraordinary Life of the Adventurer, Entrepreneur, and Diplomat Who Cofounded the Flying Tigers* (Potomac Books, Inc., 2012).

[39] In a typed Biographical sketch (1page, ca. 1961), Enders wrote the period from 1921-1937: "China (including the Manchurias & Mongolia) – U.S. Department of Commerce, 1921-1923; business, 1923-1930; Aviation Advisor to Chiang Kai-shek, 1930-31; Foreign Advisor to Panchan Lama of Tibet, 1930-1937."

[40] Enders' record typed at NARA for FOIA shows, for this period, "Dates of Service: 09/17/1941 to 09/23/1946. Decorations and Awards: Good Conduct Medal; WWII Victory Medal; Asiatic Pacific Campaign Medal; Army of Occupation Medal (Japan); American Campaign Medal; European African Middle Eastern Campaign Medal; American Defense Service Medal; Honorable Lapel Button WWII." I believe he would also be authorized to wear one Bronze Star attachment on his Asiatic Pacific Campaign Medal, because he participated in the invasion of Okinawa. From NPRC, his redacted Record of Service shows "Intelligence-MA / MIS, WD, Wash DC / 9-15-41 to 7-30-42 / M/A – Kabul, Afghanistan / 7/1/42 to 12/31/42   M/A – Afghanistan (Fr E.R.) 7/1/43 to 12/21/43   M/A – Afghanistan 1/1/44 to 3/32/44   M/O – India 4/1/44 to 5/6/44   Temporary Duty with MIS, Wash 5/7/44 to 6/30/44." Continued *infra*. In his Biographical sketch, he shows this period simply as "Military Attaché, Kabul, 1941-44." MA = Military attaché. MO = Military Observer. MIS = Military Intelligence Service. WD = War Department.

[41] Enders' obituary, probably written by his brother Robert Enders, says that he was "underground in Korea when Japan surrendered; he carried the surrender of the commander in Korea to MacArthur. Gordon's brother Robert was in the O.S.S. at that time, and he would have known of events such as this that were not in the public record.

Details of the timing of surrender in Korea are shown in https://history.army.mil/books/wwii/okinawa/; and https://history.army.mil/html/forcestruc/cbtchron/cc/027id.htm (accessed 12/1/17)

[42] The transcription of official record at NARA, obtained from FOIA, shows Dates of Service as: 9/10/1917 to 10/09/1919, 9/17/1941 to 9/23/1946, 05/17/1948 to 05/01/1961.

[43] https://www.findagrave.com/memorial/180868293 [Elizabeth Ann "Betty" Crump Enders]

[44] Charles Mosley, editor, *Burke's Peerage, Baronetage & Knightage, 107th edition, 3 volumes* (Wilmington, Delaware, U.S.A.: *Burke's Peerage* (Genealogical Books) Ltd, 2003), volume 1, pp 517-18. Sir Benjamin, the 5th Baronet, is #178966 (http://thepeerage.com/p17897.htm#i178966).

[45] Abbreviations: Transcribed from AWZ Wartime Papers by George J. Hill, February 2011. AZ = LT Albert W. Zimmermann, USNR. AWZ = Zimmermann's Wartime Papers. BB = Maj. Sir Benjamin Bromhead, OBE, IA. NB = Lady (Nancy) Bromhead. ALS – I = Autograph Letter Signed, in ink (i.e, handwritten in ink)

[46] Grindlay & Co. = Grindlay's Bank, now known as ANZ Grindlays Bank. Grindlays has gone through several name changes and owners. As described on the website of the Royal Bank of Scotland (of which it was "a past constituent") (From http://heritagearchives.rbs.com/wiki/Grindlays_Bank_plc,_London,_1828-date; accessed 2/17/2011):
"This overseas bank was established in London in 1828 as Leslie & Grindlay, agents and bankers to the British army and business community in India. It was styled Grindlay, Christian & Matthews in 1839 and Grindlay & Co. from 1843. Branch firms were opened at Calcutta in 1854 and at Bombay in 1865. From 1908 these firms became branches and were thereafter administered directly from London. Additional branches were opened at Simla (1912), Delhi (1923), Lahore (1924) and Peshawar (1926). The bank was acquired by National Provincial Bank Ltd in 1924, but continued to operate as a separate private limited company under the title of Grindlay & Co Ltd. In 1928 its balance sheet totalled almost £3 million. In 1942 it took over Thomas Cook & Son (Bankers) Ltd (est. 1924), extending its business to Burma and Ceylon. It was renamed Grindlays Bank Ltd in 1947.

"National Provincial Bank's interest was sold to National Bank of India Ltd (est. 1863) in 1948, in return for shares and a cash payment. After 1948 the two banks operated separately until merging in 1958 under the title of National Overseas & Grindlays Bank Ltd. At that time National Provincial Bank's shareholding was converted into a 9% holding in this new concern. In 1968 National Provincial Bank sold its shareholding to Lloyds Bank. The bank was renamed National & Grindlays Bank Ltd in 1959, Grindlays Bank Ltd in 1975 and Grindlays Bank plc in 1982. In 1984 the bank was acquired by Australia & New Zealand Banking Group Ltd, and was renamed ANZ Grindlays Bank in 1989."

[47] D.I.K. = Dera Ismail Khan, a city on the west bank of the Indus River in what was then the Northwest Frontier Province (now Khyber-Pakhtunkhwa Province). It is the capital of the district of the same name. I do not believe AZ ever got to D.I.K., but the use of the initials shows that he would have been familiar with the name.

[48] Ghilzai = "Ghilji or Ghilzai are one of the most famous tribes of Afghanistan. They are a large and widespread Afghan tribe, occupying the high plateaus north of Qandahar (Qalat-e-Ghilzai) and extending eastwards towards the Suleiman Mountains, westwards towards the Gul Koh range, and the North of the Kabul River. They were in power in Afghanistan at the beginning of the 18th Century and for a time even possessed the throne of Isfahan (Persia). According to the Ethnologue Data from Languages of the World, 14th Edition, the Ghilzai speakers of Pashto are 24 % of the national population of Afghanistan" (from Khyber.org, accessed 2/17/11).

[49] **Curtin Winsor**, son of James D. Winsor and Marion Harding Curtin, was b. 5 December 1905 in Ardmore, Pa.; died 12 November 1998 in Bryn Mawr, Pa. He was at Princeton University from 1925-29 and was secretary of the Class of '27. He was a member of the editorial board of the *Daily Princetonian* and of Quadrangle Club. He went to the University of Pennsylvania Law School, graduating in 1930. He served in the National Recovery Administration but returned to Philadelphia and graduated from the Curtis Institute of Music in 1942 with a degree in composition and criticism. He then went into the Navy, where he rose to the rank of lieutenant commander, USNR. He was active in business and charitable and environmental organizations in Philadelphia until his death.

He married (1), in 1937, as her second husband, the former **Elizabeth Browning Donner** (b. 12 May 1911; d. 1980), daughter of William M. and Dora B. Donner. Elizabeth had m. (1), on 16 January 1932, **Elliott Roosevelt** (1910-1990), son of the President. They had a son, William Donner Roosevelt (1932-2003), and the marriage was annulled in 1933. Winsor and Elizabeth (Donner) Curtin were divorced in 1943, and he m. (2) in 1943, Margaretta Rowland, in Washington, D.C. He m. (3) in 1959, Catherine Horst, in Moore, N.Y.; she died in 1971. He m. (4), in 1974, Eleanor Webster, in Bryn Mawr, Pa. Curtin Winsor and Elizabeth Donner had two children: Curtin Winsor Jr. (b. 28 April 1939, Philadelphia); and Joseph Donner Winsor (b.19 Sep 1941, Philadelphia; d. 2 Mar 1976, West Palm Beach, Fl.). Curtin Winsor and Eleanor Webster had a daughter, Ellen Wilson Winsor (b. 4 Oct 1977, Bryn Mawr, Pa.). His obituary also shows two other children: Karin King and Linda Edson.

[50] JICA, also referred to in lower case as Jica, was abolished after World War II. So-called "Joint Intelligence" organizations (intelligence sections and divisions of the Joint Chiefs of Staff) are now usually referred to as "Defense" or "Armed Forces" rather than "Joint"; e.g. Defense Intelligence Agency (DIA) and Armed Forces Medical Intelligence Center (AFMIC). For other examples of Defense and Armed Forces Intelligence acronyms, see *Joint Staff Officer's Guide* (Washington, D.C.: Superintendent of Documents, for National Defense

University/Armed Forces Staff College, Norfolk, Va., 1986), pp. 4-8 and 4-9; and acronyms, pp. II-38-41. Only one "Joint" intelligence acronym persisted in 1986: JIEP (Joint Intelligence Estimate for Planning).

[51] The Office of Naval Intelligence was created on 23 March 1882, by Navy Department General Order 292. The ONI was incorporated into the Office of the Chief of Naval Operations in 1915. The Near and Far East Sections were established in 1939. In March 1943, the Assistant DNI's title was changed to Deputy Director. The ONI was located in the Navy Building on New York Ave. in World War I. Its location in World War II is not given in the History of the Office of Naval Intelligence (ONI), from the Dedication Ceremony booklet, National Maritime Intelligence Center, dated 20 October 1993, updated 19 Jul 2008 (http://www.navycthistory.com/ONIHistory.txt), accessed 2/14/11. The Chief of Naval Operations was located in the Washington Navy Yard at the onset of World War II, but many of the OPNAV sections have since been transferred to the Pentagon. For example, OP-093 was at the Pentagon in 1984, near the Navy Command Center. ONI was in "Room 2646" (on the 2nd floor of a building not named) in 1926, according to Elias M. Zacharias, *Secret Missions*, p. 83. In 1935 CAPT Zacharias was chief of the FE [Far Eastern] Division.

I suppose ONI was then in the so-called "Mills Building Naval Annex" at 17th and Pennsylvania Ave., presumably where the present Mills Building (built c. 1966) is now located at 1700 Pennsylvania Ave. Trevizo's "A History of the Office of Naval Intelligence, 1882-1942" (http://odh.trevizo.org/oni.html (accessed 3/24/2011) says (p.5) that in 1903, ONI moved its Washington headquarters from the State, War and Navy Building into the Mills Building on the corner of Seventeenth Street and Pennsylvania Ave. Trevizo says (p.29) that ONI headquarters moved in April 1941 into the "new Navy Department building." This building was called "Main Navy," and it was located on Constitution Ave. at the foot of 18th Street. An image can be seen at: "Photo #: 80-G-609132 / Navy Department Building ("Main Navy"), Washington, D.C. / View of the building's main entrance, at the foot of 18th Street, seen looking south from across Constitution Avenue, NW, on 26 June 1947. ... *Official U.S. Navy Photograph, now in the collections of the National Archives.* / Online Image: 74KB; 740 x 605 pixels.

In 1945, the FE Division was located in Room 4626 of an unnamed building (letter from LT Curt Winsor to Col. W. L. Bales, 9 March 1945). I assume this was in "Main Navy," described in the previous paragraph. ONI is now located at 4251 Suitland Road, SE, Washington, DC 20395-5720, on the border of DC and Maryland.

[52] AZ arrived in Karachi on 19 July 1943 (cf. his letter #15, started in Cairo on 15 July). His reports to ONI that Winsor refers to would have been classified. Copies were therefore not sent to his wife, nor are they in his Wartime Papers. They have not been preserved in the National Archives. NARA II, RG 38, was searched to no avail.

[53] Winsor refers to the children of Elizabeth, from whom he was divorced earlier in 1943; the divorce was recorded in Brevard County, Florida. The children (Curtin, 4; and Joseph, 2; and perhaps also William, 11) will soon "leave," and the "wife here" in Washington, D.C., is his new wife, Margaretta, from whom he was divorced before 1959. Elizabeth apparently raised her three children, William D. Roosevelt, Curtin Winsor Jr., and Joseph D. Winsor. All three of these sons were involved with the Donner Foundation when they were adults. Curtin Winsor III, son of Curtin Jr., was b. 10 June 1963, and he, too, is involved with the Donner Foundation.

[54] **Morris Duane** = Morris Duane, son of Russell Duane and Mary Burnside Morris, was born on 20 Mar 1901; died on 18 Jul 1992 at Bryn Mawr Hospital, Bryn Mawr, PA, at age 91. He was an attorney. He graduated in 1923 from Harvard. He married Maud Stovell Harrison, born 10 Feb 1904 in Colorado, on 11 Jun 1927 at Church of the Redeemer, Bryn Mawr, Montgomery Co., PA; she died 23 Jan 1996 at age 91. They resided at Bryn Mawr, PA.; they had 2 children (From Google/Ancestry/ Harrison Family, 2/13/11). Other snippets suggest that he was with the firm Duane, Morris & Heckscher, Philadelphia, Pa., and that he was a Commander in the U.S. Navy in 1944. By 1943, the Zimmermanns were also attending the Church of the Redeemer, and they surely knew the Duanes. AZ and BZ are buried in the cemetery on the grounds of this church.

Henry Pemberton = Henry Rawle Pemberton, son of Henry Pemberton Jr., was b. at Philadelphia, 27 April 1898; died in Pennsylvania, 21 April 1989. He lived in Spring Lake, Monmouth Co., N.J., in 1900, and in Philadelphia in 1910, 1920, and 1930. He graduated from the Episcopal Academy of Philadelphia in 1915, where he was said to be the "most musical" member of the class. From various sites accessed on Google (2/13/11), including his passport application, his death in the Social Security Death Index (SSDI) and Edward Digby Baltzell, *Philadelphia Gentlemen: The Making of a National Upper Class* (New Brunswick, N.J.: Transaction Publishers, 2009 [Free Press, 1958]), 298. The book blurb states, "The quantitative backbone of the book is based on the 770 Philadelphians of various class and ethnic backgrounds listed in *Who's Who in America in 1940*, an index of the elite leadership structure; 226 members of this elite also came from upper-class families listed in the Social Register that year. In addition, Baltzell shows how these upper-class members dominated the financial and business power structure of the city in 1940. Thus, although he describes the upper-class style of life in Philadelphia in fascinating detail, he constantly emphasizes that it is power and influence over the whole social structure, rather than style of life per se, that is an essential quality of a properly functioning upper class."

[55] He is professing false modesty. He actually lived in a beautiful house about six blocks from the Washington National Cathedral. It is near Observatory Circle, where the Vice President now lives. This part of Massachusetts Avenue is called "Embassy Row."

[56] Jim = **James D. Winsor, Jr.** (b. 30 June 1908; d. 17 Dec 1999, Ardmore, Pa.), brother of Curtin Winsor. Sal = Sally Winsor, wife of James Winsor, Jr., and Curt's sister-in-law.

Fort Slocum = From Google (2/12/11): "now an uninhabited 80-acre property off the shore of Long Island, N.Y., known as David's Island. The installation is gone — abandoned and its crumbling buildings demolished by the U.S. Army Corps of Engineers in 2009 — but not forgotten. Established in 1867 on the site of a former Civil War hospital, Fort Slocum served as a military hospital, an artillery mortar battery and a training post. During World War II the fort was the most active recruitment center in the United States and served as a staging area for troops heading overseas during the two world wars."

[57] Google (2/12/11) for JICA and JANIS shows a few citations for JICA but only one for JANIS. It is James D. Marchio, "Days of Future Past: Joint Intelligence in World War II," *Joint Force Quarterly* (JFQ) (no. 11, Spring 1996): JANIS = JANIS (Joint Army Navy Intelligence Studies). Elsewhere, Google shows JICA = Joint Intelligence Collection Agency. There is apparently nothing else published about JANIS, and probably little or nothing about JICA. Histories of Joint Intelligence in the U.S. are indeed scarce. The Joint Chiefs of Staff had a Joint Intelligence Committee (JIC), comparable to the British JIC in London, but the Chiefs successfully resisted creation of a Joint Intelligence Agency (JIA).

Marchio says: "While lacking the impact of Ultra or Magic, joint intelligence efforts contributed to Allied operations in virtually every theater. Joint intelligence operations enhanced collection, improved production, and expedited dissemination of critical information. Nonetheless, joint intelligence efforts during the war were neither universally accepted nor appreciated. . . . JICAs were operational in four theaters: North Africa (JICANA, later renamed JICAMED), Africa-Middle East (JICAME), China-India- Burma (JICACIB, which in 1945 became only India-Burma), and China (JICA/ China). They were attached to their respective theater headquarters as separate staff sections. [i.e, JICACIB would therefore have been at Kandy] . . . JICAs performed three primary tasks. First, they collected, screened, and transmitted to Washington "all information, exclusive of combat intelligence, within the theater" desired by the War and Navy Departments. As theater collection coordinators, JICAs provided logistical support, tasking, and guidance to all human intelligence (HUMINT) sources, including OSS agents, in the JICA area of responsibility. Lastly, JICAs ensured lateral dissemination of pertinent intelligence among various agencies, military and civilian, within each theater. . . . The Joint Intelligence Study Publishing Board (JISPB), with representatives from the War Department G–2, ONI, OSS, A–2 [Army Air Corps], and Office of Chief of Engineers, was created in May 1943 when it became clear that the activities of G–2, ONI, and OSS were duplicative, particularly in preparing foreign area studies. Consequently, JISPB commissioned a series of joint Army-Navy intelligence studies (JANIS) that provided basic topographical data on likely operational areas. These studies included information from 20 government agencies and ranged from Bulgaria to Japan and Indochina. Over 2,000 copies of each JANIS study were disseminated. . . . The chief of staff, Pacific Ocean Areas, praised JANIS studies, indicating that they were indispensable references for the shore-based planner. . . . [However], as late as March 1945, joint intelligence was not being fully accepted. . . . JICAs in the Mediterranean, Africa-Middle East, India-Burma, and China theaters were deactivated between August and December 1945.

[58] "the Duke" and Briggs have not been identified. "Duke" is presumably a nickname.

[59] Winsor refers to the limerick that was popular in WWII: "There once was a lady of Madras / Who had a most wonderful ass / It was not smooth and pink / As you might think / But hairy and gray / And ate grass."

[60] Vague comments in AZ's letters to his family, and in Winsor's letters, suggest that Howard Voorhees may have been known to AZ personally before he came to Karachi. He might be the Howard Voorhees of Pennsylvania who appears in Rootsweb Search Results (nychauta/Families/Swanson - RootsWeb.com Home Page): Howard Voorhees Mar 13, 1902 married E. .... 1916 Dec 2, 1939 at Warren, PA. Their Children Dennis Bertil Aug 13 . . . Howard Voorhees Mar 13, 1902 married E. Elle Anderson Feb 5, 1908 Nov 29, 1929. I have not been able to verify this connection, nor do I have any post-war follow up on him.

[61] **Clarence Off** is later referred to as a Commander (CDR) but he also has been referred to as Lt. Col. (which would be USMC, in the equivalent rank) in a letter of AZ to his wife, although I have not been able to locate that in any letter. Intelligence officers often use assumed ranks and names for anonymity

[62] Indeed, Bullitt did lose. From Wikipedia (2/12/11): "William Christian Bullitt, Jr. (January 25, 1891 – February 15, 1967) was an American diplomat, journalist, and novelist. Although in his youth he was considered something of a radical, he later became an outspoken anticommunist." He was Scroll and Key at Yale. He married the widow of the American Communist John Reed, but divorced her after he discovered that she was involved in a lesbian affair. A close friend of FDR's, he served as the first American ambassador to Moscow, and later was ambassador to

France. He detested Sumner Welles and was a major player in the campaign to discredit him and force his resignation after his homosexuality was discovered. For this, he was never forgiven by FDR, who turned loose the forces of the Democratic party to ensure Bullitt's defeat in the mayoralty election in 1943. Bullitt then became a ferocious opponent of FDR and the Democratic party. AZ's son Warren Zimmermann was also a member of Scroll and Key at Yale, Class of 1956.

[63] CW's stay for several days in Karachi is described briefly in AZ's letters of 7 and 13 January 1944. AZ says of CW, "I think he is quite a difficult person." AZ knew that CW was his desk officer at the ONI and thus was responsible for his well-being, but he appears to have been suspicious of CW, as if Winsor was playing some sort of game with him for some purpose that AZ never could figure out. By 1945 he suspected "skullduggery" in Washington was delaying his return home, implying, but without naming CW specifically as the author of the problem.

[64] LCDR F. Howard Smith, USN, CO of NLO Karachi, was relieved under a cloud and apparently received an unsatisfactory Fitness Report, for reasons that were never made clear in AZ's correspondence folders.

Mme. Dubash = Mrs. Nadia Dubash. She appeared at the Naval Liaison Office on the night that AZ arrived in Karachi and she is mentioned in many of his letters thereafter until 28 December 1943, but never again thereafter. AZ admitted, somewhat defensively, to his wife that he referred to her as "Nadia," in response to his wife's query about his use of the given names of women in Karachi in the letters he wrote home. AZ said she was a Russian, about 50, attractive, and (on 24 October), he said that she was Smith's "girl friend." Whether she was divorced or widowed was never stated. I suspect that she was the *cause célèbre* who led to Smith being removed under a cloud with an Unsatisfactory Fitness Report, although the letters refer only to problems with the finances of the mess. AWZ's Wartime Papers do not shed light on the question of whether she was believed by ONI to be a double agent, or was suspected of this, or was just expensive and unproductive, and perhaps too close to Smith.

[65] At first glance, this would appear to be an ALS from Curt Winsor, but I believe it is a transcription made by AZ, not the original letter. See the comments about it by (presumably) Commodore Milton Miles (dated 22 August 1944, *infra*), which I also believe was transcribed by AZ. It is therefore probably not Miles' original letter. Also see CW's comments on this letter of 21 February 1944 and on Miles' letter of 22 August 1944, in CW to Col. W. L. Bales, 9 March 1945. This letter was not received by AZ, but CW apparently showed a Photostat of it to AZ when he came to Karachi, and AZ copied it at that time.

[66] Markey = **CAPT Gene Markey**, USNR. He is the Chief U.S. Naval Liaison Officer in I-B Theatre. His first name was given in a letter of AZ to BSZ (7 June 1944), who said that he had previously been married to the actress **Hedy Lamarr**. In 1945, after he was relieved in India and returned to the U.S., Markey was Director of the Office of Navy Photographic Services. From United States Government Manual / 1945 / Division of Public Inquiries / Office of War Information / Department of the Navy / Officials (http://www.ibiblio.org/hyperwar/ATO/USGM/Navy.html (accessed 2/13/11). He also had a free-wheeling career in Hollywood at that time and was often seen with **Myrna Loy**. He married her after the war, with **FADM William "Bull" Halsey** as best man.

"The colonel" = **COL W. L. Bales,** presumably USMC, Winsor's superior, whose initials appear in CW's letter of 9 March 1945. His first name and title do not appear in the letters of AZ or CW. His name appeared to be "Bates" in many handwritten letters, but it is typed "Colonel W. L. Bales" in CW's letter to him of 9 March 1945 and I have changed the spelling to Bales throughout all of the letters.

[67] **Sid. Sweet** = **Sidney S. Sweet**. From Google to ftp.resource.org/gao.gov/87-794/00004B37.pdf (accessed 2/13/11): PL 87-794: "AUGUST 13, 1964 / Office of the White House Press Secretary / THE WHITE HOUSE: The President has appointed five additional members to the Public Advisory Committee on Trade Negotiations, which advises Governor Christian A. Herter, the Special Representative for Trade Negotiations, in connection with the current Kennedy Round of trade negotiations under the auspices of the General Agreement on Tariffs and Trade (GATT). The five new appointees are: Joseph M. Baird / Dr. Persia Campbell / John M. Mathis / Sidney S. Sweet / Dr. Caroline F. Ware . . .

Sidney S. Sweet, who was born at Jersey City, New Jersey on April 9, 1913, was President of C. Tennant, Sons & Co., of New York City, traders in metals, chemicals, vegetable oils, and machinery. Before the War he taught English at a Chinese Government school partly supported by Yale and worked for the American Express Co. After service in Naval Intelligence during the War, he returned to American Express as an assistant vice president, before joining Tennants in 1946."

It is not clear why CW would have written three letters to AZ on the same day, 21 April 1944 – if indeed the year of all three was 1944.

[68] "Got rid of him" = The CO, LCDR F. Howard Smith, was recently transferred out and went to Delhi. Smith would have written a Fitness Report on all of his subordinates before he left, and they were probably not rated highly because there was ill-will between them and the CO. It is difficult to counter a poor fitness report, and even

endorsements by superior officers in Washington would not be able to remove such a tarnish on the record of AZ, Voorhees, and any other officers at NLO Karachi. It would likely cause AZ's request for promotion to be set aside.

[69] CW's advice to contact Markey came too late; Markey was already leaving for the U.S., and AZ never did get a courier run back to the States.

[70] This is a rare instance of a request for intelligence surveillance and report that is in AZ's papers. The project was a survey of the harbor at Karachi. Two copies are in the National Archives.

[71] **Marshall Green = Later an ambassador and Assistant Secretary of State** for Far Eastern and Pacific Affairs. He could be said to be the spiritual godfather of the new opera, "Nixon in China."
From Wikipedia (2/13/11): "Marshall Green (1916–1998) was a United States diplomat whose career focused on East Asia. Green was the senior American diplomat in South Korea at the time of the 1960 April Revolution, and was United States Ambassador to Indonesia at the time of the Transition to the New Order. From 1969 to 1973, he was Assistant Secretary of State for East Asian and Pacific Affairs, and, in this capacity, accompanied President of the United States Richard Nixon during President Nixon's visit to China in 1972.

[72] CW's suggestion to AZ to write his own request for promotion was probably ill-advised; AZ should have drafted the letter but he should have passed it to his superior officer to sign and forward up the chain of command. AZ would have had to get the endorsement of this officer anyway and he should have cleared it with him in advance. In the event, AZ did write a letter recommending himself, and it looks awkward. It is in his papers.

[73] **Bowen = CAPT Harold Gardiner Bowen Jr.**, USN (~1913-2000), Deputy Director of Naval Intelligence (see fn in Bowen to AZ, 20 Nov 1944). He later rose to the rank of Vice Admiral.

[74] **Gene = CAPT Gene Markey**, USNR, who was replaced by **CAPT Ransom Kirby Davis**, USN, Senior ALUSLO, I-B Theater, 1945.

[75] The Commodore was **Commodore, later RADM and (after WWII, VADM) Milton Miles.** It appears from this message that the former CO at Karachi, LCDR F. Howard Smith, is now in Ceylon.

[76] After much delay, in the end Groman did not come to Karachi, thus dashing the hopes of AZ and BSZ for him to be the replacement that would bring AZ home for good. Groman lived near them in Bryn Mawr, although they had not known each other before the war. They believed LCDR Groman and his wife maneuvered successfully to keep Groman in the U.S., leaving AZ stranded in India. However, Groman was said to have had a serious medical problem, and that appears to have been possible, given his prior service in the Far East. AZ's possible return for two months was disapproved by ONI, so this, too, was a faint hope that was dashed.

D.I.O. 3^d N.D. = District Intelligence Office, 3rd Naval District, New York City. This is the office where CAPT Vincent Astor worked before he went out on sea duty. It was the senior U.S. Naval Intelligence office on the East Coast of the U.S., and it was in close communication with the British intelligence office in NYC, which was headed by Sir William Stephenson, also known as "Intrepid."

[77] Queen's Hotel = Queen's Hotel, Kandy, Ceylon (now Sri Lanka). A Google search for Queen's Hotel in or near Colombo shows only one such hotel, and it is in Kandy. This is where SEASAC was located, so it is undoubtedly where Habecker was billeted. The hotel's website in 2011 says it was built 160 years ago by the British; i.e., about 1850. I believe AWZ wrote that he stayed there, too, but that piece of correspondence is not indexed.

[78] **Habecker = CAPT Frederick Shrom Habecker,** USN (2 April 1905-12 May 1981), U.S. Naval Academy graduate, 1927. He received the Navy Cross for service as a destroyer CO in October 1944 and the Legion of Merit for service in the Cold War. He was transferred to the Office of Naval Intelligence and was sent to India to be the Senior U.S. Naval Officer at SACSEA in January 1945. If his Legion of Merit was for intelligence work, the citation may be classified or rather vague. He retired as a Rear Admiral but is shown on one web page as "later VADM" so he apparently received a "tombstone" promotion, based on his receipt of the Navy Cross.

[79] **Linaweaver = CAPT Paul Glenwood Linaweaver, Jr.**, USN. Many websites suggest that his father was an Episcopal priest in South Carolina and his son, Paul G. III, was a Submarine Medical Officer in the Vietnam War era. He was the Senior ALUSLO, Kandy, Ceylon, 1944-45, and thus the immediate senior to AWZ. In turn, he would have been under the command of CAPT Habecker in Ceylon.

[80] **Groman = LCDR Henry Groman** first appears in AZ's correspondence on 26 Dec 1944. On 19 February AZ wrote that Groman was still expected soon to leave for Karachi, and perhaps had even left already. On 6 March, AZ wrote that he had just received a letter from Groman, saying that he would not be coming. CW wrote on 6 March that it was a surprise to him, too. Given the length of time that mail took, Groman (and perhaps CW, too) must have known that Groman would not be coming out as far back as the first week in February, but they didn't tell this to AZ.

[81] LCDR Edward F. O'Connor, USNR, who was appointed CO, NLO Karachi. His arrival on 30 March 1945 is shown in AZ's last letter to BSZ from Karachi. After several days of familiarization, he relieved AZ, and as CO, he then signed orders detaching AZ to return to the U.S. on 9 April 1945.

Appendix A

AWZ Wartime Papers

The AWZ Wartime Papers were assembled into 15 file folders in two portable file boxes (total, 1 linear foot); and the following items prepared by his wife (BSZ): a crown folio-sized leather-bound scrapbook with 36 pages, items attached recto and verso; a smaller leather-bound photograph album containing 181 photos (7 pages, maximum of 28 to the page); two reels of 16mm movie film (copied onto VCR and DVD); an envelope of approximately 250 loose photos (10x6x3in., including 130 photos taken on the NWFP trip); and a string-tied cardboard box of 119 letters (12x9x4in.) from Zimmermann to BSZ. The original handwritten letter of 25 November 1943 was in this box of letters, and also the second letter, which was started on 6 December and finished and mailed in Karachi on 15 December 1943. The letters were dated and each was given a sequential number when they were written by Zimmermann. These letters have been transcribed; they now occupy 167 pages of typescript. All envelopes were stamped "Passed by Naval Censor" and were usually initialed by the censor. Most of the postmarks were "U.S. Army Postal Service / A.P.O. 886" [APO]. The two letters about the NWFP trip were passed by the military censors without any deletions, and only a half-dozen or so words (principally, names of places on commercial postcards) were elided by censors from all of this correspondence.

Zimmermann's Navy career, in brief, was as follows: He was granted a waiver for colorblindness and was appointed LT, I-V(P), USNR, on 9 September 1942, with rank dating from 1 August 1942. He was called to active duty in October 1942. After completing basic and advanced intelligence training, his designator was changed to LT, I-V(S). He departed from New York for NLO Karachi on 1 July 1943, and he returned to the U.S. from Karachi on 1 May 1945. He was granted terminal leave in September 1945 and he was released from active duty on 30 December 1945. He was promoted to LCDR, I-V(S), in January 1947, and he resigned his commission and was honorably discharged on 29 August 1947.

The AWZ Wartime Papers were accepted by the U.S. Naval Institute in 2013, and they are now in its Special Collections in Annapolis, Maryland. The AWZ Wartime Papers consist of the following:

Box 1:
- File 1 – Brochures and Magazines
- File 2 – Bromhead letter file
- File 3 – Correspondence with family and civilian friends
- File 4 – Chitral trip
- File 5 – Humor
- File 6 – India-North West Frontier (miscellaneous, includes other items)
- File 7 – Instructional Materials
- File 8 – Loose photos in file folder
- File 9 – Maps, typed letters, notes
- File 10 – Miscellaneous correspondence
- File 11 – Navy Record documents
- Extract from File 9 of Baluchistan trip by AZ, 1944

Box 2:
- File 12 – Newspapers and clippings
- File 13 – Travel, orders, receipts
- File 14 – Socials
- File 15 – Winsor and ONI

Oversized:
- AWZ-WP A Letters to BSZ
- AWZ-WP B Loose photos in India WWII envelope
- AWZ-WP C Photo album
- AWZ-WP C Scrapbook
- AWZ-WP E Movies, 16mm, 2 reels, copied on VCR, and on DVD 4-10-11

Appendix B

Albert W. Zimmermann's Biographical Sketch, summarized from his Navy papers

- Commissioned 8 September 1942 as Lieutenant, I–V(P), USNR, Date of Rank 1 August 1942
- Information submitted / 27 January 1943:
  - born 11 June 1902, Philadelphia, Pa.
  - Resided at 400 Rose Lane North, Haverford, Pa.
  - B.S. in M.E., University of Penna. [1923]
    - Sigma Tau – Honorary Engineering Fraternity
    - President of Glee Club
    - Sphinx Honorary Senior Society
  - Married to Barbara Shoemaker, 1926. Four children
  - Member of 4 Clubs: Philadelphia Country, Merion Cricket, Fourth Street, Orpheus
  - Career: Formerly V.P., Artloom Corp. Textile Mfrs.; Partner in Ott & Zimmermann (Wool Dealers), 1937 until commissioned in 1942
- Commissioned 8 September 1942: Lieut., I-V(P), USNR 205065 (with waiver for defective color perception)
- Reported for duty, 31 October 1942
- Naval Intelligence School, Nov-Dec 1942
- Indoctrination Course, Dartmouth College, Dec 42 - Jan 43
- Promoted to Lieutenant, I-V(S), USNR
- Naval Language School, Arlington, Va., (French), 6 wks, under orders of 12 Feb to report to Dakar; cancelled 13 Apr
- Advanced Naval Intelligence School, New York, 19 Apr-1943-8 May 1943
- Awaiting orders, Washington, DC, 1-30 June 1943
- Posted to Naval Liaison Office, Karachi, India, 1 July 1943

Background for the Anglo-American Mission on the NWFP Trip, Nov-Dec 1943

- AWZ's official Navy records & orders
- AWZ's correspondence with his wife (AL & TLs)
- AWZ's Trip notes (AN), Handwritten draft report (ARd), Typed draft report (TRd), Typed report (TR)
- AWZ's Photos with captions (AC)
- AWZ's Wartime scrapbook (S)

A = Autograph (handwritten)   L = Letter   T = Typed   R = Report   N = Notes

Appendix C

Maps used by Zimmermann on his North-West Frontier Trip

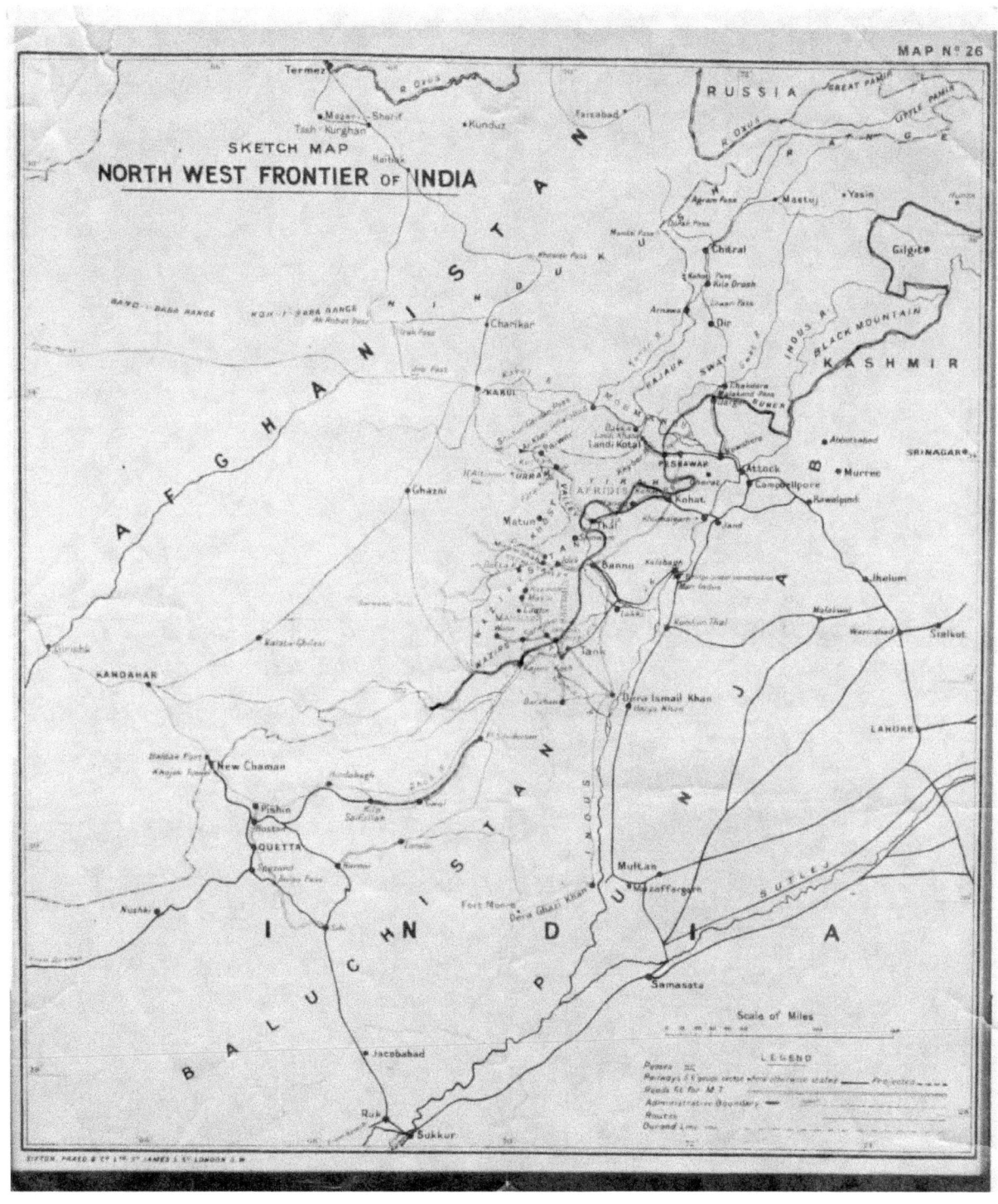

Route marked by Zimmermann, after the trip was completed

337

Peshawar and North Waziristan from Kabul map

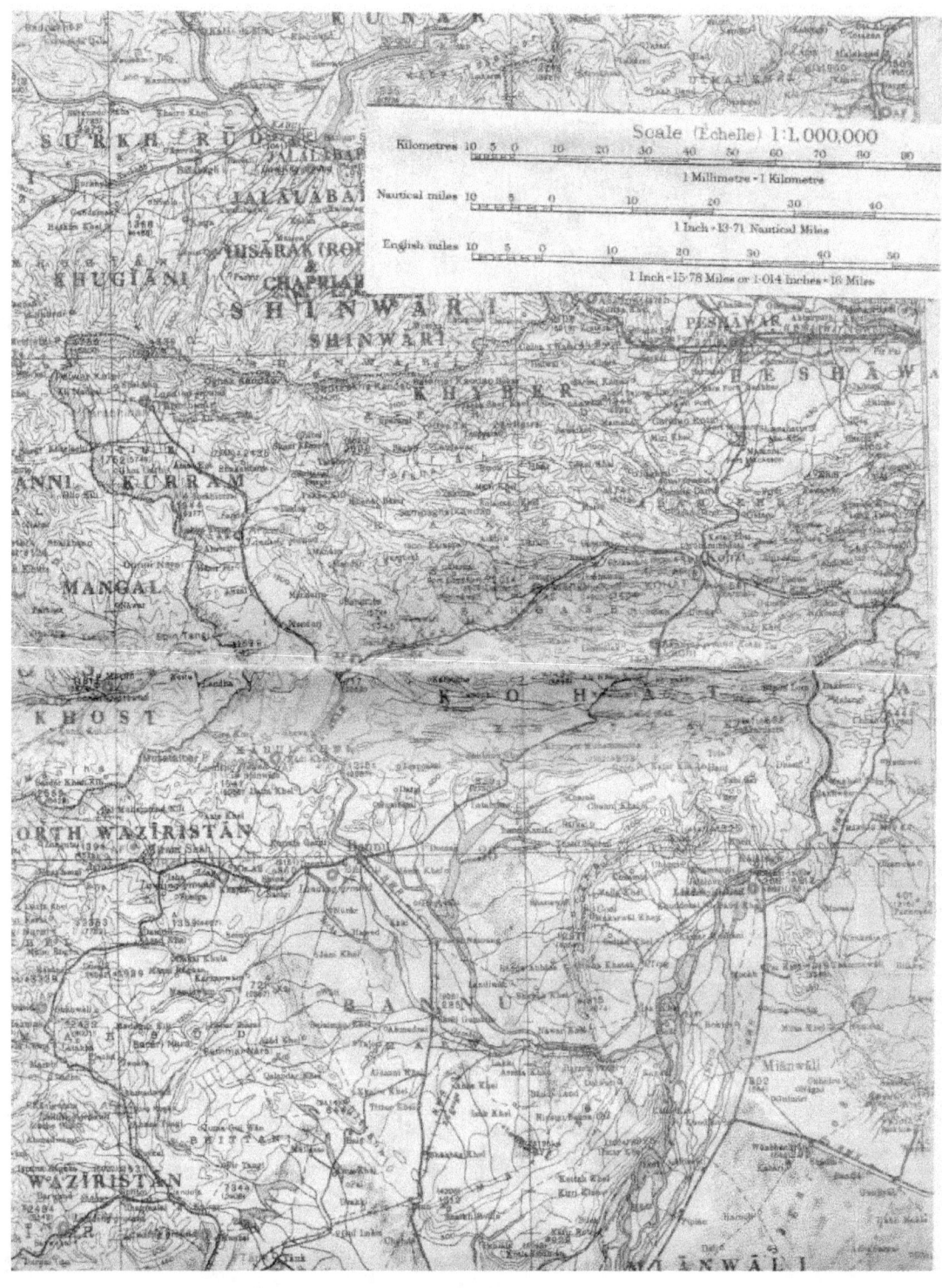

North and South Waziristan from Kabul map

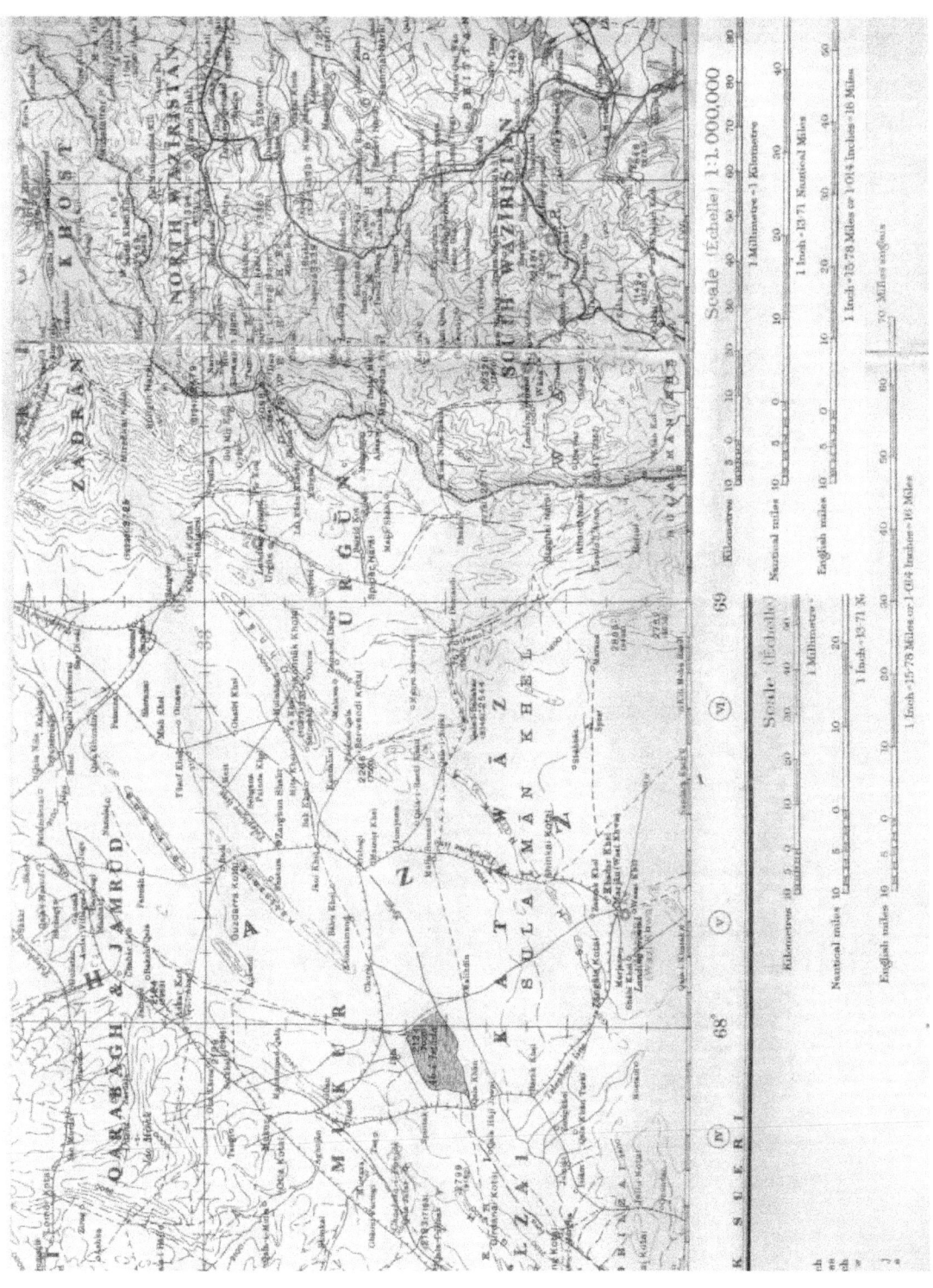

South Waziristan to Baluchistan, Sulaiman Mts map

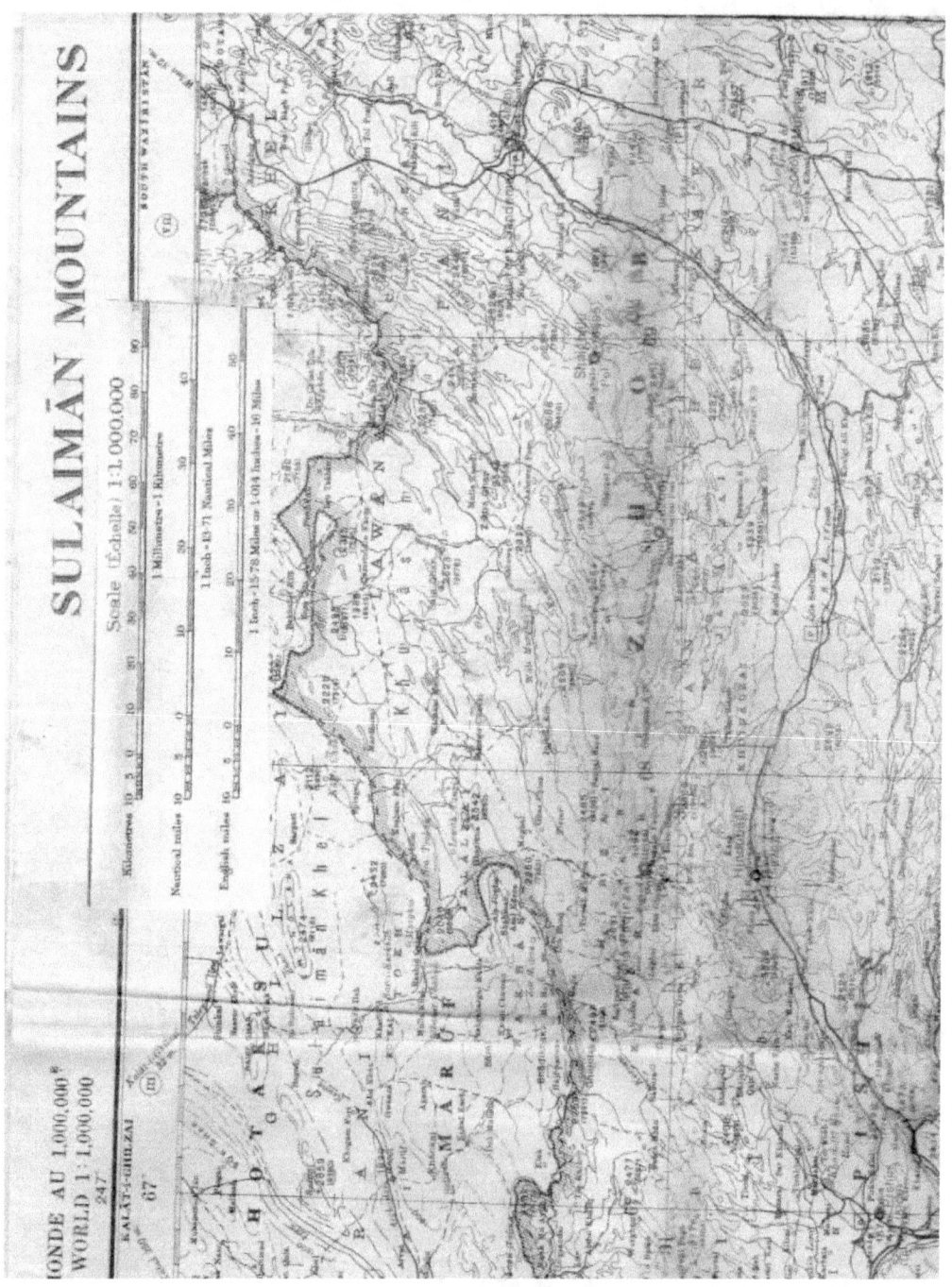

Baluchistan to Quetta on Sulaiman Mts map

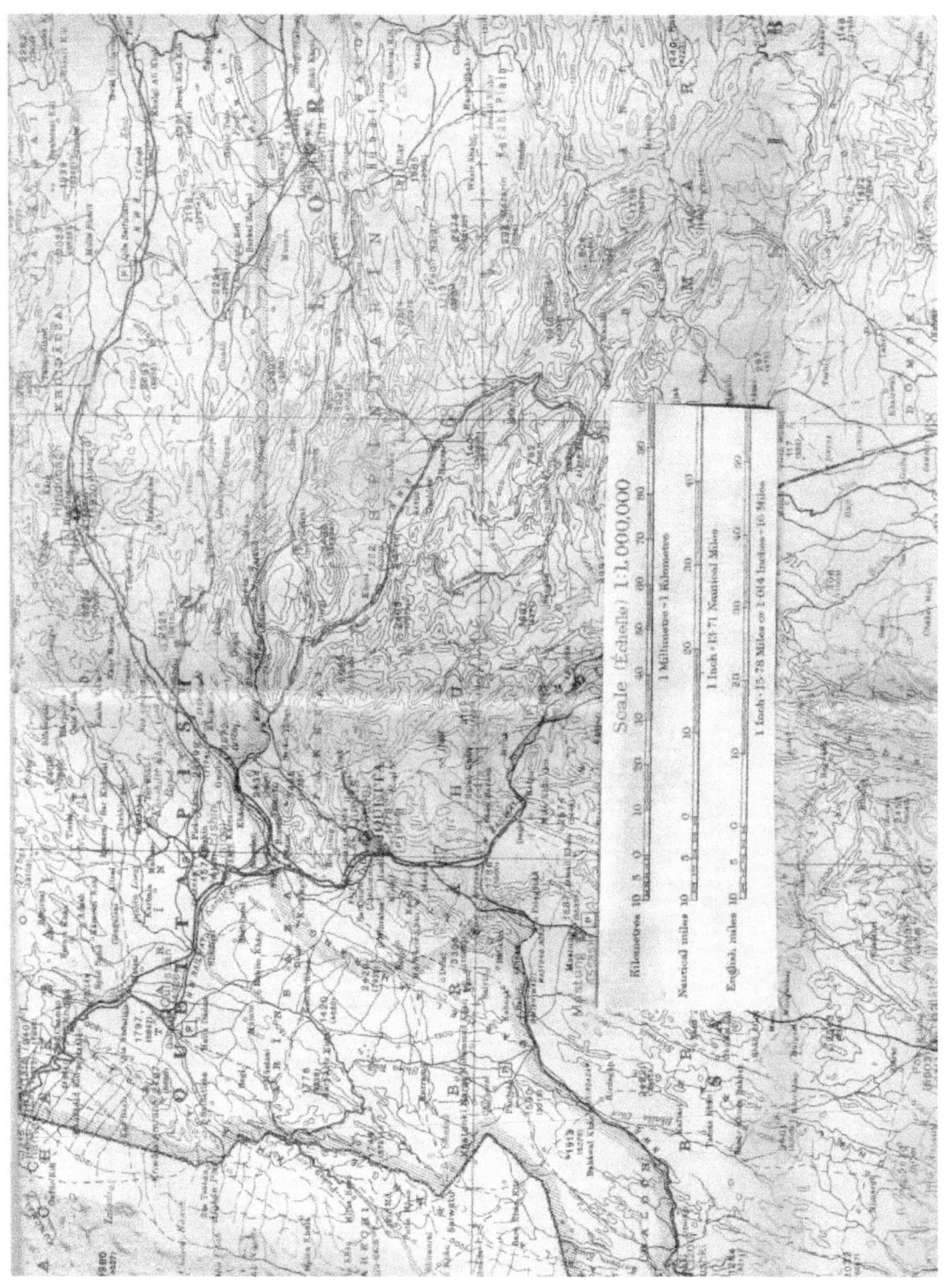

Appendix D

U.S. Naval Liaison Officer, Karachi – Intelligence Report, 5 July 1944

ISSUED BY THE INTELLIGENCE DIVISION
OFFICE OF CHIEF OF NAVAL OPERATIONS
NAVY DEPARTMENT

INTELLIGENCE REPORT

Serial 18-44
Monograph Index Guide No. 104-100

From U. S. Naval Liaison Officer at Karachi, India Date 5 July 1944

Reference

Source: Central Intelligence Officer, Karachi   Evaluation B-2

Subject: INDIA - Political Parties and Groups

BRIEF:

Submitted herewith is a transcript of a lecture delivered to British Army officers at Delhi about the first of May, turned over to this office by J. Harris, Esq., Central Intelligence Officer, Karachi. Unfortunately the lecturer could not be identified.

It is valuable in that it gives a background of the Indian political situation and defends the British for the actions they have taken in the past up to the time of Gandhi's release from confinement and probably presents the best British opinion on the Indian political situation as it stands today.

DECLASSIFIED
E.O. 13526, Sec. 3.3
NW 36725
By DM   NARA, Date 12-8-11

Distribution By Originator: JICA/CBI, ONI.

JICA/CBI #4048

A-3-0
MID (5)
OSS via 16-Z (2)
State (2)
FE-2
FE-Eile
Spare

RETAIN OR DESTROY

THE CONSTITUTIONAL ISSUE IN INDIA

I have been asked to talk to you this morning about "The Constitutional Issue in India". It is a large subject and later on I shall have to concentrate on one aspect of it. To begin with, however, what the problem itself boils down to is, I think, the answer to two questions –

(1) Who is to govern India?
(2) How is India to be governed?

By the second question I mean what the form of government is to be – not only whether it should be unitary or federal, or whether India should be divided into two or more separate countries each with its own Government. This latter question about the form of government is no less important than the former about who is to do the governing; but it is the first question that occupies men's minds almost exclusively and it is amazing how little attention is given to the second.

As to the first question – who is to govern India – taking it by and large and leaving out of account for the moment any political pronouncements or pledges made by H.M.G., there are, I suppose, three possibilities –

(1) that India should continue to be governed as at present by the British with the assistance of Indians;
(2) that India should be governed by Indians alone; and
(3) that India should be governed by Indians with the assistance of the British.

The difference between the first and last of these possibilities is that under the first the Government of India, however composed and however many Indians it might contain, would remain responsible to the British Parliment, while under the last it would <u>not</u>. Englishmen might still be members of the Government or at any rate of the Govt. services, but the Government as a whole would be responsible to an Indian electorate and control from Whitehall would cease.

Now what strikes one most is the negative character of all thought on this aspect of the problem. Everyone is agreed – and by everyone I mean the vast majority of educated Indians – that they do <u>not</u> want to be governed by the British. They reject the first alternative. It may be true – it probably is true – that 90 or 95 per cent of the actual population are not deeply interested and are probably much more concerned, if they are concerned at all, in good government than in self-government. But even 5 per cent of the population of India is an exceedingly large number and it is of course the educated and vocal section of public opinion that counts.

As soon, however, as you get to the second alternative – that Indians alone should govern themselves – unanimity ceases at once and it is here that you get the growth of parties with conflicting views as to who should – or rather should <u>not</u> – do the governing. These parties are loosely called political parties; but they are not. They are communal parties. The Hindu Mahasabha frankly claims that the Muslims are a minority and therefore should <u>not</u> be allowed an equal – much less of course a predominating – share in the government. The Muslim League say they are <u>not</u> a minority, but a separate nation, and they will <u>not</u> submit to a government in which Hindus are in majority. The Indian National Congress say that they will not do business with any other party except on the basis of their claim to represent both the Hindus and the Muslims – indeed the whole of India. This Congress claim is not accepted by the Muslims, or indeed by a good many other people like the Depressed Classes; and everybody knows that Congress is a predominantly Hindu body.

The third alternative – that the government of India should be a joint Indo-British affair responsible to the people of India and not to the people of England – has never been seriously discussed. I mention it –

(a) because it is a possibility;
(b) because I am not one of those who hold that the British connection with India is something to be ashamed of or the complete severance of which would be a good thing in itself. We may have made our mistakes but we have done immeasurably more good than harm in India/

Page 2.

(c) because it is obvious that under any scheme of self-government for India, India will, at any rate for a temporary period, be literally unable to carry on without some British assistance in certain matters. The most of important these matters is, of course, defence; and whether the assistance is given under a treaty with a Free India or in some other constitutional way, the practical result will be much the same. It is, however, fundamental that in no Dominion of the British Commonwealth can British troops come under the control of any authority other than the British Parliment; and that position would, of course, have to be maintained.

Now, let me turn to my second main question - how is India to be governed? What is to be the form of government? We have seen that while there is practical unanimity in the desire of Indians not to be governed by the British and a great deal of talk about a National Government and the national aspirations of the Indian people, there is no agreement as to who should run that Government. Is there any greater measure of agreement as to the form that the Government should take? Unfortunately no. I do not think I need bother you with any detailed account of the various possible forms of government. Nor need I trace at length how the democratic form of government, with its great principle of the responsibility of the executive to a legislature elected by the people, has gradually developed in this country, (as a result, mind you, of British rule) and displaced in men's minds the natural or traditional form of government in the East, namely, autocracy. The whole of this development has been most ably recorded by Professor Coupland in his recent books about India undertaken on behalf of Nuffield College, Oxford. The books themselves, three volumes, are fairly stiff reading, but each volume contains a most admirable summary, and those of you who are interested should make a point at least of reading these summaries. I must, however, give you a brief account of the more salient points and I cannot do better than borrow freely from one of the summaries I have just mentioned.

From the outset of British rule, India has been divided into two parts under different forms of government -
(1) British India comprising a number of Provinces (now 11), under direct British administration; and
(2) the Indian States, which, by treaty or usage, are, broadly speaking, autonomous as regards their domestic affairs but accept the suzerainty of the Crown and its control of their external relations. The government of the Indian States has maintained with modifications the oriental tradition of autocracy.

Till 1919 the Government of British India was completely controlled by the British Parliament operating since 1858 through the Secy. of State for India. It was highly centralised administration. The Provincial Governments were agents of the Central Government and under its legislative and executive control. So early as 1833 however Englishmen were contemplating the ultimate withdrawal of British rule from India and by a series of Acts in 1861, 1892 and 1909 the representative principle was recognised by the admission of Indians to the Legislative Councils in the Provinces and at the Centre. But British statesmen of all parties (and this is an important point) emphatically asserted that this development was not intended to lead to Parliamentary government as practised in Britain. Indian society, it was held, was so diversified by race, creed and custom as to preclude the normal operation of majority rule. Meanwhile, however, the Indian Nationalist movement, which had begun with the birth of the Indian National Congress in 1885, had become to regard the contititutional development of the British self-governing colonies and the democratic principle which this development contained as the model for Indian advance.

The movement was intensified during the last War and in 1916 one great obstacle to it seemed for the moment to have been overcome by an agreement between the Congress and the Muslim League which was known as the Lucknow Pact. The most important part of that Pact was the agreement of Congress to concede separate electorates for the Muslim community, that is to say, Muslim seats to be held by Muslims elected only by Muslim voters. Such an arrangement, I may remark is of course quite undemocratic. In 1917 the British Government defined its policy as "the progress realisation of responsible government in India as an integral part of the British Empire." This policy was embodied in the Constitution Act of 1919 (generally known as the Montague-Chelmsford Reforms) and the most important provisions of this act were as follows:-

(1) it established a measure of Provincial Autonomy by devolving authority in Provincial matters on to the Provincial Governments and freeing them to a large extent from Central control. It began the process of realising responsible, as distinct from representative, government in the Provinces by dividing the field of government and introducing what was known as dyarchy. Certain subjects like law and order were reserved to the control of the Governor and his Executive Councillors responsible, as before, to the Secy. of State and Parliament. The rest of the field was transferred to Indian Ministers responsible to their Provincial Legislatures. And so for the first time you got a departure from the complete control hitherto exercised by the British Parliament.

This policy also signified a definite change in the British attitude to the constitutional problem. Parliamentary government of the British type, for which dyarchy was to serve as a training. The authors of the policy did not evade or minimise the difficulties in the path, especially Hindu-Muslim antagonism, but they expressed the hope that such obstacles would be overcome by the patriotic cooperation of all communities in the common task of self-government. Separate electorates for Muslims and certain other communities were, however, retained; and, looking back, one cannot help wondering whether that was not a mistake. During the next ten years a real attempt was made to carry out this policy; but the hopes of 1919 were mostly disappointed. In the first place, the Congress, far the most powerful political organisation in India, became under Mr. Gandhi's leadership a quasi-revolutionary body, pledged to obtain complete independence by all non-violent means. It therefore rejected the Act of 1919 and refused to cooperate in working it. Secondly, partly because of the Congress attitude and partly because communal divisions, based on, and encouraged by, separate electorates, prevented the development of a true party system, dyarchy provided no effective training in parliamentary government. Thirdly, and this was the most discouraging feature, Hindu-Muslim antagonism increased. This increase was however only to be expected. The prospective removal of British control naturally led to the scramble for power after that control had been removed, and the differences between the two great communities, which had remained quiescent so long as both were governed by a neutral third party, immediately sprang to life.

The next ten years, starting with the appointment of the Simon Commission in 1927, were spent in devising further constitutional reforms in the face of unremitting opposition and two long civil disobedience campaigns by the Congress Party. The new Government of India Act was passed in 1935 and came into force in the Provinces, but not at the Centre, on April Fools Day, 1937.

The main provisions of the new Act were as follows:-
(1) It completed the development of Provincial Autonomy by liberating the Provinces entirely from Central control except for a few specific purposes. (it is true that powers of control except for a few specific purposes, have been largely restored during the war emergency, but that is only a temporary measure).
(2) It established full responsibility to Parliamentary government subject to certain safeguards in all the eleven Provinces.
(3) It was intended to establish, though in fact it has never

done so completely, the Federation of India, comprising both Provinces and States, with a Federal Central Government and Legislatures for the management of Central subjects.

(4) Dyarchy was abolished in the Provinces, but was to be reproduced at the Centre where the subjects of Foreign Affairs and Defence were to be reserved to the control of the Governor General and the other Central subjects were to be transferred to Ministers.

The Act thus confirmed and extended Parliamentary government in the Provinces and was intended to introduce it at the Centre, but it still retained separate communal electorates, both Provincial and Central, and distributed the seats on the lines of a communal award which had to be given by His Majesty's Government because the parties in India could come to no agreement among themselves. It was perhaps too late in the day to go back, but here again one may ask what was the use of constructing a democratic edifice on an entirely undemocratic foundation?

Well, the Congress won the elections in seven out of the eleven Provinces, but it was not until July, 1937, that they agreed to take office. For a brief period of about two years the Congress governed seven out of the eleven Provinces. They resigned under the orders of their High Command - the Congress Working Committee - shortly after the beginning of the war because the Congress High Command refused to support the war effort. In the remaining four Provinces non-Congress Ministries came into power, and, after various vicissitudes which I need not mention, six Provinces out of the eleven are now governed by the Governor with certain official advisers. The breakdown of the constitution in these Provinces has in fact taken them right back to the completely official or bureaucratic administration of 1860. The failure to introduce the full federal scheme at which the Act of 1935 aimed was due partly to the reluctance of the Rulers of Indian States to acquiesce in the requisite infringements of their sovereignty which the scheme entailed, and partly to the fact that both the Congress and the Muslim League repudiated the Federal scheme.

I have given you this brief historical survey because I wish to bring out two main points against its background. In the first place, during the whole of this period between 1917 and the present day, there have been only two attempts, so far as I know, by Indians themselves to put forward a detailed constructive scheme for the government of their own country - and one of these, if I may be permitted an Irish Bull, was the work of an English woman. This was Mrs. Annie Besant who, with the help of certain moderate Indian politicians, put forward a scheme for the government of India in 1924 and went so far as to draft a "Commonwealth of India Bill" to give effect to it. This scheme was never seriously considered by the Congress Party. Four years later, in 1928, a fresh scheme was put forward in what is known as the "Nehru Report". The Nehru in question was not Pandit Jawaharlal Nehru, but his father Pandit Motilal Nehru, at that time the leader of the Congress Parliamentary party. Other signatories included Sir Tej Bahadur Sapru and a prominent Muhammadan, Sir Ali Imanam. The scheme itself was based on the principle of Dominion Status within the Empire and aimed at securing a balance between the Hindu-majority provinces and the Muslim-majority Provinces. It insisted on the abolition of separate electorates. This Nehru scheme was almost immediately rejected by the Muslim League under Mr. Jinnah, and although it held the field for a short period in Congress circles, it was repudiated at the end of 1929, when Congress finally rejected all idea of remaining within the British Empire and voted for the attainment of complete independence. These, as I say, are only two constructive proposals that I am aware of. The only other proposals made by Indians which might, I suppose, be called constructive in one sense are:-

(a) The Pakistan scheme put forward by the Muslim League in 1940 and
(b) the "Quit India" scheme promoted by Gandhi in 1942.

One great defect of both these schemes (not to mention, of course, numerous other objections to them) is their careful avoidance of detail and their delightfully vague nature. In the first case, the Pakistan scheme has never gone beyond the very general principle that the Muslims are not a minority but a separate nation and have the right to govern their own homelands. None of the practical details has been worked out and no attempt has been made to meet and overcome the serious difficulties, financial and otherwise, which are involved. Professor

Coupland has made an attempt to do this in his last book about India, but I think he is the only person who has attempted the task. His own feeling, with which I personally agree, is that any partition of India would be extremely difficult in practice, as well as most regrettable in principle. And he suggests that the only way to avoid it is to devise some scheme - by geographical regrouping or otherwise - which will bring into existence a Central Government in which the two great communities will have an equal share. It is of course Hindu control at the centre - and particularly Hindu control of the Army - that the Muslims are most afraid of.

In the second case, - Gandhi's Quit India Movement - there was even less attempt to work out the conception. Indeed, in Gandhi's own words, it was a case of leaving India to anarchy - or to God.

The second main point I wish to bring out is that, in spite of the talk about democracy and parliamentary institutions, in spite too of the discussion of fine ideas like Federation, and in spite of the provisions made in the various Government of India Acts to bring these ideas into practice, there has, in fact, been no real attempt in India up to the present day to work democratic institutions as they should be worked, nor has there been the slightest readiness among the various classes and communities in India to cooperate in the working out of any large conception such as Federation, or to give up one iota of their communal demands in order to bring into existence a truly National Government. You may say that I have just told you that full Parliamentary Government was established in the Provinces in 1937. That may be so in form; but let us see what has happened in practice. The Congress Ministers in the seven Congress Provinces remained throughout under the autocratic control of the Congress Parliamentary Board, which in fact took its orders from the Congress Working Committee, which again, in fact, took its orders from a single man, Mr. Gandhi. It may be said that any government which governs with the support of the majority of the people is a democratic government and that the Congress Ministries had such support. This, however, could equally be said of Hitler's government, but no one, I think, will maintain that the Government of Germany is a democracy. Again, the Congress Ministries made no attempt to regard the other Parties in their Provincial Legislatures as a legitimate opposition.

Their main opponents were the Muslim League; but instead of admitting them as a constitutional opposition party, or even trying to form some kind of coalition with them in order to carry on the Government more successfully, their one object was to absorb the Muslim League, and indeed all minority Parties, into the Congress organization itself. This ill-conceived attempt failed; and the only result was a vigorous campaign against the Congress Governments by the Muslim League under Mr. Jinnah's leadership and, eventually a great increase in the strength and influence of the League. Jinnah himself finally declared that democracy based on a majority rule could not work in India and, when the Congress Governments resigned, he ordered the observance of a "Day of Deliverance" and said they must never come back.

But you will say what about the non-Congress Governments? Have not they governed democratically? I am afraid that the general answer here too is "no". The only Province having a stable Ministry was the Punjab which has continued successfully in power since 1937. The success of this particular Ministry seems to me to be the one bright spot in an otherwise murky landscape, but even this success has been partly due to certain special circumstances. In the first place, personalities. It has had exceptionally good men such as the late Sir Sikandar Hyat Khan. In the second place, it has throughout taken the form of a coalition in which both Muslims and Hindus are represented. In the third place, the Province, mainly owing to the war, has enjoyed a period of exceptional prosperity. In the other non-Congress Provinces - Bengal, Assam, Sind, and more recently, Orissa and the North-West Frontier Province - government, so far as Ministers are concerned, has consisted largely in countering and defeating the personal intrigues of their political opponents; and they have been so busy trying to keep their own seats in the saddle that they have not been able to spare but little attention to guiding the horse. There is, indeed much to be said for another suggestion made by Professor Coupland that democratic institutions in India, if adopted at all, should follow the Swiss model, under which the Ministry, although democratically elected and given a limited lease of life, has complete security of tenure and cannot be turned out by an adverse vote of the Legislature during that limited period.

So far as the Centre is concerned, there is, of course, no attempt at real parliamentary govt., although this is often misunderstood, especially perhaps in England or America. We have an elected Legislature and we play at parliamentary methods. But the Government of India – the Viceroy's Council – are merely a set of individuals chosen to represent not the same, but different interests; they are not connected by any common programme beyond, of course, winning the war; they have no elected party behind them; they have no Party Press even; and they enter the Legislature in a permanent minority. They are often defeated and, as things are, they can always be defeated. But under the Consitution they cannot resign and appeal to the country, nor can they say to the opposition – "If you defeat us you must yourself accept responsibility for carrying on the Government". The result is, and always has been, a direct encouragement to responsible destructive criticism, since the opposition know very well that in no circumstances can they be called upon to accept the onerous task of governing the country.

What we want to know is what is going to happen now or in the near future. Is it not a fact that H.M.G. through Sir Stafford Cripps has offered to accept any scheme for the government of India that Indians may themselves put forward? Does not this offer still stand, and does it not, in fact, amount to an offer of independence if Indians so wish? If so, why do Indians go on speaking as if the struggle for freedom, as they put it, were still to be fought against an obstructive Government? On the other hand, how do Government expect Indians to put forward a scheme for the government of India while their leaders are kept in jail and cannot even consult with one another?"

Well, of course, those are the burning questions of the day and my previous remarks have been intended to lead up to answering them. It is true that the Cripps offer stands. As soon as this war has been won, Indians can settle for themselves how they are to be governed and by whom. They could start doing so now, but so great are their internal differences and so remote are the chances of their being able to work the democratic principles of political freedom for which they have been clamouring that an atmosphere of frustration has settled down over the whole political scene and the instinctive reaction, born of the teachings of the last 25 years, is simply to lash out at Government and lay the blame on them. If you ask me who is to blame, if Government are not, my answer would be – one man – Gandhi. I am sorry to say this, but I believe it to be true. Gandhi may be a great man, I think he must be a great man to have acquired the amazing influence over millions both in Indian and abroad which he undoubtedly possesses. He may be a great social reformer a notable ascetic, even a saint, but as a political leader, in my opinion, he goes right down to the bottom of the class. It seems fitting that I give you the reasons for my opinion.

(1) even from his own point of view and that of his countless followers Gandhi has failed to deliver the goods. None of his movements, based as they have been on his faith in the magical power of soul force and his creed of non-violent passive resistance, has ever succeeded in attaining, even partially, the objects at which they were aimed.

(2) On the other hand, the only results of his policy of non-cooperation have been violence, suffering, loss of life, damage to property and, above all, ill-will. And he must know it. Listen to what he said when he pleaded guilty at his first important trial in 1922:—

"As a man of responsibility, a man having received a fair share of education, having a fair share of experience of this world I should know the consequences of every one of my acts. I knew this. I knew that I was playing with fire. I ran the risk and, if I were set free, I would still do the same."

He has done the same again and again. If he were released tomorrow there is no guarantee that he would not try to do it once more.

(3) Gandhi's object has always been to lower the authority of the State. He has glorified jail going and disobedience to the law; and no Govt. can stand by and allow its laws to be broken with impunity. That surely cannot be good training for any people about to embark on the task of governing themselves.

It has been well-put in this way -
> "Blind to the lessons of history, Gandhi had taught
> men, while exalting God, to despise Caesar. But in
> India, Caesar is one, while God is worshipped in many
> forms, whose adherents dwell in mutual toleration only
> through Caesar's constraint. Inevitably, therefore
> Gandhi's doctrine has brought not peace, but a sword
> to his luckless country".

(4) His whole teaching has always been negative and obstructive rather than positive and constructive. He has thereby encouraged those traits in the oriental character which have always lent themselves to putting talk before action and to preferring bargaining and debate to the acceptance of responsibility. Again, surely, the worst possible training for self-government and surely responsible for the present atmosphere of frustration.

(5) Above all, I believe, his idea of non-cooperation is all wrong. It is bound, human nature being what it is, to breed ill-will, and it entirely fails to mobilise the enormous forces of good will which are there for the asking and which, I profoundly believe, will always in the long run exercise a more powerful influence, in all human relations, whether between man and man or between nation and nation. A quarter of a century of non-cooperation has produced lamentable results. Is it not time to give co-operation a chance?

There are of course many even among Gandhi's own adherents who profoundly disagree with his methods and with many of his decisions. I would bet, for instance, that most Congressmen would now admit to themselves - to take only the most recent cases - that it was a bad mistake -
(a) to make the Congress Ministries resign in 1939,
(b) to reject the Cripps offer in 1942, and
(c) to launch the "Quit India" mass movement.

And yet there were all mistakes made by the very same leaders whom they now want back. On every occasion - and there have been several - on which Gandhi has for the time being left the centre of the stage, there has always been a party of no less patriotic Indians who have shown themselves ready to cooperate instead of to non-cooperate. The tragedy is that these people, and I believe their numbers are considerable, have never been sufficiently sure of themselves and have never been allowed to go very far. Why? Because you have in Gandhi a man who has set the standard for all political activities in this country, who commands the unquestioning obedience of millions, to whom practically every politician has surrendered his conscience, who must not be offended, and without whom nothing can be done. To my mind also, if people would only see it, there was never a better opportunity for doing so than today. Never surely was there a moment that called most urgently for constructive thought and action. Never surely, with Congress leaders removed from the scene, was there a better chance for India's statesmen - and India's soldiers too - to step forward boldly and put an end to the blighting and deadening influence which has clogged the wheels of national progress for so long. And yet what do we find? The whole situation is very clearly xxxxxxxxxx exemplified by three things that happened during the last session of the Legislative Assembly in Delhi.

(1) At the beginning of the session a member gave notice of a Resolution that the Assembly would appoint a committee to draft a new constitution for India. Notice was, however, sent in of so many amendments from all sides of the House, showing that there was no hope of agreement, that the mover took fright and did not even move his Resolution.

(2) Later on another Resolution was moved recommending the release of political prisoners, which means, of course, Gandhi and his colleagues, <u>with a view to removing the present deadlock and furthering the interest of the war effort</u>. Congress policy, of course, has always been a mixture of opportunism and of "heads I win, tails you lose". Whenever they failed in a particular project, as they have failed in their "Quit India" movement and as they have failed in every other movement, their one idea is to rehabilitate themselves and get back into a position in which they can start the same kind of game again. And yet on this occasion in the Assembly you heard Member after Member -

Page 8.

some of them strongly disagreeing with the policy of the Congress - getting up and demanding the release of the very people who had themselves brought about the so-called deadlock when they were at liberty, and whose whole war record showed that they never had the slightest intention of supporting the war except on their own terms - that is to say as a means of extracting the last ounce of their own political demands, and (incidentally) the right to negotiate with Japan. The Home Member in a speech that I wish you could all have heard, gave chapter and verse, by quoting Congress declarations and writings, and to show that in 1939, 1940, 1941 and 1942 the Congress, under Gandhi's leadership, had scornfully and summarily rejected every attempt made by H.M.G. and the Viceroy to reach a political settlement. He also gave their disgraceful war record step by step and rightly denounced them in the following passage:-

> "If we are told that the war effort cannot be prosecuted or that the constitutional problem cannot be solved without such aid as the Congress have shown themselves minded to give in the past, I would assert that in time of war the affairs of the country must be in the hands of people who will not desert their constituents, will not refuse to share responsibility of the war administration, will not panic in the face of danger, and will not exploit that danger to their own advantage; and before we can feel confident in people who have done all these things, we are at least justified in asking for some definite assurance that there will be no more obstruction of the war effort, no more attempts to set the country in a turmoil and, most of all, that they are unequivocally in favour of using all the resources of India for the prosecution of the war against the Axis powers, and in particular, Japan, until victory is won."

(3) Finally all parties in the Assembly combined to defeat Government over the Finance Bill. Thus the session began with a complete failure to combine for a constructive purpose. It went on to demonstrate that no one is yet strong enough to stand up against Gandhi. It ended with a complete readiness to combine for a purely destructive purpose.

There you have it in a nutshell.

What then is to be done? If Government are convinced, as I think they are, that the unconditional freedom of the Congress leaders would be prejudicial to the war effort at this juncture; if further they believe, as I think they do, that the activities of these leaders have never been in the best interests of India herself, you will I hope agree that Government are fully justified in rejecting a policy of appeasement towards them. But is this sufficient? Is it wise for Government simply to sit back and do nothing in the political sphere? Many of you will no doubt say that we have had quite enough to do in winning the war and tackling the various administrative and economic problems which the war has produced. Others may say that a larger measure of popular support would certainly make it easier to deal with some of these problems, especially perhaps the food problem, and that something ought to be done to secure that support. I would agree with this latter view to a certain limited extent. Government do I think sometimes almost go out of their way to encourage the impression that they have no friends and that they are content to act as a pure bureaucracy. There are quite a large number of people in India who are prepared to support Government if they are given a chance to do so. Something more might be done - this is only my own personal opinion - to give such people the chance especially in the Provinces where there are no Ministries. But that by itself is not going to solve the problem. Nor do I think that any fresh move by H.M.G. or the Government of India in the constitutional field - any fresh offer or anything of that sort - would do any good at this moment. It would only be regarded as another Aunt Sally to be bombarded by critical coconuts. The moment must come - and in my opinion has already arrived - when Indians will have to realise that they must do something for themselves and not simply criticise every suggestion made by Government. Therefore to my mind the only solution lies in trying, and going on trying, to divert men's minds from the political frustration of the moment and get them to concentrate on the practical problems of the future. This is mainly a propaganda problem. And the way I would tackle it would be to start controversies - in the press, on the air, through pamphlets, and in every possible way - o the form of the future Government of India. If you could get people discussing that sort of thing - on the basis of course that they were going to govern themselves - I believe that gradually you would get less and less dispute about who is going to do the governing, and certainly less and less distrust of British motives. The great misunderstanding has always been that the British Government are the opponents of Indian nationalism as such. The root cause of most of the political unrest in this country is the sense of inferiority which Indians feel because they think that Great Britain will not trust Indians to govern themselves and do not want them to do so. They do not realize that there has been a very big change in public opinion about India. I believe it was very largely for that reason that Cripps offer was rejected. It was too good to be true. There must be a catch somewhere. If that misunderstanding could only be removed; if it could be brought home to the people of India that further talk about "the struggle for freedom" is mere heroics, because freedom has already been granted and the people of England wish them to enjoy it, if they could only combine for a constructive purpose and get rid of Gandhism; and if they would only realize that Government have never opposed national progress in India, but only the illegal methods followed by a particular party with the object of obtaining a short cut to the sort of power they want for themselves, then I believe the present sense of frustration might disappear and Indians might wake up one fine day to realize that they could get on much more quickly and more happily, without the particular old men whom they have regarded as indispensable for so long. Whether anything of this kind will happen I cannot say. I am afraid it is not likely. All I do say is that if it does not there can be no solution to the present so called deadlock that will lead to any permanent or fruitful results. I will conclude with a quotation which I think ought to be printed every day in large letters in every newspaper in India until everyone knows it by heart. "A genuinely national Government cannot be imposed; it arises from the sinking of differences among sections of the people for the achievement, through an agency which the whole nation can trust, of a purpose known to be desired by the whole nation."

Appendix E

Colonel Harry Reginald Antony Streather, OBE

Telephone conversation
20 January 2010
George J. Hill (H) and "Tony" Streather (S)

About 3 p.m. EST

The list of climbers on K2 in 1953 in the front pages of Houston and Bates, *K2: The Savage Mountain* shows his name as H. R. A. Streather. An article in *The Sunday Times* of 2007 says that Streather lived in Hindon, Wiltshire. I therefore put his name as H. Streather and this town into a search on phone book u.k. and got:

> H Streather
> Apple Tree Cottage
> High Street
> Hindon, Salisbury, Wiltshire SP3 6DR
> Phone 01747 820514

The international operator intercepted my call and said to use the following number:
011-44-1747-820514

S answered on the fourth ring. I introduced myself very briefly, told him about the trip to Chitral, Peshawar, and Quetta by Zimmermann, Gordon Enders, and Sir Benjamin Bromhead in Nov.-Dec. 1943, adding that Zimmermann was chief of Naval Intelligence in Karachi and Enders was the U.S. military attaché in Kabul. I said that I knew of S through the report of the first ascent of Tirich Mir in Chitral that was mentioned by Ed Viesteurs in his 2009 book on K2. Viesturs said that S was in the Chitral Scouts at that time (1950). This was only seven years after the 1943 trip to Chitral by my father-in-law, and I wondered if he might know anything about the 1943 trip. I said I went to the web and read about his climb on Haramosh in 1957 and re-read portions of Houston and Bates' *K2* that included descriptions of him and his tape-recorded comments.

H: I know exactly what the travelers did, and when. They left Peshawar, drove through Swat, met the Wali and Waliahad – took pictures of them – to Dir, met the Nawab, and made the first traverse of the Lowari Pass in a motor vehicle and were met in Chitral by the mehtir. Returned directly to Peshawar, were interrogated by the Viceroy at Governor Cunningham's garden party, went to the Khyber Pass, spent a few more days near Peshawar, and then took a winding drive through the Waziristans to Quetta where they had a picnic with Gov Hay and his family. Along the way, climbed Peiwar Khotal and looked into Afghanistan. I think that was about 20-25 miles from Tora Bora mountains. I believe the trip was Gordon Enders' idea, and that Bromhead and Cunningham accepted it, but the Intelligence Bureau at Quetta thought a Naval Intelligence officer from Karachi should be added to the trip. Why, I did not know, but I thought it may have been to have a counterbalance to the flamboyant Enders, and that the chargé in Kabul, Charlie Thayer, may have suggested Zimmermann, who was a friend of his brother George Thayer in Pennsylvania. Enders thought of himself as a "Kim," having grown up in the Himalayas near Almora (S corrected my pronunciation of this town).

My major questions for the book are:
  Whose idea was this trip?
    I thought it was Enders', and that he suggested it to Bromhead.
  What was the purpose of the trip?
    The official orders did not give a purpose. The letter from IB [Intelligence Bureau] Quetta said that the Americans were to be shown what the problems were with the locals along the border and what the British were doing about that. I thought there was more to it than that.
  Who authorized the trip?
    I now think the mission must have been approved locally, by Joint [i.e, British and American] intelligence officers in India [i.e., Quetta and Karachi], and by the governor of the N.W. Frontier Province. And although [probably] some British and Americans at HQ in Delhi and [certainly] the U.S. Minister in Kabul knew that the trip was going to take place, it was neither initiated or approved at a higher level (i.e., in Cairo, London, or Washington).
  Who knew of the trip, either before or afterwards? [I didn't get into that with S. I did not
    mention the names or titles of many people who I believe heard about the trip, except the Viceroy, Lord Wavell; Governor Cunningham in Peshawar; Governor Hay in Quetta; and Minister Engert and Chargé Thayer in Kabul]
  What has become of the reports of the trip? [I said that is still ongoing research.
    But since the travelers found nothing surprising, and nothing bad happened to them, I think their reports were simply filed away and forgotten]

H: Did S know about the trip? When did he come to Chitral? Did he know any of the people who were involved in it? What was his background and what else has he done in life?

S: My full name is Colonel Harry Reginald Antony Streather, OBE. I was born in 1926 and I was educated in England. Went to India after the War, about 1944, and was commissioned in the Indian Army in 1945. I arrived in Chitral in about 1949 or 1950, shortly before the Norwegian team arrived to climb Tirich Mir. [He confirmed that he was then a captain in the Chitral Scouts and was assigned as transportation officer for the climbers]. Never heard about the trip that was taken in 1943, and I didn't know any of the people you mentioned, although I heard about Bromhead many times – he was somewhat of a character. He had left before I got to the North-West Frontier. I didn't know Cunningham or Caroe [who followed Cunningham as governor of the N.W. Frontier Province]. Dundas was governor when I was there, and I was his ADC. The trip in 1943 – It was all part of the Great Game.

[Wikipedia says that Sir Ambrose Dundas Flux Dundas, KCIE, CSI (1899 -1973) served as the last British governor of the North-West Frontier Province of Pakistan. He entered office on 9 April 1948 and left office 16 July 1949]

H: The travelers went through Swat, where my father-in-law took pictures of the Wali and the Waliahad, and then to Dir, where the ruler was …

S: He was the Nawab.

H: Yes, and then they were the first to go over the Lowari Pass in a motor vehicle.

S: The airstrip at Drosh was terrible, just a tiny strip. There was one that was a bit better in Chitral.

H: Was the mehtir who was killed in a plane crash on the pass trying to get into Drosh?

S: Don't know.

H: The Russians had come into Chitral previously but the travelers in 1943 found no sign of them.

S: Yes, the Russians got interested in this area again a bit later, sometime after 1943. I still speak Pastun, and I offered to be a translator for the forces who are working in that area now, but no one wants to put a retired colonel there now.

H: Was I right in thinking that Peiwar Khotal, on the border, and the Kurram Valley, where the travelers went in 1943, are close to Tora Bora?

S: Yes, I have been in one of the Tora Bora caves.

H: I have read about your climbs on Tirich Mir in 1950 [he was on the team that made the second ascent, one day after the first ascent was made by one member of the team, solo], on K2 in 1953, and on Haramosh in 1957. Did you do other climbs?

S: Yes, I climbed about 20 peaks over 20,000 feet – many were unnamed – and my OBE was for climbing. Most of the climbs were in Pakistan. I had previously been made MBE and that was for a more interesting or more important piece of work, peace-keeping in Cyprus. But the MBE was superseded by the OBE, so it doesn't appear after my name now.

H: Do you have e-mail?

S: No, but my grandchildren do.

H: I then gave him my e-address, name and phone number, and said that if he planned to come to America, or if he ever did come, please let me connect him with the New York Section of the American Alpine Club.

S: I am planning to come to the American Alpine Club in Aspen in August, will be with Dee Molenaar [one of the others on the American 1953 K2 expedition]. Don't expect to come to New York. Hope to fly directly to Denver. I doubt that I would or will come to New York, but I will remember your invitation.

H: The map of England that appeared on the computer screen when I found your address and phone number shows that you live near Frome and Trowbridge, a bit east of Taunton. My wife and I are descendants of a man named Trowbridge who came to American from Taunton, and we have visited the churches in Taunton and Exeter that were important in his life and the lives of his wife and her father and grandfather. The Trowbridges originally came from the town that is now called Trowbridge, near you, so I can visualize the part of England where you live. I hope we may have a chance to meet some day. Please let me know if you come to New York. Thanks for talking with me. Good-bye.

Glossary
Slang, Foreign Words, Abbreviations, Acronyms, and Military Rank Equivalents

Slang and Foreign Words
aur billi = Urdu, for a high-ranking person
baksheesh = tip or bribe
burrah = high level gentleman
chablis = sandals
dog colla*r* = white uniform blouse with standup collar
Festung Europa = Fortress Europe (Hitler's term)
geedunk = standard expression in Navy and Marine Corps for standard GI, snack bar quality.
gyppy tummy = indigestion and diarrhea in the Mediterranean and South Asia
mehter (now usually spelled mehtar) = ruler of Chitral
memsahib = high level English lady
mullik (now usually spelled malik, and various others) = Pashtun religious leader
nawab = ruler, as in the state of Dir
phut = ended (derived from the German *kaput*)
pukka = upscale (Indian)
sahib = Englishman
salaam = gesture of respect
scuttlebutt = rumor
shriner = a Masonic Order
squeeze = Westerner's term for the Chinese custom of bribing an official
tiffin = afternoon tea with many delicacies
wali = ruler of Swat
wog = derogatory expression for non-white residents of the Middle East and Asia

Abbreviations
AZ or AWZ = Albert Walter Zimmermann, known as "Al."
  From 1942-1945, he was a Navy Lieutenant
BSZ = Barbara (née Shoemaker) Zimmermann, his wife
Family members include the four children of AWZ and BSZ, identified as follows:
  Barbara Warren "Babs" Zimmermann (BWZ), b. 1927
  Helene "Nanie" or "Lanie" Zimmermann (HZ), b. 1929; later Helene Zimmermann Hill, Ph.D.
  Warren "Warr" Zimmermann (WZ), b. 1934; later Ambassador Warren Zimmermann
  Albert Walter "Albie" Zimmermann, Jr. (AWZJr.), b. 1937; later Albert W. Zimmermann, M.D.

Acronyms
AAF or USAAF = U.S. Army Air Force
ADC = aide de camp
ALS-P = Autographed Letter, Signed, handwritten in Pencil. ALS-I is written in Ink.
ALUSLO = Cable address for Naval Liaison Office
ALUSLO = U.S Navy Liaison Officer (used for all NLOs in India-Burma Theatre)
AMBICA = Code word for NLO mailing address in Karachi
APO = U.S. Army Post Office
ARC = American Red Cross
ATS = Air Transport System
AUS = Army of the United States (also USA)
CBE = Officer of the Order of the British Empire

CBI = China-Burma-India Theatre
CIA = Central Intelligence Agency (U.S.)
CO = Commanding Officer
CPO = Chief Petty Officer (U.S.)
DC = District Commissioner (British)
DNI = Director of Naval Intelligence (U.S.)
EFM = Fleet Mail
ENIAC = Electronic Numerical Integrator and Computer
FANY = First Aid Nursing Yeomanry (UK), called "Fanny"
FBI = Federal Bureau of Investigation (U.S.)
FC = Flight Lieutenant Commander (British)
FDR = Franklin Delano Roosevelt, President
FPO = Fleet Post Office
GI = Government Issue (also slang for U.S. Army)
HE = His Excellency
HH = His Highness
HMIS – His Majesty's Indian Ship
HQ = Headquarters
IA = Indian Army
IBC = India-Burma-Ceylon Theatre
IBT = India-Burma Theatre
JICAME = Joint Intelligence Center Africa – Middle East
JICAME = Joint Intelligence Collection Agency – Middle East
JICSEAC = Joint Intelligence Center South East Asia Command
MI5 = Secret Service (British Domestic Intelligence Service)
MI6 = Secret Intelligence Service (British Foreign Intelligence Service)
MO = medical officer
MO = Military Observer
MP = Military Police (U.S. Army)
NLO = Naval Liaison Office
NOIC or Noic = Naval Officer in Charge (British)
NWFP = North West Frontier Province, India (now Khyber-Pakhtunkhwa Province, Pakistan)
ONI = Office of Naval Intelligence (U.S.)
OSS = Office of Strategic Services (U.S.)
PA = Political Agent (British)
PO = Political Officer (British)
RAF = Royal Air Force
RHIP = Rank Hath Its Privilege
RN = Royal Navy (British)
RNA = Ribonucleic acid
SACO = Sino-American Cooperation Organization
SACSEA = Supreme Allied Commander South East Asia (Lord Mountbatten)
SATO = Sino-American Special Technical Cooperative Organization
SEAC = South-East Asia Command (facetiously, "Save England's Asia Colonies")
SEAT = South-East Asia Theater
SOS = Services of Supply (U.S. Army)
UK = United Kingdom (Great Britain)
US = Unserviceable (slang)
US AAF = U.S. Army Air Force
USAF = U.S. Armed Forces
USMC = U.S. Marine Corps

USN = U.S. Navy (also U.S. Naval Officer)
USNLO = U.S. Navy Liaison Office
USNR = U.S. Navy Reserve
WAVE = Women Accepted for Volunteer Emergency Service (U.S.)
WAAF = Women's Auxiliary Air Force (British)
WAC= Women's Army Corps (U.S.)
WOG = "Worthy Oriental Gentleman" (dismissive slang for men of Egypt, Middle East, and India)
WREN or Wren = Women Reserve in the English Navy
WSA = War Shipping Administration

U.S. Military Ranks (in WWII)

NAVY			ARMY and MARINE CORPS	
O-1	ENS	Ensign	2LT	2nd Lieutenant
O-2	LTJG	Lieutenant, junior grade	1LT	1st Lieutenant
O-3	LT	Lieutenant	CPT	Captain
O-4	LCDR	Lieutenant Commander	MAJ	Major
O-5	CDR	Commander	LTCOL	Lieutenant Colonel
O-6	CAPT	Captain	COL	Colonel
O-7			BG	Brigadier General (1 star)
O-8	RADM	Rear Admiral	MG	Major General (2 stars)
O-9	VADM	Vice Admiral	LTGEN	Lieutenant General (3 stars)
O-10	ADM	Admiral	GEN	General (4 stars)
O-11	FADM	Fleet Admiral	GEN ARMY	General of the Army (5 stars)

In WWII, "Commodore" was usually a courtesy title given to an officer who commanded two or more ships. This could even be a LCDR, who was responsible for two or three landing crafts, each of which might have a LT or LTJG as its skipper. The title was occasionally used to denote a one-star flag officer such as Commodore Milton Miles, who was later promoted to two stars and in retirement became a VADM. Throughout WWII, most officers promoted from Captain to flag rank were called Rear Admiral and wore two stars. This custom was in contrast to officers promoted from colonel, because they only acquired one star, with the title Brigadier General. This difference rankled the Army. When the Department of Defense Officer Personnel Manpower Act (DOPMA) was enacted in 1980, the Navy instituted a one-star rank and the title "Commodore Admiral" was briefly used to denote a one-star flag officer. It was soon changed to be Rear Admiral (Lower Half), with the acronym RDML. The acronym RADM now denotes a two-star Rear Admiral (Upper Half).

British Military Ranks generally corresponded to those of the United States, but the abbreviations differed slightly; in the U.K, these titles were hyphenated: Lieut.-Col. and Lieut.-Cdr. The counterpart of Brigadier General in the U.K. was Brigadier, and the counterpart of the 5-star rank of General of the Army was Field Marshal in the U.K.

Some British acronyms:
CBE = Commander of the Order of the British Empire
CH = Order of the Companions of Honor
CIE = Companion of the Indian Empire
CSI = Companion of the Order of the Star of India
FRS = Fellow of the Royal Society
MC = Military Cross
MP = Member of Parliament
OBE = Officer of the Order of the British Empire
OM = Order of Merit
VC = Victoria Cross

Acknowledgements

I gratefully acknowledge my father-in-law, Lieutenant Albert W. Zimmermann, USNR, who served in India in World War II, and his wife, Barbara (*née* Shoemaker) Zimmermann, who could "only stand and wait" until the war was over. These were dangerous times, and no one was really safe, especially overseas. They believed in what they were doing – he for the Navy and for the country – and she taking care of their home and their young family, and volunteering in a canteen for servicemen. Their letters show their anxiety about the future, and their frustration with the uncertainty of it all. Both Al and Barbara developed peptic ulcers before the war was over. But she saved the crucial documents that made it possible to construct this book, in files of letters, and a scrapbook, and a photo album. They were put away after the war, in about 1947, and were not examined again for sixty years.

I also acknowledge the previous work of others on the Great Game, and especially Rudyard Kipling, whose book, *Kim*, popularized this expression for the contest in Central Asia between Britain and Russia. The history of the Great Game has been admirably summarized by Peter Hopkirk, and many other authors. I have cited a few of them, and I apologize for not including more.

A brief account of this trip, accompanied by three photos, was published in *Appalachia* in 2008. I thank the Appalachian Mountain Club for the opportunity to introduce this story to the public, incomplete as it was at that time. Correspondence from readers helped me to find other sources that have enriched my understanding of the NWFP Trip and the travelers. The U.S. Naval Institute encouraged me to continue in this research.

To Al and Barbara Zimmermann's children, I give special thanks: Barbara Warren "Babs" Zimmermann Johnson, her husband Melvin Thornton Johnson; Ambassador Warren Zimmermann and his wife Corinne "Teeny" (née) Chubb; Dr. Albert W. Zimmermann Jr., and his wife Lenore Marie (née) Lisbinski. The wartime papers of Al Zimmermann were filed and preserved by Barbara (née Shoemaker) Zimmermann, and then by her daughter, Babs. Without them, this book could never have been written.

To my parents, Gerald L. Hill, Field Director, American Red Cross, 303rd Infantry Regiment, 97th Division, 3rd Army; and Essie Mae (née) Thompson, Master Sergeant, Civil Air Patrol; and my brother, Lt. Col. Thomas D. Hill, USAF, DFC.

To the friends of Al and Barbara Zimmermann who contributed to this story, and to the next generation in those families: Mrs. Jack (Amelie née Sexias) Kane and her daughter, Sheila Kane; Jack Thayer and his daughter, "Dodie" Thayer; 1st Lt. Clarence Lewis, USMC, OSS, his wife Mrs. Georgiana (née Wetherill) "Sam" Lewis, and her daughter, Susan (née Lewis) Lillien; Virginia "Peachy" Durand of the OSS, and her mother, known only to the Zimmermann children as "Mrs. Bitz Durand"; the Lincolns: "Daddy Joe" and his wife, their son Lt. Col. J. Freeman Lincoln, OSS, and his wife; and their daughters, Anne and Crosby.

To the Enders family and their friends: Dr. Gertrude "Trudy" (née Enders) Huntington, and her brother, Dr. Allen Coffin Enders. To the family of Elizabeth (née Crump) Enders, especially Patty Hoenigman. To Maynard Creel, for his research on Gordon Enders and other World War I aviators, and for his personal recollections of Gordon and Betty Enders. To David M. Hovde, Purdue University, and Sarah Uschak, Office of Alumni Relations, and The College of Wooster, for their help in locating records of Gordon Enders. To the descendants of Cornelius and Sara Engert: Roderick K. Engert, OSS; and his daughter, Jane Morrison Engert. To Colonel "Tony" Streather, OBE, who served in Pakistan in 1950 and recalled his experiences there and in Afghanistan.

To the archivists of the Franklin Delano Roosevelt Library; Amanda A. Hegge of the Patrick J. Hurley Collection, Oklahoma University Library; the New York Public Library; Georgetown University Special Collections; and Nate Patch, Military Archive Section, National Archives and Records Administration (NARA II).

To veterans of the OSS who inspired me, although at the time I first knew them, I never knew about their wartime service: They include my teachers at Yale, Frederick G. Kilgour and S. Dillon Ripley; and my advisor and chairman of the History Department at Yale, Professor Hajo Holborn.

And to the memory of World War II veterans including: B-17 Bombardier Lieut. Marc Pitts and B-17 and XB-49 pilot Major Daniel H. Forbes; my cousin, Marine Corps pilot John Paul Thompson; and his father, CPT Manly Grant Thompson, AUS.

For inspiration, encouragement, and assignments, I thank the following individuals, listed alphabetically, and others who must be nameless: Joseph Alsop; CAPT James Bates, MSC, USN; LTGEN Quinn H. Becker, USA MC; VADM John D. Bulkley, USN; Ambassador Ellsworth Bunker; RADM James P. Carey, USNR; Mr. and Mrs. Stuart (Mary Bishop "Bish") Coan; LCDR Charles Crandall, M.D.; VADM Donald L. Custis, MC, USN; VADM Dwight Dickinson, MC, USN; Professor William Drucker, M.D.; ADM Noel A. M. Gayler, USN; RADM Ben Eiseman, MC, USNR; LT Alvin Friedland, M.D.; RADM Donald A. Hagen, MC, USN; CAPT Arthur Halstead, DC, USNR; Heinrich Harrar; MG Carl Hughes, USA MC; CAPT John Hurley, MC, USNR; Hon. Joe Foss, MoH; Hon. Donald M. Kerr, Ph.D.; Vernon Knight, M.D.; RADM Robert Laning, MC, USN; Ambassador Samuel W. Lewis; Ambassador William "Butts" Macomber; Rev. Burton MacLean; CAPT John Proctor, USNR; Bill Putnam; Col. Norman Rich, USA MC; Professor Charles Rob, FRCS; RADM Victor C. See Jr., USN; Col. Kenneth G. Swan, USAR MC; LTGEN Alexander M. Sloan, USAF MC; Hon. Robert Ten Broeck Stevens; RADM David Sullins, MSC, USNR; LTGEN Paul K. Carlson, USAF MC; MG David Trump, USAF MC; BG Hoyt Vandenberg Jr., USAF; BG J. Lionel Villavicencio; VADM James B. Wilson, USN; GEN Frederick Weyand, USA; GEN William Westmoreland, USA; RADM Winston Weese, MC, USNR; Denis B. Woodfield, D.Phil. (Oxon.); LT John Howard Woodin, SC, USNR; LTGEN John M. Wright Jr, USA (ret); and VADM James A. Zimble, MC, USN.

For opportunities and trust: the Laboratory of Clinical Investigation, National Institute of Allergy and Infectious Diseases, National Institutes of Health (LCI-NIAID-NIH); Armed Forces Medical Intelligence Center (AFMIC); Navy Medical Research and Development Command (NMRDC); Naval Medical Research Institute (NMRI); Office of the Surgeon General of the Navy (OPNAV 093), Plans and Policies Division, and my Reserve Unit, NR OPNAV 093 106; Navy Command Center; National Naval Medical Center, Bethesda, Md.; Naval Hospital Keflavik; Uniformed Services University of the Health Sciences and my Reserve Unit, NR USUHS NH BETHESDA 106; Naval Reserve Construction Brigade (the Seabees), 7th Regiment, and 21st and 23rd Battalions; Commander Naval Forces Vietnam (COMNAVFORV); Military Assistance Command Vietnam (MACV); Naval Medical Research Unit (NAMRU) Cairo; National Reconnaissance Office (NRO); 4th Marine Division and staff of NATO Operation Alloy Express (1982); U.S. Army Airborne School, Ft. Benning, Georgia; USMC Force Recon, Quantico and Camp Pendleton; Medical History Society of New Jersey; Dunworkin Club of Montclair, N.J.; New Jersey Medical Club; Baronial Order of the Magna Charta; and Northern New Jersey Chapter, Military Officers Association of America.

I thank the anonymous author of a Naval Medical Instruction for Medical Officers, from whom I learned how to write a "Report on Ports and Countries Visited," and to submit it as a Naval Medical Intelligence Report. In this NAVMED Instruction, I learned that "medical intelligence is real intelligence." And I thus began my study of military intelligence.

I am very grateful to Richard "Rick" Russell, Director of the Naval Institute Press. He decided to accept the manuscript of *Proceed to Peshawar* for publication and he guided the great team that made it into a book. They include Adam C. Kane, Adam Nettina, Claire Noble, Marlena Montagna, Judy Heise, and my wonderful copy editor, Alison Hope. I thank those who read the manuscript and wrote comments for the cover: Admiral Harry D. Train II, Rear Admiral Joseph Callo, John E. "Jed" Williamson, and Barnet Schecter. The thoughtful review by Michael Swisher was very helpful, as was the comment I received from Miss Romilly Greer Leeper, daughter of Lt.-Col. William C. Leeper, who knew the Bromhead family in England after WWII. Her mother, Dorothea (Lloyd) Leeper, is mentioned in *Proceed to Peshawar*; Mrs. Leeper met LT Albert Zimmermann in Peshawar in 1943.

Many individuals and organizations have helped to introduce *Proceed to Peshawar* to the world, beginning with a pre-publication lecture arranged by Dr. and Mrs. Charles Waldron to speak at the Great Decisions Seminar, Fearington, N.C. Early in 2014, soon after the book was published, the Naval Institute Press arranged for me to speak at the 8-Bells Lecture Series of the Naval War College in Newport, R.I.; this lecture was recorded, and it can be viewed on the YouTube website. I am also grateful to the organizers of the Virginia Book Fair in Charlottesville, Va., who gave me the opportunity to participate in a panel discussion with several other speakers on recent events in Asia, and to Dr. and Mrs. John and Jacqueline Gergen of the Pantops Mountain Lecture Committee for sponsoring my talk. I gave a lecture about *Proceed to Peshawar* at the Harvard Club of New York City, and I later was invited to participate in the Club's Annual Book Fair. I grateful for the opportunity to speak about *Proceed to Peshawar* to the New Jersey Society, Sons of the American Revolution; to the New York Commandery, Naval Order of the United States; to the North Jersey Council, Navy League of the United States; and to the New Jersey Society of Mayflower Descendants. Thanks to Governor General John Bourne, I was awarded Honorary Membership in the Descendants of the American Colonists after I spoke about *Proceed to Peshawar* to that organization at its annual meeting in Washington, D.C. *Proceed to Peshawar* was featured in the Bulletin of the Army and Navy Club, Washington, D.C. Through the courtesy of Carol Schlitt, Esq., of West Orange, N.J., I presented my discussion of the additional photos and movies taken by LT Albert Zimmermann in 1943-45. I am especially honored that *Proceed to Peshawar* received a Finalist Medal in the 2015 Next Generation Indie Book Awards ceremony at the Harvard Club of New York City.

And most of all, I thank my daughters, Sarah and Lana Hill, and my wife, Helene "Lanie" Zimmermann Hill, Ph.D., who have been patient and helpful throughout this long project.

Bibliography

A Partial Bibliography for Gordon Enders is in Chapter 6

Archival Sources

"Astor, Vincent" Folder, in Small Collections, FDR Library, Hyde Park New York.

"Cutting, C. Suydam" – a thin file of letters in President's Personal File, box 4556-4612, PPF 4596. FDR Library, Hyde Park, N.Y.

Donovan, William J. – President's Secretary's File – Box 153. FDR Library, Hyde Park, N.Y.

Cornelius Van Engert Papers, Special Collections, Georgetown University. Index at http://library.georgetown.edu/dept/speccoll/cl169.htm, accessioned 6/12/09.

"Engert, Cornelius Van H." A thin file folder in the box for Official File, File 1922, FDR Library, Hyde Park, N.Y.

Hurley, Patrick J. Papers. University of Oklahoma Libraries, Western History Collections (Hurley Papers). Finding Aid, 676 pp. Includes 2 memoranda by Gordon B. Enders, letter of Elizabeth C. Enders to General Hurley, a newspaper story by GBE, and two newspaper stories about GBE (*infra*).

"Engert, Sara" Folder in Vertical Files, FDR Library, Hyde Park, N.Y.

Franklin Delano Roosevelt Presidential Library, Hyde Park, N.Y. (FDR Library)

Foreign Relations of the United States. http://digicoll.library.wisc.edu/FRUS/

National Archives and Records Administration, Washington, DC (NARA), and NARA II (Archives at College Park, MD), Record Group 38 (RG38) – Office of Naval Archives Records, December 7, 1941ff. The records of RG38 prior to Pearl Harbor are said to be at NARA, in Washington, DC. This was confirmed by both the archivists in the Naval History Section at NARA, and in the Military History Section at NARA II.

"Tolstoy, Ilia" Folder in Vertical Files, FDR Library, Hyde Park, N.Y.

[Zimmermann]. FOIA: Freedom of Information Act. Request from Department of the Navy by Helene Z. Hill, 4 August 2009, for any memoranda, orders, reports, and endorsements that relate to the trip of her father, LT Albert W. Zimmermann, I-V(S), USNR, from Karachi, India, to Peshawar, North-West Frontier Province, and "other such places" from 12 November 1943 to 15 December 1943.

Books

Adams, Henry H. *Witness to Power: The Life of Fleet Admiral William D. Leahy.* Annapolis, Md.: Naval Institute Press, 1985.

Aldrich, Richard J. *American Intelligence and the British Raj: The OSS, the SSU and India, 1942-1947.* Routledge, 1998.

Amis, Kingsley. *Rudyard Kipling and His World*. London: Thames and Hudson, 1975.

Andrew, Christopher. *For the President's Eyes Only: Secret Intelligence and the American Presidency from Washington to Bush*. HarperPerennial, 1996.

_____. *Defend the Realm: The Authorized History of MI5*. New York, N.Y.: Vintage Books/Random House, 2009.

Bachman, Bruce M. *An Honorable Profession: The Life and Times of One of America's Most Able Seamen: Rear Adm. John Duncan Bulkeley, USN*. New York City, N.Y.: Vantage Press, 1989.

Baltzell Edward Digby. *Philadelphia Gentlemen: The Making of a National Upper Class*. New Brunswick, N.J.: Transaction Publishers, 2009 [Free Press, 1958].

Beesley, Patrick. *Very Special Admiral: The Life of Admiral J. H. Godfrey, CB*. London: Hamish Hamilton, 1980.

Brobst, Peter John. *The Future of the Great Game: Sir Olaf Caroe, India's Independence, and the Defense of Asia*. Akron, Ohio: The University of Akron Press, 2005.

Brown, Anthony Cave. *The Last Hero: Wild Bill Donovan*. New York City: Vintage Books/Random House, 1982.

_____. *The Secret Servant: The Life of Sir Stewart Menzies. Churchill's Spymaster*. London: Michael Joseph, Published by the Penguin Group, [1987] 1988.

Bryant, Arthur. *The Turn of the Tide: A History of the War Years Based on the Diaries of Field-Marshal Lord Alanbrooke, Chief of the Imperial General Staff, 1939-1943*. Garden City, N.Y.: Doubleday, 1957.

Bryant, Arthur. *Triumph in the West: A History of the War Years Based on the Diaries of Field-Marshal Lord Alanbrooke, Chief of the Imperial General Staff*. Garden City, N.Y.: Doubleday, 1959.

Buhite, Russell D. *Patrick J. Hurley and American Foreign Policy*. Cornell University Press, 1973.

Carlson, Elliot. *Joe Rochefort's War: The Odyssey of the Codebreaker Who Outwitted Yamamoto at Midway*. Annapolis, Md.: Naval Institute Press, 2011.

Carroza, Anthony R. *William D. Pawley: The Extraordinary Life of the Adventurer, Entrepreneur, and Diplomat Who Cofounded the Flying Tigers*. Potomac Books, Inc., 2012.

Chua, Amy. *Day of Empire: How Hyperpowers Rise to Global Dominance – And Why They Fall*. New York City, N.Y.: Doubleday, 2007.

Churchill, Sir Winston. *The Story of the Malakand Field Force: An Episode of Frontier War*. Seven Treasures Publications, 2009 [Churchill, Winston. *Malakand Field Force*. London: Longmans, Green and Co., 1898].

_____. *A History of the English Speaking Peoples*. v.4. *The Great Democracies*. New York: Dodd, Mead & Company, 1958.

_____. *The Second World War*. 5 vols. Vol. 4: *The Hinge of Fate*. Vol. 5: *Closing the Ring*. Boston, Mass.: Houghton Mifflin Company, 1981.

Coll, Steve. *Ghost Wars: The Secret History of the CIA, Afghanistan, and Bin Laden, from the Soviet Invasion to September 10, 2001*. New York City, N.Y.: Penguin Group/Penguin Press, [2004] 2005.

Connell, John. *Wavell: Scholar and Soldier, to June 1941*. London: Collins, 1964.

Courcy, Anne de. *The Viceroy's Daughters: The Lives of the Curzon Sisters*. New York City, N.Y.: HarperCollins/Perennial, [2000] 2002.

Crile, George. *Charlie Wilson's War: The Extraordinary Story of How the Wildest Man in Congress and a Rouge CIA Agent Changed the History of Our Times*. New York City, N.Y.: Grove Press, 2003.

Curzon, George N. *Persia and the Persian Question*. London: Frank Cass & Co., Ltd. [1892] 1966. 2 vols.

_____. *The Pamirs and the Source of the Oxus*. London: The Royal Geographical Society, 1896.

_____. *Russia in Central Asia in 1889 and the Anglo-Russian Question*. London: Frank Cass & Co., Ltd [1889] 1967.

Cutting, Suydam. *The Fire Ox and Other Years*. New York City, N.Y.: Charles Scribner's Sons, 1940.

Dorril, Stephen. *MI6: Inside the Covert World of Her Majesty's Secret Intelligence Service*. New York City, N.Y.: Simon & Schuster/A Touchstone Book, [2000] 2002.

Dorwart, Jeffery M. *Conflict of Duty: The U.S. Navy's Intelligence Dilemma, 1919-1945*. Annapolis, Md.: Naval Institute Press, 1983.

Drea, Edward J. *MacArthur's ULTRA: Codebreaking and the War Against Japan, 1942-1945*. Lawrence, Kansas: University Press of Kansas, 1992.

Dulles, Allen. *The Secret Surrender*. New York City, N.Y.: Harper & Row, Publishers, 1966.

_____. *The Craft of Intelligence: America's Legendary Spy Master on the Fundamentals of Intelligence Gathering for a Free World*. Guilford, Conn.: The Lyons Press, an imprint of The Globe Pequot Press, 2006.

Eisenhower, David. *Eisenhower at War, 1943-1945*. New York: Random House, 1986.

Elizabeth Crump Enders, *Swinging Lanterns*. New York, N.Y.: D. Appleton and Company, 1923.

_____, *Temple Bells and Silver Sails*. New York, N.Y.: D. Appleton and Company, 1925.

Enders, Gordon Bandy, and Edward Anthony. *Nowhere Else in the World*. New York: Farrar & Rinehart, 1935.

Enders, Gordon Bandy. *Foreign Devil: An American Kim in Modern Asia*. New York: Simon and Schuster, 1942.

Engert, Jane Morrison. *Tales from the Embassy: The Extraordinary World of C. Van H. Engert.* Westminster, Md. Heritage Books/Eagle Editions, 2006.

Ewans, Sir Martin. *Afghanistan: A Short History of Its People and Politics.* Perennial/HarperCollins Publishers, [2001] 2002.

Filkins, Dexter. *The Forever War.* New York: Alfred A. Knopf, Borzoi Book, 2008.

Fort, Adrian. *Archibald Wavell: The Life and Times of an Imperial Servant.* London: Jonathan Cape, 2009.

Fox, Ernest F. *Travels in Afghanistan.* New York, N.Y.: The Macmillan Company, 1943.

_____. *By Compass Alone* (Dorrance & Company, Philadelphia, PA 1971).

Fraser, George MacDonald (ed). *Flashman in the Great Game: From the Flashman Papers 1856-1858.* [1975] New York: Penguin, 1989 [fiction].

Gilmour, David. *The Long Recessional: The Imperial Life of Rudyard Kipling.* New York: Farrar, Straus and Giroux, 2002.

Goodson, Larry P. *Afghanistan's Endless War: State Failure, Regional Politics, and the Rise of the Taliban.* Seattle, Washington: University of Washington Press, 2001.

Grose, Peter. *Gentleman Spy: The Life of Allen Dulles.* Amherst, Mass.: The University of Massachusetts Press, [1994] 1996.

Goradia, Nayana. *Lord Curzon: The Last of the British Moghuls.* Delhi: Oxford University Press, 1997.

Green, Peter. *Alexander of Macedon, 356-323 B.C.: A Historical Biography.* [1st ed., 1970; 2nd ed., 1974; this is a reprint of the 2nd ed.] Berkeley, Calif.: University of California Press, 1991.

Harlan, J[osiah]. *A Memoir of India and Avghanistaun With Observations on the Present Exciting and Critical State and Future Prospects of Those Countries. Comprising Remarks on the Massacre of the British Army in Cabul, British Policy in India, A Detailed Descriptive Character of Dost Mahomed and His Court, Etc.* Philadelphia, Pa.: J. Dobson, 1842.

Harris, Charles H., III, and Louis R. Sadler. *The Archaeologist Was a Spy: Sylvanus G. Morley and the Office of Naval Intelligence.* Albuquerque, N.M.: University of New Mexico Press, 2003.

Hay, William Rupert. *Two Years in Kurdistan. Experiences of a Political Officer, 1918-1920.* London: Sidgwick and Jackson, 1921.

Hersh, Burton. *The Old Boys: The American Elite and the Origins of the CIA.* New York City, N.Y.: Charles Scribner's Sons, 1992.

Hill, George J. *Proceed to Peshawar. The Story of a U.S. Navy Intelligence Mission on the Afghan Border, 1943.* Annapolis, Md.: Naval Institute Press, 2013.

_____. *Quakers and Puritans: The Shoemaker, Warren and Allied Families. Ancestors and Descendants of William Toy Shoemaker and Mabel Warren, Who Were Married in Philadelphia in 1895*. Berwyn Heights, Md.: Heritage Books, 2015.

_____. *American Dreams: Ancestors and Descendants of John Zimmerman and Eva Katherine Kellenbenz, Who Were Married in Philadelphia in 1885*. Berwyn Heights, Md.: Heritage Books, 2016.

_____. *Four Families: A Tetralogy. Readers Guide to* Western Pilgrims, Quakers and Puritans, Fundy to Chesapeake, and American Dreams. Berwyn Heights, Md.: Heritage Books, 2017.

Holt, Thaddeus. *The Deceivers: Allied Military Deception in the Second World War*. Skyhorse Publishing Inc., 2007.

Hopkirk, Peter. *Trespassers on the Roof of the World: The Secret Exploration of Tibet*. [1982] New York: Kodansha International, 1995.

_____. *The Great Game: The Struggle for Empire in Central Asia*. New York, N.Y.: Kodansha Globe, 1992.

_____. *Setting the East Ablaze: Lenin's Dream of an Empire in Asia*. [1984] New York, N.Y.: Kodansha International, Inc., 1995.

_____. *Quest for Kim: In Search of Kipling's Great Game* [1996] Ann Arbor, Mich.: University of Michigan Press, 1999.

Hotchkiss, Henry. *A Treatise on Aviation Law*. 1938.

Houston, Charles S., and Robert H. Bates. *K2: The Savage Mountain*. Seattle, Wash.: The Mountaineers, [1954] 1979.

Hyde, H. Montgomery. *Room 3603: The Incredible True Story of Secret Intelligence Operations During World War II*. New York City, N.Y.: The Lyons Press, [1962] 2001. Foreword by Ian Fleming.

Isaacson, Walter, and Evan Thomas. *The Wise Men: Six Friends and the World They Made. Acheson, Bohlen, Harriman, Kennan, Lovett, McCloy*. New York, N.Y.: Simon & Schuster Inc. / A Touchstone Book, [1986] 1988.

Jablonsky, David. *Churchill, the Great Game and Total War*. New York, N.Y.: Routledge, 1991.

Jeffers, H. Paul. *In the Rough Rider's Shadow: The Story of a War Hero – Theodore Roosevelt Jr*. New York City, N.Y.: Presidio Press Book / Ballantine Publishing Group, Random House, Inc., 2003.

Jeffery, Keith. *The Secret History of MI6*. New York City, N.Y.: The Penguin Press, 2010.

Jeffreys, Alan, and Duncan Anderson. *British Army in the Far East 1941-1945*. Oxford: Osprey Publishing, 2005.

Joint Staff Officer's Guide. Washington, D.C.: Superintendent of Documents, for National Defense University/Armed Forces Staff College, Norfolk, Va., 1986.

Jones, Seth G. *In the Graveyard of Empires: America's War in Afghanistan*. New York City, N.Y.: W. W. Norton, [2009] 2010.

Kaplan, Robert D. *Monsoon: The Indian Ocean and the Future of American Power*. New York, N.Y.: Random House, 2010.

Kiernan, Frances. *The Last Mrs. Astor: A New York Story*. New York City, N.Y.: W. W. Norton & Company / Norton paperback [2007] 2008.

Kipling, Rudyard. *The Writings in Prose and Verse of Rudyard Kipling*. New York: Charles Scribner's Sons, 1897-8. 12 vols. *Under the Deodar, The Story of the Gadsbys, Wee Willie Winkie* (vol. 6). *Verses, 1889-1896* (vol. 11). [copy in the library of Albert W. Zimmermann]

_____. *Kim*. [1901] Mineola, N.Y.: Dover Publications, Inc., 2005.

_____. *Kipling Stories: Twenty-eight Exciting Tales by the Master Storyteller*. New York City, N.Y.: Platt & Munk, 1960.

Kleveman, Lutz. *The New Great Game: Blood and Oil in Central Asia*. New York City, N.Y.: Grove Press, 2003.

Knauth, Arnold W., Henry Hotchkiss, and Emory H. Niles. *United States Aviation Reports*. 1942.

Leider, Emily W. *Myrna Loy: The Only Good Girl in Hollywood*. Berkeley, Calif.: University of California Press, 2011.

Lewin, Ronald. *The Chief: Field Marshal Lord Wavell, Commander-in-Chief and Viceroy, 1939-1947*. New York City, N.Y.: Farrar, Straus and Giroux (1980).

Liddell, Guy Maynard and Nigel West. *The Guy Liddell Diaries: MI5's Director of Counter-Espionage in World War II*. Taylor & Francis, 2005.

Lohbeck, Don. *Patrick J. Hurley*. Chicago: Henry Regnery Company, 1956.

Lukacs, John (ed.) *Through the History of the Cold War: The Correspondence of George F. Kennan and John Lukacs*. Philadelphia, Pa.: The University of Pennsylvania Press, 2010.

Manchester, William. *The Last Lion: Winston Spencer Churchill. Visions of Glory, 1874-1932*. Boston, Mass.: Little, Brown and Company, 1983.

Markey, Gene. *His Majesty's Pyjamas*. New York, N.Y.: Covici-Friede-Publishers, 1934.

_____. *Women, Women, Everywhere*. New York, N.Y.: Bobbs-Merrill, 1964.

Marston, Daniel. *Phoenix from the Ashes: The Indian Army in the Burma Campaign*. Westport, Conn.: Prager Publishers, 2003.

McIntosh, Elizabeth P. *Sisterhood of Spies: The Women of the OSS*. Annapolis, Md.: Naval Institute Press, 1998.

Meyer, Karl Ernest, and Shareen Blair Brysac. *Tournament of Shadows: The Great Game and the Race for Empire in Central Asia.* New York: Basic Books, 1999.

Miles, Milton E. *A Different Kind of War: The Unknown Story of the U.S. Navy's Guerilla Forces in World War II China.* Garden City, N.Y: Doubleday & Co., 1967.

Mitchell, Norval. *Sir George Cunningham: A Memoir.* Blackwood, 1968.

Moran, Lord (Charles Wilson). *Churchill: Taken From the Diaries of Lord Moran. The Struggle for Survival, 1940-1965.* Boston, Mass.: Houghton Mifflin Co., 1966.

Mosley, Charles, editor. *Burke's Peerage, Baronetage & Knightage,* 107th edition, 3 volumes. Wilmington, Delaware: Burke's Peerage, Genealogical Books, Ltd. 2003.

[Mountbatten] Ziegler, Philip. *Mountbatten: A Biography.* New York: Alfred A. Knopf, 1985.

Morison, Samuel Eliot. *The Two Ocean War: A Short History of the United States Navy in the Second World War.* Boston, Mass.: Little, Brown and Co., 1963.

Osborne, Frances. *The Bolter.* New York City, N.Y.: Vintage Books, [2008] 2010.

Packard, Wyman H. *Century of U.S. Naval Intelligence.* Naval Historical Center, 1996

Page, Bruce, David Leitch, and Phillip Knightley. *Philby: The Spy Who Betrayed a Generation.* London: Andre Deutsch, 1968. Introduction by John Le Carré.

Persico, Joseph. *Roosevelt's Secret War: FDR and World War II Espionage.* New York City, N.Y.: Random House, 2002.

Philby, Kim. *My Silent War: The Autobiography of a Spy.* New York City, N.Y.: Modern Library, [1968] 2002. By H. A. R. Philby. Introduction by Philip Knightley. Foreword by Graham Greene.

Powers, Thomas. *Intelligence Wars: American Secret History from Hitler to Al-Qaeda.* New York City, N.Y.: New York Review of Books, 2004.

Prescott, Kenneth W. (CAPT, USNR, ret). *A PT Skipper in the South Pacific: A Naval Officer's Memoir of Service on PTs and a PT Boat Tender.* Bennington, Vt.: Miriam Press, 2009.

Rasanayagam, Angelo. *Afghanistan: A Modern History.* London: I. B. Tauris [2003] 2005.

Rashid, Ahmed. *Taliban: Militant Islam, Oil and Fundamentalism in Central Asia.* New Haven, Conn.: Yale University Press, [2000] 2001.

Renault, Mary. *The Persian Boy.* [1972] New York: Bantam Books, 1974.

Rhodes, Richard. *Hedy's Folly: The Life and Breakthrough Inventions of Hedy Lamarr, the Most Beautiful Woman in the World.* New York, N.Y.: Doubleday, 2011.

Ricketts, Harry. *Rudyard Kipling: A Life.* New York City, N.Y.: Carroll & Graf Publishers, Inc., [2000] 2001.

Rolandshay, The Rt. Hon., the Earl of. *The Life of Lord Curzon: Being the Authorized Biography of George Nathaniel, Marquess Curzon of Kedleston, K.G.* London: Ernest Benn, Ltd., 1927.

Ross, Nancy Wilson. *Buddhism: A Way of Life and Thought.* New York City, N.Y.: Vintage Books / Random House, [1980] 1981.

Smith, Richard Harris. *OSS: The Secret History of America's First Central Intelligence Agency.* Guilford, Conn.: The Lyons Press, an imprint of The Globe Pequot Press, [1972] 2005.

Soong, T. V. *Reports of Cases Heard and Determined in the Appellate Division of the Supreme Court of New York.* Vol. 250. Banks & Bros., 1937.

Stevenson, William. *A Man Called Intrepid: The Secret War.* Guilford, Conn.: The Lyons Press / Globe Pequot Press, [1976] 2000.

_____. *Intrepid's Last Case.* New York City, N.Y.: Ballantine Books / Random House, [1983] 1984.

Sutton, Antony G. *America's Secret Establishment: An Introduction to the Order of Skull & Bones.* Walterville, Ore.: Trine Day, [1983] 2002.

Swayne-Thomas, April. *Indian Summer: A Mem-sahib in India and Sind.* London: New English Library, Barnard's Inn, Holborn, 1981.

Taylor, Edmond. *Richer by Asia.* New York City, N.Y.: [Houghton Mifflin Company, 1947] Time Life Books, 1964.

Thayer, Charles W. *Bears in the Caviar.* Philadelphia, Penna.: J. B. Lippincott Co., [1950] 1951.

_____. *Hands Across the Caviar.* Philadelphia, Penna.: J. B. Lippincott Company, 1952.

_____. *Diplomat.* New York City, N.Y.: Harper & Brothers, 1959.

_____. *Guerrilla.* New York City, N.Y.: Harper & Row, 1963.

Thomas, Evan. *The Very Best Men: The Daring Early Years of the CIA.* New York City, N.Y.: Simon & Schuster, [1991] 2006.

Thomas, Lowell. *Beyond Khyber Pass into Forbidden Afghanistan.* New York City, N.Y.: Grosset & Dunlap, by arrangement with The Century Company, revised ed., 1925. Illustrated with photographs by Harry A. Chase, F.R.G.S. and the author.

Thompson, Nicholas. *The Hawk and the Dove: Paul Nitze, George Kennan and the History of the Cold War.* New York: Henry Holt/Picador, 2009.

Treanor, Tom. *One Damn Thing After Another.* Garden City, N.Y.: Doubleday, Doran & Co., 1944.

Tung, Rosemary Jones. *A Portrait of Lost Tibet.* New York City, N.Y.: Holt, Rinehart and Winston, 1980.

Viesturs, Ed, and David Roberts. *K2: Life and Death on the World's Most Dangerous Mountain.* New York: Broadway Books, 2009.

Wake, Jehanne. *Kleinwort Benson: The History of Two Families in Banking.* Oxford: Oxford University Press, 1997.

Warner, Philip. *Auchinleck: The Lonely Soldier.* Barnsley, South Yorkshire, England: Pen & Sword Books, Ltd., [1981] 2006.

Wavell, A. P. [Archibald Percivall] [Field Marshall Viscount Wavell]. *Other Men's Flowers: An Anthology of Verse.* New York City, N.Y.: G. P. Putnam's Sons, 1945. Book Club edition. Foreword by Wavell, April 1943. The first English edition was probably 1943 or 1944.

_____, and Penderel Moon. *Wavell: The Viceroy's Journal.* Oxford, 1973.

Wedemeyer, Albert C. *Wedemeyer Reports!* New York City, N.Y.: Henry Holt & Company, 1958.

Weiner, Tim. *Legacy of Ashes: The History of the CIA.* New York City, N.Y.: Random House/Anchor Books, [2007] 2008.

Wolpert, Stanley. *Shameful Flight: The Last Years of the British Empire in India.* Oxford, 2009.

Young, Desmond. *Rommel.* London: Collins, [1950] 1954.

Yu, Maochun. *OSS in China: Prelude to Cold War.* New Haven, Conn.: Yale University Press, 1997.

_____. *The Dragon's War: Allied Operations and the Fate of China, 1937-1947.* Annapolis, Md.: Naval Institute Press, 2006.

Zacharias, Ellis M. *Secret Missions: The Story of an Intelligence Officer.* Annapolis, Md.: Naval Institute Press, [1946] 2003.

_____. *Behind Closed Doors: The Secret History of the Cold War.* New York City, N.Y.: G. P. Putnam's Sons, 1950.

[Zacharias] Wilhelm, Maria. *The Man Who Watched the Rising Sun: The Story of Admiral Ellis M. Zacharias.* New York City, N.Y.: Franklin Watts, Inc., 1967.

Zimmermann, Barbara S. *Mutterings.* Wynnewood, Pa.: Livingston Publishing Co., 1969.

Zimmermann, Warren. *First Great Triumph: How Five Americans Made Their Country a World Power.* New York, N.Y.: Farrar, Straus and Giroux, 2002.

Book Chapters

Cutting, Col. C[harles]. Suydam. "Cheetah Hunting," pp. 167-172, in Plimpton, George (ed. and Introduction). *As Told at the Explorers Club: More Than Fifty Gripping Tales of Adventure.* Guilford, Conn.: The Lyons Press / Globe Pequot Press, [2003] 2005.

Wake, Jehanne. "Benson, Sir Reginald Lindsay [Rex] (1889-1968)," in *Dictionary of National Biography.* Vol. 5. (Oxford: Oxford University Press, 2004), 194-6.

Newspapers, Magazines, and Periodicals

[Aldrich] "Malcolm Aldrich, 86; Headed a Foundation. Obituary, *New York Times* (3 August 1986).

[Benson] "Sir Rex Benson, Banker, 79, Dies: Was Military Attaché Here from 1941 to 1944" *New York Times* (28 September 1968).

Benson, Sir Reginald. Knighthood awarded. *The London Gazette* (18 July 1958), 4514.

[Brady] "James Cox Brady Marries in London; Son of Late Financier Wed Miss Helen McMahon in Westminster Cathedral. Heir's Third Marriage. Bride Won Honor as Typical American Girl of Today in The Times Contest in 1913" *New York Times* (3 October 1920).

[Brady] "Business & Finance: Brady Estate" *Time* (7 April 1930).

[Child] Thomas, Louisa. "Cloak and Dagger Was Her Bread and Butter: A Group Portrait of Idealists, Including Julia and Paul Child, Who Served in the OSS During World War II" *New York Times Book Review* (3 April 2011), 21. Review of Jennet Conant, *A Covert Affair: Julia Child and Paul Child in the OSS* (New York: Simon & Schuster, 2011).

Chivers, C. J. "U.S. Courting a Somewhat Skittish Friend in Central Asia" *New York Times* (21 June 2007), A3. Dateline: Ashgabat, Turkmenistan.

Dorwart, Jeffery M. "The Roosevelt-Astor Espionage Ring" *Quarterly Journal of the New York State Historical Association* 62 (No. 3, July 1981): 307-22.

Enders, Elizabeth Crump. "Three Vignettes: Women in the East." *Journal of the American Asiatic Association* (June 1934): 386-8.

_____. "East Meets West in Japan's Women." *Independent Woman*. 15-16 (April 1936): 102-4, 130.

Enders, Gordon B. "Prohibition in Old India" (with decoration by Herb Roth) *Asia: The American Magazine on the Orient* 20 (No. 10, November 1920): 1003-4.

_____. "The Last Theocrat: The Panchan Lama of Tibet" *World Youth* (30 November 1935), 7. Copy in Patrick J. Hurley Collection, Western History Collection, Oklahoma University Libraries, Box 493C, Folder 4. With photo of the Panchan Lama.

_____. "Night Raiders in China," *Liberty* (3, 10, 17, 24, 31 July 1937) [five parts], 7-10, 28-35, 26-32, 40-4, 50-2.

_____. "The Nomad Woman" *Colliers* (7 October 1950), 16, 62.

[Enders] Anon., "Flying Gold Out of Tibet: Planes Invade Land of the Lamas" *Modern Mechanix Hobbies Inventions* (November 1936), 76-7, 120.

[Enders] *Purdue Exponent* issues in which Gordon Enders is mentioned (pages copied and sent from David M. Hovde, Purdue University hovde@purdue.edu, subsequent to inquiry in e-correspondence by George Hill, 19 May 2009).

[Enders, Gordon B.] Hoy, William. "A Yankee Adventurer and the 'Living Buddha'." *Chinese Digest* (28 February 1936), 11, 14.

_____. Gordon B. Enders, Obituary. "Former Adviser to Chiang Dies." *Albuquerque Journal* (September 1978).

Fisher, John. "India's Treasures Helped the Allies." *National Geographic* 89 (April 1946), 501-22.

Fox, Jean. "Gordon Enders, Confidant of Chinese Leaders, Visits Shore: Eminent Author Is Foreign Counsellor to Grand Lama of Tibet." *Winnetka [Ill.] Talk* (21 January 1937), 27. In Hurley Collection, University of Oklahoma Libraries, Box 493C, Folder 4.

Fox, Margalit. "George MacDonald Fraser, Author of Flashman Novels, Dies at 82." *New York Times* (3 January 2008), C12. Fraser died at his home on the Isle of Man on 2 January 2008.

Friedberg, Aaron L. "The New Great Game." Review of Robert D. Kaplan, *Monsoon* (New York City, N.Y.: Random House, 2010), in *New York Times Book Review* (21 November 2010), 21.

Gall, Carlotta. "A Taliban Leader in Pakistan Says He Would Aid bin Laden." *New York Times* (22 April 2007), A6.

Graufner, Steffen, and Kathrin Münzel. "Pamir: Oxus Snow Lake (ca 5,300m), new discovery; Peak 5,588m, attempt; Pamir-i-Wakhan Range, first traverse of central section." *The American Alpine Journal* 51 (Issue 83, 2009): 257-9.

Ernest O. Hauser, "A Reporter at Large: Pathans Behind the Rocks." *New Yorker* 90 (30 September 1944), 169-222.

Heilbrunn, Jacob. "The Dishonorable Schoolboys: Charles Cumming's Thriller Posits a Sixth Man Among Britain's Notorious Cambridge Spies." *The New York Times Book Review* (20 March 2011), 9.

Kostecka, Daniel J. "A Bogus Asian Pearl." U. S. Naval Institute, *Proceedings* (April 011), 48-52.

Johnston, David. "Search for Al Qaeda Leader Focuses on Pakistan Border Area." *The New York Times* (8 March 2003), A12.

La Guardia, Anton, and Ben Aris. "Putin Concedes New Moves in the Great Game." *Telegraph.Co.Uk* (24 September 2001).

[Lincoln, Anne] *Life Magazine*, 5 March 1945, 118-121

Macintyre, Ben. "Smiley's People: The Facts Behind the Myth. What MI5 Is and Was." *The New York Times Book Review* (31 January 2010), 15.

Marchio, James D. "Days of Future Past: Joint Intelligence in World War II." *Joint Force Quarterly* (no. 11, Spring 1996), 5, 116-122.

Masood, Salman, and Carlotta Gall. "Pakistani and Afghan President Discuss Border Woes." *New York Times* (27 December 2007).

Muir, Peter, and Frances Muir. "India Mosaic." *National Geographic* 89 (April 1946), 442-70.

"National Geographic Maps: The Complete Collection." 8 CD-ROM set (Washington, D.C.: National Geographic, 2001). "South Asia, Afghanistan, and Mynamar" (1997); and "Caspian Region" (1999).

[Roosevelt, Kermit] Kinkead, Gwen. "Kermit Roosevelt: Brief life of a Harvard Conspirator: 1916-2000." *Harvard Magazine* (Jan-Feb 2000), 30-1.

Stafford, David. "School for Spies." *World War II* (March-April 2010), 36-43.

Steinberg, Saul. [Cartoons from China] *New Yorker* (15 January 1944).

[Streather, Tony] In "Friends Who Died at the Top." *[London] The Sunday Times* (2 September 2007).

Taylor, Mark. "What happened to the Sandia bighorn herd? Gordon Enders Focused on Sheep." *Albuquerque Journal* (22 December 1986). In Hurley Collection, University of Oklahoma Libraries, Box 493C, Folder 4.

Thomas, Louisa. "Cloak and Dagger Was Her Bread and Butter: A Group Portrait of Idealists, Including Julia and Paul Child, Who Served in the OSS During World War II." *New York Times Book Review* (3 April 2011), 21. Review of Jennet Conant, *A Covert Affair: Julia Child and Paul Child in the OSS* (New York: Simon & Schuster, 2011).

Tolstoy, Ilia. "Across Tibet from India to China." *National Geographic* 89 (August 1946), 169-222.

[Traub] Khalidi, Rashid. *Sowing Crisis: The Cold War and American Dominance in the Middle East*. Beacon Press, c.2008. Reviewed by James Traub in *The New York Times Sunday Book Review* (15 March 2009), 6.

Williams, Maynard Owen. "South of Khyber Pass." *National Geographic* 89 (April 1946), 471-500.

[Zacharias] Pfeiffer, David A. "Sage Prophet or Loose Cannon? Skilled Intelligence Officer in World War II Foresaw Japan's Plans, but Annoyed Navy Brass." *Prologue Magazine* [National Archives] 40 (No. 2, Summer 2008), 11 pp.
http://www.archives.gov/publications/prologue/2008/summer/zacharias.html (Accessed 4/16/10).

Zimmermann, Barbara S. "This War – And Brave Little Women." *Vogue* (1 March 1944), 137, 140.

Other Media

In Harm's Way, movie, 1965, on DVD. Hollywood, Calif: Paramount Home Entertainment, 2005.

Letters and Personal Communications

Brady, Nicholas F., letter to George J. Hill, 14 October 2009, re Charles Suydam Cutting.

Enders, Elizabeth C[rump]. Letter to Brigadier General [sic] Patrick J. Hurley, Washington, D.C., 22 April 1944, TLS, 2 pp. In Hurley Collection, University of Oklahoma Libraries, Box 84, Folder 4.

[Enders, Gordon B.] Letter from Sarah Uschak, Office of Alumni Relations, The College of Wooster, Wooster, OH 44691, to George Hill, 8 Jun 2009.

Enders, Gordon B. Letter to Department of State, 30 March 1943, re Dodge "carry all" for the King of Afghanistan, State Department Archives, Microfilm Publication #1219, Afghanistan, 1940-45, Box 4.

Engert, Cornelius Van H., letter to NLO Karachi, 2 December 1942, discourages the sending of a Navy lieutenant to Karachi on a "special mission." State Department Archives, Microfilm Publication #1219, Afghanistan, 1940-45, Box 5.

____. Letter to Department of State, 4 November 1943, re brigands operating in South Afghanistan. Ibid, Box 5.

____. Letter to Department of State, 15 July 1944, re rebel leader Abdurrahman, known as "Pak." Ibid, Box 4.

____, Letter to Department of State, 13 Sep 1944, re Mazrak Khan, known as "Mazrak." Ibid, Box 4.

[Kane, CDR (select) John] Telephone conversation by George Hill with his daughter, Amelie, 26 May 2009.

Lomazow, Steven. "FDR's Deadly Secret: A Cancer on the Presidency." Lecture to Medical History Society of New Jersey, Princeton, N.J., 20 May 2009.

Thayer, Charles W. Letter to Department of State, 20 February 1943, "Some Observations regarding Motor Roads through Afghanistan and Central Asia to Eastern China." In State Department Archives, Microfilm Publication #1219, Afghanistan, 1940-45, Box 5.

____. Letter to Department of State, August 1943, August 1943, "Afghan Reactions to Recent Developments in Sicily, Italy." Ibid., Box 5.

____. Letter to Department of State, 30 November 1943, with Dispatch No. 348 of 8 December 1943, "Observations of the Italian Minister, Mr. Quaroni, on Afghan-Axis-Allied Relations, 1939-1943." Ibid, Box 5.

State Department to Thayer, at Am Embassy, London, 23 December 1943, re shotgun shells for King of Afghanistan. In State Department Archives, Microfilm Publication #1219, Afghanistan, 1940-45, Box 4.

Manuscripts, Memoranda, and Reports

Bohlen, Charles E. Memorandum for the Record, 22 August 1944, recommending rejection of request for exchange of Afghan and American officers for training purposes, enclosing comment from Major Charles W. Thayer, formerly of the U.S. Legation, Kabul. State Department Archives, Microfilm Publication #1219, Afghanistan 1940-45, Box 5

Enders, Gordon B. "Afghanistan's Strategic Geography." 8 pages (4 double-sided)
RG 165, Entry 7 Afghanistan 5000-6400 Box 4 NND 745008, Geography (5900), 5930- Report 107-A, mimeographed, 6 September 1943.

____. RG 165, Entry 7. 128 Kabul. Memorandum for the Minister, Report No. 103, 4 September 1943; 8 November 1943, addendum to 22 Oct and 4 Nov, ref M.A. Report # 103 of 1 Sept 1943; Memorandum for the Minister, 4 Nov 1943.

_____. Memorandum, "Afghanistan in General" to General Patrick Hurley, 13 January 1944, Kabul (5 pp., typed), from Hurley Collection, Box 83, Folder 7.

_____. "Report of Military Attaché, Kabul, Afghanistan, on General Hurley's Visit to Afghanistan," 1/18/44 (5 pp., typed), in Hurley Collection, Box 493C, Folder 4.

Zimmermann, Albert W. "Intelligence Report" on "INDIA – Political Parties and Groups," a 9-page transcript of a talk given to British Army officers by a "lecturer [who] could not be identified," entitled "The Constitutional Issue in India." This report was sent by AWZ to JICA/CBI with many copies made for others who are listed on the cover page, and marked RETAIN OR DESTROY. It was still classified as SECRET in 2011, but it was declassified under FOIA on 8 December 2011 (see Appendix D for a copy). From NARA II, RG 38, ALUSLO Report files 1944-45, Box 25, ALUSNO Karachi, Reports 1944-5.

_____. (prepared by Howard Voorhees), 12 July 1944. USNLO Karachi to JICA/CBI and DNI. Subj.: INDIA – Progress on Improvement of Karachi facilities. "Anglo-American shipping and ports mission report of progress on recommendations during the months of April and May 1944." From NARA II, RG 38, ALUSLO Report files 1944-45, Box 25, ALUSNO Karachi, Reports 1944-5.

_____. USNLO Karachi to CNO (DNI), 1 Sep 1944 / Serial 960 / CONFIDENTIAL Subj.: Landing Beaches in the vicinity of Karachi, Study of. From NARA II, RG 38, Records of the Office of the Chief of Naval Operations / Correspondence with Naval Attaches, Observers, and Liaison Officers, 1930-48, 370, 15/1/2, Box 8 (Hobart, Australia to Lima, Peru, includes Karachi, India).

_____. From NLO Karachi to JICA/CBI and DNI, 13 October 1944, Serial 23-44. Subj.: LIFE RAFT – Washed ashore at From U.S.N.L.O. Karachi to ONI (FE) via JICA/CBI (2 pp.). Subj: LIFE RAFT – Washed ashore at $25^0 11'N - 66^0 45'E$. From NARA II, RG 38, ALUSLO Report files 1944-45, Box 25, ALUSNO Karachi, Reports 1944-5, file ALUSLO Karachi 21-44.

_____. "Facilities of Port of Karachi, India," c. 31 October 1944. 72 blue mimeographed pages, unbound, enclosed with cover letter from Zimmermann and Voorhees to Joint Transportation Committee, Delhi, through JICA/CBI Naval Section, with copies to 13 other addressees, including FE-4. In NARA II, RG 38, ALUSLO Report files 1944-45, Box 25: ALUSNO Karachi, Reports 1944-5, file ALUSLO 1-S-44. Also black mimeographed, 42 bound pages, in RG 165, Entry 77 India 4730-5900 Box 1629, "Port Summary of Karachi, India," 15 May 1944, from Op-16-FE.

_____. From NLO Karachi to Chief of Naval Operations (Director of Naval Intelligence), 25 Nov 1944, Serial 1068. CONFIDENTIAL [declassified]. Subj: Records of Mess at the Office of U.S. Naval Liaison Officer, Karachi – Retention of. From NARA II, RG38, Records of the Office of the Chief of Naval Operations / Correspondence with Naval Attaches, Observers, and Liaison Officers, 1930-48, 370, 15/1/2, Box 8 (Hobart, Australia to Lima, Peru, includes Karachi, India).

_____. From NLO Karachi to Distribution List, 1 March 45, Serial No. 2106. Subj: Monthly summary of intelligence reports, with enclosure (copy) showing reports submitted in February 1945. The enclosure showed four reports were prepared, and two more were being undertaken, as directed in Op-16-FE/A3-1/JICA CONFIDENTIAL No. 0354016 of 6 Feb 1944. From NARA II, RG38, JICA Administrative and Serial Files 1943-45, 12 W3 24 (17) C-F 370-15/5/3-6. Boxes 24-25 Serials JICA/CBI 1943-44.

Internet Sources

Benson, Lt.-Col. Sir Reginald "Rex" Lindsay, DSO, MVO, MC from Google search: AIM25 Archives in London.

Benson, Lt.-Col. Sir Reginald, in "Philip Morris's predecessor, Benson & Hedges" http://www.smokershistory.com/Benson.html.

"CBI Order of Battle: United States Army Forces, China, Burma and India (HQ USAF CBI)" from "Order of Battle of the United States Army Ground Forces in World War II," Office of the Chief of Military History, 1959, accessed from http://www.cbi-history.com/part_xxii_hq.html (12/3/2010).

Chambers, John Whiteclay, II. *OSS Training in the National Parks and Service Abroad in World War II.* National Park Service, 2008. From http://www.nps.gov/history/history/online_books/oss/ (accessed 3/8/11).

Crosby, Bing. Quotation from http://www.finadeath.com (accessed 11/1/2011) quotation.

[Enders, Gordon Bandy]. Find A Grave. Gordon Bandy Edwards. Birth May 7, 1897. Death Sept. 2, 1978.

Engert, Cornelius Van. Department of State, Office of the Historian.

D'Aoust, Gerald. "The Holy Tibetan Dogs," Part 2: *"Lhasa Apsos ... is the story of Gordon Enders who got to see the Pan-Chen Lama's 'Holy Dogs' one century or so ago."* From www.dharmapalalhasa.org/holytibetdogs2.php (accessed 23 April 2007).

[Di Gemma, Joseph Paul] Various items about Joseph Di Gemma, USCG Combat Artist, who was at NLO Karachi from 13-16 January 1945.

Dorwart, Jeffery M. "The Roosevelt-Astor Espionage Ring." *Quarterly Journal of the New York State Historical Association* 62 (No. 3, July 1981): 307-22. (at http://cryptome.info/fdr-astor.htm, 7 pp. (accessed 4/27/2009).

Dyer, George C. [VADM, USN]. The Amphibians Came to Conquer: The Story of Admiral Richmond Kelly Turner. Viewed at http://www.ibiblio.org/hyperwar/USN/ACTC/actc-4.html (4 January 1911).

Finnegan, John Patrick, and Romana Danysh. *Military Intelligence.* In Army Lineage Series. Washington, D.C.: U.S. Government Printing Office, 1998. Foreword, Contents, and Chapters 5 and 6 downloaded from http://www.history.army.mil/books/lineage/mi/, 9/5/2009.

[Godfrey, ADM John Henry, C.B., R.N.] Google search, 4 January 2011.

[Hind] Major-General Neville Godfray Hind, C.S.I., MC. In "The London Gazette" (27 June 1952), 3521 [found and printed from a Google search, January 2011]

[Hunter, Gen. John H. "Nonnie"] Notes re Gen. John H. "Nonnie" Hunter from internet sources.

Khan, Mohammad Afzal "A Short History of Chitral and Kafiristan," www.anusha.com/chitralh.htm, and www.site-shara.net/_chitral (accessed 13 August 2003).

[Kipling] Lewis, Lisa. "Kipling's Biographers," http://www.Kipling.org.uk/rg_biogs.htm, 16 January 2008 (accessed 11/1/2011).

Macy, Clarence Edward. Entry in Political Graveyard.com.

[Markey] thePeerage.com: Gene Markey, M, #298074, and Melinda Markey.

[Markey] University of Virginia School of Medicine: www.medicine.virginia.edu/research/research-centers/cell-signaling/welcome-information/history-page.

[Markey] Find A Grave Memorial# 11486250.

[Markey] *The New York Times*. http://movies.nytimes.com/movie/53745/Wee-Willie-Winkie/details. (2008-12-10): *Wee Willie Winkie: Hollywood's Version of a Highland Regiment on the NW Frontier*, Soldiers of the Queen (Journal of the Victorian Military Society).

[Office of Naval Intelligence, Op-16]. Information from various sources about ONI, aka Op-16 in the Office of the Chief of Naval Operations].
For the history of ONI in World War II, see: http://www.oni.navy.mil/This-is-ONI/Proud-History (accessed 1-30-18)

Reeve, Walter. "Peshawar Remembered." From http://www.khyberlodge.co.uk/about-khyber-mainmenu-26/peshawar-remembered-mainmenu-43.html (accessed 1/13/2011).

Rugby, birth of. Quotation from http://www.turtleheadrugby.com (accessed 11/1/2011),

Smith, Francis Howard. From the Internet, various sources.

Thayer, Charles Wheeler. Biographical outline from Political Graveyard.

_____. Department of State, Office of the Historian.

_____. Biography from *Contemporary Authors Online – 2006*.

Trevizo, _____. "A History of the Office of Naval Intelligence, 1882-1942" (http://odh.trevizo.org/oni.html (accessed 3/24/2011)

Valero, Larry A. "The American Joint Intelligence Committee and Estimates of the Soviet Union, 1945-1947: An Impressive Record. 21 pp. Center for the Study of Intelligence, Central Intelligence Agency. At https://www.cia.gov/index.html, accessed 9/5/2009

Wedemeyer, Albert Coady. On line Pacific War Encyclopedia entry.

[Wheeler] Rear Admiral Charles Julian Wheeler, USN. Information from various internet sources.

[Wheeler] Lt. Gen. Raymond Albert Wheeler, USA. Information from various internet sources.

[Winsor] Information from various internet sources re Curtin Winsor, Curtin Winsor Jr., and Curtin Winsor III.

Zacharias, Ellis Mark. Arlington National Cemetery Website. http://www.arlingtoncemetery.net/emzacharias.htm (accessed 4/16/2010).

Reviews of *Proceed to Peshawar*

C. Naseer Aman, *Pakistan Link*, http://pakistanlink.org/Opinion/2015/Dec15/11/04.HTM (accessed 1-29-18)

Robert F. Baumann, *Military Review* (1 February 2015), 127-8.

James A. Cox, *Midwest Book Review* (December 2018) [on-line].

Roger D. Cunningham, *Journal of America's Military Past* (Spring/Summer, 2014), 113-5.

Charles C. Kolb, *Naval Historical Foundation* (Book Review, posted 10 February 2014).

Emma Reid, *Canadian Naval Review* 11 (no. 2, 2015): 41-2.

Ronald F. Rosner, *Asian Affairs: Journal of the Royal Society of Asian Affairs* (March 2014): 157-8.

Michael Scott Swisher, *Bulletin of the Founders and Patriots of America* 54 (no. 194, Spring 2015): 14, 22, 30.

List of Illustrations

* Albert Walter Zimmermann, by Bachrach, c.1923	ii
* LT Albert Zimmermann, c. 1943-5, on the bank of the Indus River	iii
* LT Albert Zimmermann, USNR, in dress blues, c. 1942	v
U.S. Naval Liaison Office, Karachi, Ingle Road street scene, c. 1944	v
* Albert W. Zimmermann, in civilian suit, 1924	vi
* Barbara Shoemaker, in formal dress, 1924	vi
Finalist Certificate for *Proceed to Peshawar*, Indie Book Awards, 2015	xi
Albert W. Zimmermann, as a child	xii
Barbara Shoemaker, as a child	xii
FDR with Vincent Astor	2
John Zimmermann and Eva Kellenbenz and their five surviving children in 1907	3
Zimmermann family roadster; Albert in right front seat; and home in Philadelphia	3
Art Loom Factory building, Philadelphia. John Zimmermann's mill	3
Home of Barbara Shoemaker, daughter of Dr. William T. Shoemaker, in Philadelphia	4
Albert W. Zimmermann, Diploma, U. Penn., B.S. (Mechanical Engineering), 1923	5
Albert and Barbara (Shoemaker) Zimmermann's wedding party in 1926	5
Home of Albert and Barbara Zimmermann, 400 North Rose Lane, Haverford, Pa.	6
Children of Albert and Barbara Zimmermann: Albert Jr., Barbara, Helene, Warren	6
German Bund guard at St. Paul's Church, Philadelphia (c.1937), taken by AWZ	7
Nazi troops marching in Stuttgart, Germany, from Zimmermann's movie (1936)	7
Albert W. Zimmermann's commission as a Lieutenant, USNR, 8 September 1942	8
* LT Albert Zimmermann (two photos in dress blue uniform)	9
LT Zimmermann's Basic Orientation Class, Washington Navy Yard, October 1942	9
RADM Harry Train and RADM Ellis Zacharias	10
"Main Navy" (U.S. Naval Headquarters, Washington, DC). Location of ONI in 1941	10
Example of LT Zimmermann's handwriting, 30 December 1943, Peiwar Khotal, Waziristan	16
Zimmermann between a French colonial officer and a man in turban in Algiers	19
Newspaper clipping, showing Viceroy of India at Garden Party in Karachi, March 1944	88
"Chippy" the monkey at NLO, Karachi	91
Field Marsal Claude Auchinleck "teaching a sepoy how to shoot" in July 1944	113
Kandy, Ceylon, entrance to SEAC Headquarters, July 1944	113
Kandy, Ceylon – veranda at OSS Headquarters, with "Peachie" Durand, July 1944	119
"The Thing" (a life raft, washed overboard from a Liberty ship), Baluchistan, Sept 1944	124
Iranian Consul's "At Home" reception, with Zimmermann in dress whites, Nov 1944	133
Typed notes on Chitral Trip (5 pp.)	135-40
* Three U.S. officers in summer whites, outside of NLO Karachi, November 1944	150
"Tiffin" (large lunch) at NLO Karachi, showing 13 at table, probably December 1944	150
* Cartoon drawing of Zimmermann, and two others by Joseph Di Gemma, USCG	154
Five Naval Officers at NLO Karachi. LTs Zimmermann and Voorhees, and 3 LTJGs	162
Mahan Kishin, Zimmermann's "bearer"	163
"Our bungalow": Naval Liaison Office, 225 Ingle Road, Karachi, India	163
Group picture, U.S. officers of Services of Supply with British officers, c. 1944-5	163
Group picture, British officers, with some U.S., including Zimmermann, c. 1944-5	164
* Zimmermann, informal, in khakis, "on the bank of the Indus River"	164
Naval Liaison Office, Karachi – LCDR Smith and Hashim Ali Shah, head watchman	166
LTJG Phil Halla (drawing by Di Gemma) and Sen. Richard Russell (photo)	167
April Swayne-Thomas, self-portrait, and 11 of her paintings from *Indian Summer*	168-70
Lady Vere Birdwood, "Abode of Peace," photo of page in *Illustrated Weekly of India*	171
Cornelius Van H. Engert	172
Charles Thayer	172

---

* From original photos in AWZ Wartime Papers, previously published in *Proceed to Peshawar*.

Gene Markey	172
Sir George Cunningham	172
Ambassador Clarence E. Gauss	173
Brigadier General John D. Warden (drawing by Di Gemma)	173
Winston Churchill (two photos: as a young officer, and as an elder statesman)	173
Khyber Pass painting	174
Bolan Pass (painting and photograph)	174
Peshawar: Dean's Hotel, and Peshawar Club	175
Peshawar: Services Club (two photos)	176
Governor's House and Garden, Peshawar (two photos)	177
Commodore Milton "Mary" Miles, USN, with Ambassador Patrick Hurley	178
Field Marshals Bernard Montgomery, Archibald Wavell, and Claude Auchinleck	178
Letter from IB Quetta to J.R. Harris, 26 October 1943	180
Zimmermann's orders on 8 November 1943 to "Proceed to Peshawar"	181
Attock, India, crossing the Indus River	182
Attock, India	183
* Peshawar railroad station, Bromhead meeting Zimmermann	184
Shabkadar (two photos)	185
Shabkadar (two photos of bazaar)	186
Shabkadar bazaar street	187
Halki Gandao, promontory	187
Halki Gandao, khadassars	188
* Kasai Munidi reception committee	188
Malik of Kasai Munidi	189
Yusef Khel – Hill 19408	189
* Yusef Khel – two photos of lunch with the malik and his staff	190
* Yusef Khel – two photos, showing Shazada, the malik, and his bodyguards	191
Malakand – two photos showing the Swat River basin, where "Gunga Din" fought	192
Malakand – two photos, showing valley of Swat River	193
Saidu – two photos, showing the Wali of Swat and his palace	194
* Saidu – two photos, showing the wali's palace	195
Swat, Malakand, and Zimmermann's notes for 16 November 1943 – three photos	196
* Swat River hydroelectric works – two photos	197
Malakand, the political agent's home – two photos	198
* Malakand, the political agent's home; and Enders in the jeep, Ma Kabul – two photos	199
Malakand and Bat Khel – two photos	200
* Punjkara River Valley – two photos – and Zimmermann's notes for 17 November	201
Dir State – terraced valley and rest house – two photos	202
Dir bazaar – two photos	203
Dir – two photos of the departure	204
* Dir to Gujar – two photos of the difficult road	205
Gujar – two photos of the levy post, and Zimmermann's notes for 18 November	206
Approaching the top of the Lowari Pass – two photos	207
* Jeep and its passengers at the summit of the Lowari Pass – two photos	208
* The summit of the Lowari Pass, and starting descent – two photos	209
* Descending from the pass into Chitral – two photos	210
Descending, with ice on the road and "tough going" – two photos	211
Descent, showing Enders alone in the jeep – two photos	212
Jeep is off the road – two photos	213
Ziarat levy post – two photos	214
Difficult road to Ziarat – two photos	215
* Tirich Mir at dawn, 20 November 1943	216
Tirich Mir, south face, from Wikipedia	217
Tirich Mir and the palace of the mehtar (two photos) from Zimmermann's movies; his notes	218
Guards at Khyber Pass, looking into Afghanistan at "Frontier of India" (postcard)	220
Peshawar Club, Dance Card, 27 November 1943	221

North West Frontier of India, sketch map, marked by Zimmermann to show their route	221
Kohat, group photo: Enders, District Commissioner Sheiku, and his brother, Mir Ali	222
* Kohat – District Commissioner's House	228
* Peiwar Khotal, Waziristan	224
Zimmermann's note on 30 Nov 1943: *"potentially most powerful position in the world"*	224
* Peiwar Khotal, Waziristan (2 photos)	225
Peiwar Khotal, Waziristan (2 photos)	226
* Parachinar Plain from Peiwar Khotal	227
Thal – Indian Army Post	228
Spinwam – Tochi Scout outpost	228
Dasali – two photos	229
Dasali – two photos	230
Dasali – two photos	231
Dasali to Miram Shah – two photos	232
* Miram Shah – two photos	233
Gilzai Tribes near Bannu	234
Jandola – two photos	235
Jandola	236
Village near Laaha	236
Laaha	237
* South Waziristan Khassadars	237
South Waziristan Khassadars	238
Koniguram	238
Koniguram	239
South Waziristan	239
Wana	240
Mahsud Khassadar	240
Zhob Militia Post on border of Baluchistan	241
Sambaza – Zhob Militia Post	241
Sambaza	242
Tora Gharu Piquet near Sambaza	242
Fort Sandeman (2 photos)	243
Fort Sandeman (2 photos)	244
Fort Sandeman (2 photos)	245
Hindubagh Gulzai camp (2 photos)	246
Hindubagh Gulzai camp (2 photos)	247
Hindubagh Gulzai children (2 photos)	248
Sheep grazing outside of Quetta, Baluchistan	249
Shepherd & his sheep near Quetta (2 photos)	250
Fat tail sheep	251
Temporary Government House, Quetta	251
Tents for visitors to Governor, Quetta	252
Gathering for chikhor shoot	252
Unloading lunch	253
* Roasting mutton for lunch – Gov's youngest child	253
Looking up the valley	254
A Muslim says his prayers	254
The beaters have their lunch	255
* Lunch on the chikhor shoot	255
* Leaving Quetta for Karachi. Maj. Alston, Mrs. Wood, Benjie, Tommy, Enders	256
Endorsement of Zimmermann's Orders on return to Karachi	256
Vale of Swat (4 images)	257
Kohat (2 images)	257
Dir River Valley	258
Hunzakut children of Chitral	258
Lowari Pass (2 images)	259

Khyber Pass (photo from *Illustrated Weekly of India*, 1944)	260
Albert Zimmermann, in his office, Bala Cynwyd, Pennsylvania	261
Albert W. Zimmermann, in about 1950	262
Barbara (Shoemaker) Zimmermann, in about 1950	263
Al and Barbara Zimmermann, at home	264
Descendants of Al and Barbara Zimmermann: Their Children	264
Their twelve grandchildren	265
Their children and spouses, and some of their grandchildren and great-grandchildren	265
Al and Barbara Zimmermann, graves, Bryn Mawr, Pa.	266
Gordon Enders' letter to Al Zimmermann, 22 December 1943	276
Col. Gordon Bandy Enders - tombstone illustration	283
Elisabeth Crump Enders – tombstone illustration	283
Enders family in Iowa	284
Enders family in India	284
Gordon Bandy Enders – as a boy, in WWI uniform, and as a dapper young man	285
Gordon Enders – five images, one with the Panchan Lama – and "Golden Passport"	286
Enders' citation for Legion of Merit	287
Enders' curriculum vitae, typed by him in about 1961	288
Enders' passport application with photo, 5 April 1917	289
Enders' service record	290-1
Gordon Enders' obituary, at College of Wooster	291
Elizabeth Crump Enders, two photos: in WWI uniform, and publicity photo	292
Lt.-Col. Rex Benson	292
Major General Patrick Hurley	292
Major Ernest F. Fox	293
General Archibald Wavell and Mrs. Wavell	293
Field Marshal Wavell	293
Sir Benjamin "Benjie" Bromhead, photo by Zimmermann, November 1943	297
Telegram from Bromhead to Zimmermann, 22 September 1944	301
Decorations to which Sir Benjamin Bromhead was entitled to wear (5 images)	303-4
Thurlby Hall, the seat of the Bromhead family (exterior)	304
Thurlby Hall (grounds and interior, 4 images)	305
Benjamin Bromhead's schools: Wellington College and Sandhurst (2 images)	306
LT Curtin Winsor to LT Albert Zimmermann (Dear Al, from Curt), 12 September 1944	318
Curtin Winsor, his wife Elizabeth (Donner) (Roosevelt) Winsor, and their home in Washington	319
Social cards: Acceptance and Invitation from Governor Dow; Invitation from Iranian Consul	320
Acceptance by Captain Burrows Sloan, Jr.	321
Letter and photograph of C. Rollins Hanlon, M.D., F.A.C.S. (then LT, MC, USNR)	322

Index to Chapters 1-6

9th Panchan Lama (see Panchan Lama)
14th Dalai Lama (see Dalai Lama)
27th Division (see Army, 27th Division)
223d Bomb Squadron, 277
Abadan, 142
Abyssinia, 116
Academy Award, 30, 105
Acorn Club, 4, 76
Adams, Jeanne Hope, 116
Aden, Yemen, 81, 310
ADC (aide de camp), 33, 96, 98, 102-3, 118, 133, 153, 273
Aedes, 44
A.E.F. (American Expeditionary Force), 277
Afghan War, Third, 224
Afghanistan / Afghan, 60-5, 68, 108, 114, 174, 179, 184, 220, 224-8, 234-6, 246-7, 268-74, 279-81
Africa / African, 10, 18-29, 41, 58, 127, 151, 157, 261, 268, 281
Afridi, 68, 220
Afridi Lushkar, 220
Agra, India, 57, 109, 112
Ahmedabad, 14, 56
Air Force, U.S. Army (US AAF), 44-5, 114, 141, 222, 285
Air Service Journal, 277
Akron, Ohio, 37
Akyab, India, 316-7
Albrecht [jewelers], 155
Aldrich, Malcolm Pratt, 12, 120-1
Aldrich, Winthrop, 12
Alexander the Great, 60, 63, 68, 258
Alexandria, Egypt, 12, 25-6
Alexandria, Virginia, 106
Algeria, 12
Algiers, Algeria, 12, 17, 18, 21, 23-6, 34, 39, 43-5, 53
Ali Baba, 122
All Black Magic, 160
All Star game, 126
Allah, 274
Almorah, India, 277
Alps, 53
Alston, Major ____, 256
ALUSLO (see Naval Liaison Officer)
America (see United States and various units)
American College of Surgeons, 148, 322
American Gun Mystery, 77
American Red Cross, 27, 67, 76, 147-8
Ames, ____, 18
amethyst, 23, 36

amoebic dysentery (see dysentery), 14, 33, 110, 127, 130 (bacillary), 145, 158
An Honorable Profession, 89
Anad Villa, 270
Anderson, Andy, 70
Anfa Hotel, Casablanca, 18
Angus and Indie ____, 116, 120
anna (Indian currency), 127
Annenberg, Walter, 86
Anon, LCDR, David and Martha, 127
Anopheles, 44
antafagasta, 84
Anthony, Edward, 278
antimony, 203
Antheil, George, 103
Aquacade, 94
Arab / Arabic, 277
Arabian Sea, 30, 35, 43, 45, 54, 76
Argenti and Co., 103
Argentina, 59, 65, 297, 301
Armistice, 278
Army, British, 27, 42, 59, 63-4, 76, 93, 100, 121, 132, 159, 182, 185, 270-3, 278, 298-306 (implied), 311
Army, Indian, 64-5, 98-100, 219, 228, 229, 240, 248-9, 260, 272-3, 297
Army, Japanese, 278, 281
Army, Korean, 282
Army, United States, 2, 12-15, 20, 24-5, 32-47, 59, 69, 74, 90-1, 99, 102, 108, 116, 118, 141-8, 151, 154. 158, 182, 268, 270, 277-93, 302, 315, 317
Army, 2d U.S., 277, 281-2, 285
Army, 10th U.S., 281
Army, U.S., 27th Division, 281
Army, U.S., XVI Corps, 282
Artloom Corporation, 1, 3
Asia, 14, 60, 74, 87, 92, 110, 141, 268, 278, 302-3, 307
Asphalt, 142
Assam, 142
Astaire, Fred, 117
Astor, John Jacob, 2, 61
Astor, Vincent, 2, 61, 121
atabrine, 14, 29,
Athens, Greece, 131, 148
athlete's foot, 130
Atlantic Ocean, 2, 17, 89, 148, 301
Atlantic City Beauty Pageant, 148
Attock, India, 182-3
Auchinleck, General Sir Claude, 62, 132, 142, 178, 189, 192

aur billi, 108
Australia, 44, 89, 100, 158, 261, 313
Axis, 13, 24, 135
B-24 "Liberator", 12, 27
B-26, 282
B-29, 309
Babetz, Dr. Irma, 60
Babson, LT Arthur, USNR (ONI), 57, 312, 314
Baden-Württemberg, 1
Badoglio, Marshal, 42
Bagararh, 31
Bahadur, 94, 108, 272
Baker, LTJG Paul, 141, 162
baksheesh, 59
Bala Cynwyd, Pa., 261
Baldwin School, 53, 101, 105, 110
Bales, W. L., Colonel, USMC, 74, 157, 312-7
Baltazzi, CAPT ____ (ONI), 143, 312, 314
Baltimore, Md., 14, 56
Baluch Regiment, 98
Baluchistan, 64, 78, 98, 124-5, 241, 245, 249, 252, 268, 297
Bamboo, Mr., 278
Bangalore, India, 57, 69, 111
Bannu, 65, 227, 234-6
Banville, Louise, 82
Baralochi, 59
barber, 81
Barlow, ____, RAF officer, 31, 70, 86
Barnes [prob mis-named; could be Bales], Col. ____ (ONI), 312
Barnes, LCDR Hampy, 111-2, 120, 123-4
Barstow, Brigadier (?) ____, 222,
Barton, ____, Mr. and Mrs., 98
Base Hospital No. 8, 278
Base Hospital No. 9, 278
Base Hospital No. 38, Cairo, 25
baseball, 12, 37, 46, 48, 53, 70, 102, 110, 121-2, 129
Batdorf, ____, 134
Battle of the Bulge, 120, 148, 151
Bay of Pigs, 282
Bayfield, Captain ____ (RN), 101-2
Bales, Col. ____, USMC, ONI, 307, 311
bear (animal) [possible Russian], 68
Beard, Mr. and Mrs. Jack, 54, 78, 96
Bears in the Caviar, 68
Beaver Camp, 34
Behts, Captain ____, 204
Beiderbecke, Bix, 100
Beijing, 278
Belic, George, 50, 72
Belgium, 282
Benghazi, Libya, 44
Bennett, Barbara, 152

Bennett, Constance, 44
Bennett, Joan, 44
Benson, Lt. Col. Rex (later Sir Rex), 61, 272, 279, 292
Bergman, Ingrid, 18
Best, Janet, 116
billet doux, 147
Bingo, 108
Birdwood, Lady Elizabeth Vere, 100-1, 132, 171
Birdwood, Lt.-Col. Christopher, 100-1
Birdwood, William, Field Marshal, 100-1
Bitulithic, 142
black market, 206
Blackwell, John, and Pauline, 93, 100, 108, 118, 121, 129-34, 150
Blalock, Dr. Alfred, 148
Blynn, Bryce, COL, 151
Brittany, 277
Boenning, Henry Dorr, Jr., 151
Boericke, Dorothy "Dottie", 58, 156
Bogart, Humphrey, 18
Bohlen, Amb. Charles "Chip", 61
Boles, Dr. ____, 81
Bolshies, 135
BOMBA, 42
Bombay Company, 69, 80, 101
Bombay, India, 14, 36, 38, 57, 65, 69, 73, 80-1, 88, 103, 105, 111, 125, 130, 134, 141, 144-5, 153, 299, 308, 310, 314,
Bomber Pilot, 277
Bond, Mr. and Mrs. Arlo, 13, 38, 43, 51, 53, 55, 67-8, 81, 94, 160
Boone, Col. ____ (ONI, A4B, 16F), 316
Border, The (as in Kipling's *Kim*; also see Durand Line and Afghan border), 135, 179, 268, 279
Boston, Mass., 145
Botwood, Newfoundland, 12, 16-17
Boulter, Mr. and Mrs. ____, 222
Bouvier, "Black Jack", 4
Boy Scout, 274
Bowen, CAPT Harold Gardiner Jr., USN (ONI), 314
Bradley, ____, 95
Briggs, CAPT H. M., USN, 319
Britain / British (also see England), 12-14, 18, 23, 27-33, 42, 45-6, 51-2, 55-6, 59-72, 75-87, 91-8, 100-2, 107, 121-7, 132, 148, 153, 160, 182, 185, 192, 199, 220, 222, 227-8, 234, 270-2, 279-82, 296-306 (implied)
British Overseas Airways Corp., 14, 98
Bromhead, Anne Kathleen G. (b. 1942), 297
Bromhead, Benjamin (d. 1782), 296-3
Bromhead, Col. Sir Benjamin Parnell, 4th Bt. (d. 1935), 100-1, 296-9

Bromhead, Col. Boardman (d. 1804), 296, 299
Bromhead, Diana Jane Gonville (b. 1940) 297
Bromhead, Maj. Sir Edmund, 3d Bt. (d. 1870) 296-9
Bromhead, Edward, father of 1st Bt., 296, 299
Bromhead, Edward Gonville (d. 1910), father of Sir Benjamin, 5th Bt., 296-9
Bromhead, Sir Edward ffrench, 2d Bt., 296-9
Bromhead, Emily (née Hosking), 297
Bromhead, Sir Gonville, 1st Bt. (d. 1822), 296
Bromhead, Janetta, 100-1, 296-9
Bromhead, John Desmond G. (b. 1943), 297, 301
Bromhead, Lady Nancy (née Lough), wife of Sir Benjamin, 5th Bt., 60, 296-302
Bromhead, Sir Benjamin ("Benjie"), 5th Baronet, 14, 16, 59-64, 67, 93. 100-1, 124, 129, 132, 208-9, 225-6, 229, 255-6, 267, 296-306
   genealogy, 296-306
   letters, 299-301
   additional notes, 302
   medals, decorations, images, 303-6
Broudhead, Squadron Leader "Flash", 77
Brown, H. Tatnall, Jr., 92,
Browning, LT Powell (aka Paulus), 13, 32, 36-7, 42, 92, 95, 119, 155, 308
Bruce-Steer, Col. ___, 252
Brundage, Story and Rose, 129, 146
Bryn Mawr Hospital, 100
Bryn Mawr, Pa., 17, 61, 80, 100, 101, 110, 149, 261
Buddha / Buddhist, 58, 82, 88, 277
buddist (see Buddhist)
Buenos Aires, Argentina, 59, 60, 297
Bulgaria, 121
Bulkeley, John D., VADM, 89
Bullitt, William, 309
Buna (synthetic rubber), 93
Bund, German (see German Bund)
bungalow, 13, 37, 84, 95, 100, 158-9, 163, 169, 269
Bunker, ___ (RN), 77
Bureau of Naval Personnel (BUPERS), 311
burial / buried, 36, 61, 65, 69, 101, 103, 157, 249, 261-6, 269-70, 278, 282, 296
Burma, 11, 25, 42, 45, 59, 84, 100, 102, 110, 118, 141-2, 280, 282, 307, 314
Burmah Shell, 100, 102, 118
Burns, LTJG Harmon, USNR, 13, 31-2, 36-7, 57, 66, 308
Busleby, ___, 141
Buzby, George, 50
C'est impossible, 121
C 53 transport plane of US AAF, 28
Cabin in the Sky, 269

Cadillac, 148
Cairo, Egypt, 12, 17, 18, 21, 23-4
Calcutta, 97, 103, 105, 109, 111, 148, 160, 269, 279, 316-7, 322
Caledonia Camp, Botwood, 17
California, 38, 61, 81, 96-7, 124,
Callahan, LT ___, USNR, 13, 32, 36-7, 66, 70, 81, 162
Calumet Farms, 25
camel, 19, 43, 54, 61, 122-6, 149, 168, 185, 201, 227, 249, 269-71
Cameroon, 83
Camp Russell B. Huckstep, 25
Camp Tecumseh, 99
Canada / Canadian, 95, 142
canteen, 20, 48-9, 95, 105, 132
Cargill, Mr. and Mrs. ___, 155
Carmarkar, CDR ___, 108
carmeline, 131
carriage ghāree, 37
carrier pigeon, 122
Carrington, CPT ___ (MC, USA), 141
Casablanca, 18
Casablanca Conference, 18
Casablanca, Morocco, 12, 18, 19, 21, 32, 39
Castle, LT ___, USNR (ONI), 43, 152, 311 317
Castro, Fidel, 282
Cat = Catalina (Consolidated PBY), 114
Catholic, 123, 147
censor, 16, 20-3, 42, 49, 310 (also, implied for all letters printed on pp. 11-178)
Central Intelligence Agency (CIA), 74, 282
Central Intelligence Officer (CIO, British), 67, 85, 107, 180
C'est en guerre, 16
Ceylon (see Colombo and Kandy)
chablis (sandals), 157
chaga (aka *chagra*), 68
chagra (Chitrali robe), 63
Chakdarra, 200
Chandra Arya, Prof. Ramesh, 157
Chandri Chau, 271
Chanti, 277
chapatti, 190, 245, 274
Charleston, S.C., 51
Charlottesville, Va., 281
Chateauroux, France, 278
Chatham Hall, Chatham, Va., 45, 49, 78, 88, 96-7, 101, 130
Chatham, Mass., 43
Chatrubhuj, R.F., 52
Chawda, Mr. ___, 55
chemin de fer, 20
Chengdu, China, 277
Chennault, General Claire, 142

Chestnut Hill, Pennsylvania, 22
Chevalier of the Legion of Honor, 278
Chiang Kai-shek, Generalissimo, and Mde., 60, 142, 278
Chicago, Ill., 281
chicors aka *chikhor*, 63, 216, 219, 252-5
chifferobe, 90
Child, Paul, 82, 92, 119 (also see McWilliams, Julia)
children of AWZ and BSZ (see Zimmermann)
China / Chinese, 11-13, 26, 39-45, 59, 64, 67, 74, 79, 82, 92, 110, 141-2, 150, 156, 178, 222, 268-82 (implied), 280, 312-3
China-Burma-India Theater (CBI), 11, 25, 44-5, 107, 133, 156, 179, 280, 312-3
Chipman, ___, CDR, 12
Chippy, 90-1
Chitral / Chitrali, 62-8, 78, 88, 135, 204-21, 236, 257-9, 267, 276, 280, 299, 300, 302
Christmas, 47-8, 52, 54, 68-77, 81-3, 94, 97, 128, 130-4, 141-8, 152-6, 270, 299, 314
chrome, 249
Chunking, China, 26, 31, 74, 115, 142, 315
Church Farm School, 4
Church of the Redeemer, 3, 61, 261, 266
Churchill, Winston, 18, 61-2, 127, 173, 175, 192, 197
CIA (see Central Intelligence Agency)
Clee, Charles Beaupre Bell, Esq., and Mary, 83, 109, 116, 127, 144
Clei (see Clee)
Clement, Browning, 151
Clifton, district in Karachi, 84
CNO (Chief of Naval Operations), 316
Coast Guard, U.S., 97, 120, 153-4
Cobbet, Col. ___ and Mrs., 102, 104
cobra, 41, 84
codene (codeine), 124
Coffin, "Tince", 66
Cole, Capt. ___, 235
Collet, Roger (aka Collett), 96, 133
Collins, CAPT ___ (ONI), 313
Colombo, Ceylon, 14, 55-8, 68-9, 76, 79, 81, 104, 106, 109-12, 117, 123, 152, 309, 316-7
Colonial Dames, National Society of, 4
CBE, 63, 298
Commanding Officer (CO), 11-14, 31-2, 87, 157, 161, 309
Commonwealth Fund, 120
Communist, 282
Congress Party, 56
Congress, U.S., 38
Consolidated PBY, 114
Constantine, Tunisia, 12, 17, 43
Contax, 68
Cooks Travel Agency, 59

Cornell Medical School, 278
Cornell University, 70
Corsair bombers, 278
Cotswold Corner, 1, 4
Coughlan, Raymond and Frances, 69, 80, 84, 146, 150
Coupe, Captain ___, 108, 154
Covington, CDR ___ (ONI), 313
courier, 14, 56, 74, 92, 117, 125, 131-3, 144, 147-8, 155-7, 161, 309-17
Coward, Noel, 96, 98
Craft, Marie, 83, 94
Craft, Millie, 83, 94
Craft, William "Bill", 73
cremate, 65, 249
Crete, 23
cricket, 1, 4, 23-4, 37, 151
Crump, Elizabeth (see Elizabeth Enders)
Crump, Samuel, father of Elizabeth, 278
Crump, Samuel, Jr., 278
Cuba, 142, 282
Cullen, Bill, 102, 116, 160
Cunningham, Sir George, 14, 60-3, 135, 172, 177-8, 185, 272-3, 279
Curren, LCDR ___, 111, 119, 311
D-Day, 312
D'Elseux, Dr. ___, 100
daffodils, 96
Dalai Lama, 14th, 278
Dalmia Cement Factory, 157
Darjeeling, India, 43
Dartmouth College, 10, 317
Darya Ismail Khan, 98, 301
Dasali (aka Dosali), 229-33
Datta Kheyl, 234
Daur, 227
Davis, CAPT Ransom, Senior Naval Officer, India-Burma-Ceylon, 151-2, 156, 158-60, 314-7
Dawson, LCDR ___, 111, 119
DC 3 (C-47), 29
Dead Sea, 29
Dean's Hotel (aka Deans), 175, 271, 302
Delaware Water Gap, 131
Delaplaine, ___ (ONI), 314
Delhi, India (sometimes abbreviated from New Delhi), 12, 14, 36, 45, 61, 69, 81-2, 87-92, 99, 109-13, 124-6, 141-4, 152, 157-160, 178, 268-80, 299, 301, 307-9, 313, 316-7
Denderius, Petsa, 102
dengue, 14, 44, 111, 118, 125 (AWZ's), 129, 316
Denning, Major ___, 227, 234
Department of Commerce, U.S., 278
Department of State, 279

385

Dera Khan, probably Dera Ismail Khan, 98
Dewey, Thomas, 311
Di Gemma, Joseph P., USCG, 154, 167, 173
DiPinna, 78
Dickson, Abbott, 103
D.I.K. (see Darya Ismail Khan)
Dilowa, Prince, 278
Dimitrom, 31
Dinar Kheyl, 234
Dinshaw, Aspey, & children, 43, 50-1, 78, 80
DIO (see District Intelligence Officer)
Diplomat, 68
Dir state, 61-3, 199-214, 219, 258-9
Director of Naval Intelligence (DNI), 10, 90, 123-6, 133, 156
District Intelligence Officer (DIO), 3d Naval District (Philadelphia), 315
Dixon, Fitz Eugene, Jr., 71
Division, 27th, U.S. Army, 281
Dodge, 272
Dodgers, Brooklyn (baseball team), 90
donkey, 60, 62, 185, 192, 203-5, 249, 253, 256n, 270, 273
Donald, Kenneth (PA, Kurram), 222
Donner Foundations, 66
Donner, Elizabeth, wife of (1) Elliott Roosevelt and (2) Curtin Winsor (q.v.)
Donner, William H., 66
Donovan, MG William "Bill", 12, 74, 92
Dosali (aka Dasali), 64, 228-34
Double Indemnity, 156
Dow, Lady ____, 13, 41, 97, 133, 320
Dow, Sir Hugh, Governor, 13, 31, 33, 37, 41, 47, 51-2, 82-6, 96-7, 107-9, 126, 133, 142, 320
dowry, 55
Drexel Building, 38, 71
Dreyfus, Amb. Louis Goethe, 280
Drosh (aka Drash), 62, 204-10, 215-6, 218-19
Dubash, Mme. Nadia, 13, 31-2, 47, 53, 58, 67, 308
DuBois, Cora, 82
Duke of Acosta, 116
Duncan, Acting P.A. "Pat", 234-6
Durand, "Bitz", 17, 105, 119
Durand, Virginia "Peachy", 17, 81-2, 119, 153, 315
Durand Line, 60, 184, 280
Durbin, Deanna, 30
dysentery, 14, 32, 110, 126, 130, 145, 158
earthquake, 251
Easthampton, Long Island, N.Y., 82, 92-4, 100, 104-28, 134
Eastman (see Kodak and Kodachrome)
Eastwick, Joe, 92, 108
Ecker, Mr. ____, 36, 39

Eddy, Col. William A. "Bill", 22, 277, 279
Eddy, Edward, 289
Edwards, Dixon "Dick", 134, 145, 158
Egan, CDR ____ (ONI), 317
Egypt, 12, 17-18, 21-30, 156, 279
Eighth Army (UK), 42
Eisenhower, Dwight, General, 42, 282
El-Alamein, 18
elephant, 36, 58, 75, 82, 88, 93-4, 106, 111
Elphinstone St., 84
Emmanuel, Victor, King, 93
Enders, Elizabeth "Betty" (née Crump), 1st wife of Gordon Enders, 268-75, 28 (implied) 283, 292
Enders, Elizabeth "Liz" (Schwalbe) (Garrahan), 2d wife of Gordon Enders, 283
Enders, Emmanuel Allen, father of Gordon, 277
Enders, Frances Marie (née Seibert), mother of Gordon, 277
Enders, Major Gordon Bandy, 14-16, 60-7, 81, 100, 114, 123, 134, 141, 175, 178-9, 184, 199, 208-9, 218-29, 235, 248, 255-6, 267-293, 299, 300, 302
  letters to Betty, 268-75
  correspondence with AWZ, 276
  biography, 277-81
  images, 282-95
Enders, Miriam (married Stockwell), 277
Enders, Robert "Bob", 277
Engert, Hon. Cornelius Van H., 61, 172, 280
Engert, Jane Morrison, 61
Engert, Roderick, 61
England / English (also see Britain), 2, 4, 31, 37, 39. 46-7, 55, 58-60, 67, 70, 83-4, 89, 106, 108, 126, 134, 144, 183, 270-4, 295-7, 300-5, 311-2
ENIAC, 151
Episcopal / Episcopalian, 120, 261
Epsom salts, 130
Essex, Page County, Iowa, 277
Etawah, India, 269
Euphrates River, 46
Europe / European, 2, 26-7, 89, 107, 120, 130, 134, 144
Evans, Mrs. ____, 39
Evening Bulletin [Philadelphia], 156
Everett, Mass., 38
Fagin, Col. ____, 59, 182
falcon, 63, 201, 219
Fanny aka FANY; First Aid Nursing Yeomanry (UK), 102
Faqir of Ipi, 135, 227, 295
Farbos restaurant, Calcutta, 269
Farmington, Conn., 145
Farrell, Dr. ____, 48

Fatehpur-Haswa, 269
Faye, Robert (ONI), 314
faux-pas, 45
FBI (see Federal Bureau of Investigation)
FDR (see Roosevelt, Franklin D.)
Federal Bureau of Investigation (FBI), 1, 7
Feninaro, ____ (OSS), 83
Fernandez, 150
Festung Europa, 121
fez (Egyptian men's hat), 19, 22, 26, 51, 275
film, 4, 23-4, 30, 50, 53, 55-6, 74, 78, 83-6,
  101, 103, 110-111, 219
Finland, 80, 88
Fischer, ____ (OSS), 127
Fisher's Road, Haverford, Pa., 110, 148
Flecks, 75
Fligg, James and Lillian, 3, 52-3
flu epidemic, 77
Flukes, 82
Flying Tigers, 34
Foreign Devil, 60, 64, 220, 269, 280
Foreign Service (US), 44
Fort Benning, Ga., 282
Fort Logan National Cemetery, Santa Fe,
  N.M., 36
Fort Meade (aka Fort George Gordon Meade),
  Maryland, 281-2
Fort Riley, Kansas, 282
Fort Sandeman, 58, 65, 240-51
Fort Slocum, Long Island, 308
Fourth Street Club, 1, 85
Fox, Maj. Ernest F., 280
Foynes, co. Limerick, 12, 17
Frame, Rosamond, 82
France / French, 10, 18-20, 26, 36, 39, 59, 71,
  82, 89, 92, 107, 119-20, 160, 277-8, 285,
  302
Francis, Lt. Col. ____, 16, 222-6
Frankford, Pa., 53
Freeman's friends (code word for OSS), 14,
  39, 50-1, 78, 83, 85, 127
French Club, 71
French Foreign Legion, 277
French Without Tears, 160
Friday Evening Club, 71
Frontier Force, 272
G-2, 2, 12, 24 (also see Enders, Gordon, and
  U.S. Army Intelligence, MIS)
Galle F'ace Hotel, 111
Ganesh, 58
Ganges River, 84
Garden of Eden, 29-30, 45
Garden Party, 63-4, 84-5, 88, 135, 177-8,
  219-21, 302
Garrett, Maj. Peter, 244-5
Garson, Greer, 105

Gauss, Amb. Clarence E., 14, 44, 141, 173
gazelle, 27
Geldard, Mrs. Ruth (aka Geldart), 86, 98
General Hospital 38, 12, 25
Genoa, Italy, 42
Georgetown University, 37
German Bund, 1, 7
Germany / German, 1, 2, 7, 10, 13, 18, 27, 34,
  38, 80, 89, 104, 119, 120, 135, 148, 155-6,
  224, 277, 279, 282
Germantown Academy, 151
Germantown, Penna., 72
Getchell, ___, 78
Gezira Club, Cairo, 23, 25
Ghandi, Mohatma, 12, 56
ghat, 79, 122
Ghazi ud-Din, 216
Ghiza, 28
Gilgit, 62, 203
Gilzai, 234, 248, 301
Giraud, General Henri Honoré, 19
Goa, India, 36, 58
God Save the King, 52
Godfrey, VADM J. H., RN, 109, 125-6
Gokalda, Thakirdas, 144
Golden Passport of Tibet, 278
Goldsborough, LCDR ___, 14, 56-7
Goldstraw, Blossom, 144
golf, 1, 4, 23, 55, 86, 92, 95-98, 101-6,
  119-23, 131, 134, 155
Government House, Karachi, 84, 96-8, 118-20
Government House, Peshawar, 64
Graduate Hospital, 4, 48
Grand Trunk Road, 269
Grant, John and his wife Joy, 117, 147
grave, 61, 65, 79, 249, 261
Great Game, 280
Greatbatch, Mrs. Betty, 47-8, 58, 76, 94
Greece / Greek, 23, 31, 76, 95, 102-3, 131,
  144, 148-9, 258
Green, CAPT ____, USN, ALUSLO, 317
Green, Ann Carter, 92, 105-6
Green, LTJG Marshall (ONI), 312
Greene, Julian, 105-6
Greene, Louis Storrow, 105-6
Grindlay & Co., Ltd., 298
Groman, Henry, LCDR, 110, 148-51, 156-60,
  315-6
Gross Süssen, 1
Grout, Pamela, 116
Groves, General Leslie, 151
Guam, 279
Guerrilla, 68
Gujar, 205-6, 219
Gul Kutch, 234, 245
Gulzai, 246-8

Gulph Mills Golf Club
Gunga Din
Gurkha, 274
Gussenstadt, 1
Gustav, King of Sweden, 93
Gymkhana Club, 31-2, 69, 72, 76-7, 88, 94, 116, 151
gyppy tummy, 30
Habanniyah, Iraq, 21
Habecker, CAPT Frederick Shrom Habecker, USN ____ (ONI), 151-2, 307, 316
Haddon, BG Julian B., 14, 44-5, 67, 108, 133, 163
Haiti, 282
Halazone, 54
Halki Gandao, 187-8
Halla [aka Haller], LT Phil, USCG, 94, 97, 120, 122, 125, 145, 150, 153, 156, 160, 166-7, 316
Halliburton, Richard, 41
Halsey, FADM William, USN, 25
Hands Across the Caviar, 68
Hanlon, C. Rollins, LTJG, 148, 267, 322
Hanounum (or Hanoum), Camp, 38, 43
Hanover, New Hampshire
   U.S. Naval Training School, 319
Hanuman, 270
Hardy, Andy, 55
Harkness, Edward S., 120
Harris, John R., Esq., 14, 45, 74, 76, 84-5, 107, 116, 131, 163, 180, 267
Hartwell, Janet, 130
Hassett, "Doc", 300
Hatoun, M., 22
Haverford School, 78
Haverford, Pennsylvania, 1, 3, 6, 11, 15, 26, 29, 33, 49-135 (implied), 141-61, 261, 264
Hawaii, 81, 280-1
Hawk's Bay (aka Hawke's Bay), 129, 145
hawking, 63, 216, 218
Hay Adams Hotel, 155
Hay, Lt. Col. William Rupert, 65
Haymes, Dick, "In My Arms", 52
Haynes boys, 120-1
Haynes, Justice, "investigative work for," in Bombay, 134
Heidenheim, 1
Heliopolis, Egypt, 25
Henreid, Paul, in *Casablanca*, 18
Henson, ____ (ONI), 17
Hershey Bar, 33
Higgins, LT ____, (ONI), 317
Hightstown, N.J., 86
Hill 19408, 189
Hilley, Petty Officer ____, 37
Himalaya Mountains, 63, 97, 142, 149, 279

Hind, Brig. Neville and Marguerite, 13, 32-3, 85, 103, 118, 132, 152-3
Hindu, 12, 13, 38, 44, 50, 55-6, 58, 65, 79, 84, 122, 135, 156, 161, 227-8, 249, 269-70, 277
Hindustan / Hindustani, 62, 269, 80, 107-8, 269, 271
Hindu Kush Mountains, 63, 208
Hindubagh, 246-9
Hiroshima, 281
Hirst, Bill, 21
Hirst, Mrs. ____ (British), 121
Hitler, Adolph, 107, 112, 114, 121
Hobbs, Brig. Mervin, 231, 233
Holden, Robert, 36
Hollywood, California, 25, 104-5
Holmes, photographers, 299
Homestead, The, 49-52
Hong Kong, 41
Hope, Bob, 27
Hot Springs, Va., 49-51, 57, 59, 66, 150
Hotchkiss, Henry, LT, USNR, 12, 22, 25
Hsu Chih-chuan, 278
Humayon, Emperor, 99
Hungary / Hungarian, 60
Hunneman, 38
Hunter, ____, Major (British), 38
Hunter, "Nonnie" or "Lanie", 59, 283
Huntingdon, ____ (ONI), 316
Huntington, Gertrude (Enders) [Ph.D.], 268, 284-5
Hunza, 216
Hunzakut, 258
Hurley, MG Patrick, 44, 61, 141-2, 177-8, 280
Husky, Operation, 26
Hyderabad, India, 14, 56
IB Quetta, 14, 45, 123, 179, 267
ibex, 62, 197-8
Iblanke, 229, 233
In My Arms (and quoted by AWZ), 11, 52, 95
India / Indian, 11-13, 15-261 (implied), 267-84 (implied), 297-308 (implied), 311-14 (implied)
India-Burma Theatre, 42, 110, 307
Indian Army (specific), 64-5, 98, 100, 185, 219, 228-9, 240, 249, 297 [also see many postnominals: IA]
Indian Civil Service, 127, 153
Indian Naval Training Schools, 94
Indiana, 268, 271, 279
Indus River, 13, 56, 98, 182-3
infantile paralysis, 50
Ingle Road, Karachi, 11-166 (implied), 300-1
Intelligence Bureau Karachi (see Harris, J. R., Central Intelligence Officer, Karachi)

Intelligence Bureau Quetta (IB Quetta), 14, 45, 123, 179, 267
invitations, 13, 31, 37, 63, 79, 82, 102, 109, 125, 133, 148-9, 153, 157, 160, 329-30
Iowa, 60, 276, 284, 292
Ipi, 135, 227
Iran / Iranis, 61-2, 107, 133 268, 273, 279-81, 320
Iraq / Iraqis, 36, 44, 122, 297, 303
Ireland / Irish, 11, 18-9, 36, 71, 303
Isaac, Mohammed, 185
It Started with Eve, 30
Italy / Italians, 13, 18, 27, 34, 42-3, 93, 116, 151, 269, 281
ivory, 58, 68, 82, 93-4, 121
J Z Sons (John Zimmermann's Sons), 92
J. E. Caldwell and Company, 73, 155
jade, 106
Jalalabad, 220, 279
Jamrud, 65, 220
Jandola, 64, 234-40
Jane Eyre, 130
Janson, Lt. Col. ____, 225, 235-6
Japan / Japanese / Jap, 2, 13, 38, 74, 104, 111, 142, 268, 277-81
Jarvis, James, USNR, 159
jeep, 26, 52, 64, 122, 129, 199, 203, 210, 219-20, 234, 245, 279-80, 299
Jefferson Medical College, 25
Jenkintown, Pennsylvania, 96
Jerusalem, 120
Jew, 39, 61, 103
JICA (Joint Intelligence Collection Agency), 307-9, 314 (also see regional JICAs)
JICA-Chunking, 315
JICA-CBI (Joint Intelligence Collection Agency-China Burma India), 107, 134, 156, 312-3
JICA-FE (Joint Intelligence Collection Agency-Far East), 307-16
JICA-ME (Joint Intelligence Collection Agency-Middle East), 25
JICA-SEAC (Joint Intelligence Collection Agency-South East Asia Command), 92
jirga, 236
John Zimmermann's Sons, 92
Johns Hopkins University, 148
Johnson, Harry, 91
Johnson, Melvin, 265
Johnson, Walter, 72, 149,
Jones, Dr. ____, 47, 58
Joyce, Mr. and Mrs. ____, 273
K-2 (mountain in Pakistan), 63
Kabul, Afghanistan, 13, 14, 36, 60-2, 67, 123, 141, 178, 185, 199, 222, 268-9, 273, 279-80
Kabul River, 234, 271-2

Khan, Sawar, 185
Kaiser, Henry, 148
kala azar, 44
Kali, 88
kamakazi, 281
Kandy, Ceylon, 14, 17, 81-2, 92, 98, 110-3, 119, 280, 307,
Kane, CDR John and Amelie, 1, 2, 12, 25, 36, 39-40, 47, 54, 58, 73, 105, 127, 130, 151, 155, 261
Kane, Sheila and Suzanne, 127, 131
Kaniguram, 236
Karachi, India (now Pakistan), 10-15, 17-18, 21, 25, 30-167 (implied), 174, 178-9, 182-3, 255-6, 261, 267-8, 273, 276, 299-301, 308-22 (specific or implied)
[Karachi] Boat Club, 13, 37, 55, 67, 76, 121, 131, 134, 144
Karachi Club, 13, 50, 52, 80
Karachi Club Annexe, 144
karez, 249
Karlachi, 16, 223
Karoti, 249
Kasai Munidi, 188-9
Kashmir, India, 53-4, 58, 68, 100, 131, 156, 301
Keating, Lt. Col. Geoffrey, 245, 248
Kheyl, 234, 248 (also see Datta Kheyl)
Kellenbenz, Evakaterina, 1, 3
Kelley, Anna (née Zimmermann), 43, 69, 75, 92, 120, 132
Kelley, Richard Carlyle, 43, 69, 75, 120
Kelley, Richard C., Jr., CPT, 43, 69, 81, 92, 116-20, 132
Kenilworth, Ill., 148
Kennedy, Jacqueline, 4
Kennedy, John Fitzgerald, President, 4, 282
khan (title = ruler), 234,
khassadar, 187-8, 200, 232, 237-8, 240
Khotal, 58 (also see Peiwar Khotal and Landi Khotal)
Khowar, 68
Khyber Pass, 60, 63-5, 68, 131, 170, 174, 179, 184, 212-21, 259, 272-4, 279-80, 302
kike, 95 (derogatory, for Jewish)
Killarney Hotel, 101
Kim, 60, 62, 280
Kim (as name or nickname), 62, 280
King, Bill, and his wife Nell, 119
King, FADM Ernest J. (CNO), 316
Kipling, Rudyard, 280
Kirk, Ambassador Alexander C., 23-4
Klein-Eislingen, 1
Knott, Dodie, 109
Knox, Frank, 8
Kodachrome, 78, 91, 99, 117, 131, 153

Kodak (see Kodachrome)
Kohat, 222-3, 257, 272, 300
Koniguram, 235-40
Korea, 61, 280-2
　U.S. Army in, 281
Ku Hu-ming, 278
kullah (Afghan cap), 68
kukri, 275
Kung, Prince, 278
Kunming, 142
Kuntz, John J. "Jack", 92, 101
Kurram, 58, 64, 222, 224
Kurram Militia, 64, 224-5
Kurram River, 64, 224
Kushwakht, 236
La Rochelle, France, 278
Laaha, 236-7
Ladd, CDR ___ (ALSULO), 313
Ladha, 236
Lafayette, Indiana, 268, 271, 279
Lahore, India, 31-2, 59, 182, 329
Lama, 60, 121, 278, 286
Laman, Eric, 103
Lamarr, Hedy, 25, 103
Lamboit, Paul (FC, RAF), 28-31
Laminou, Eric
Lamour, Dorothy, 27
Lamsbury, Stuart, 144
Landi Khana, 220
Landi Khotal, 220
Langlands, ___, Brigadier (?), 45
lapis lazuli, 268
Laughton, Charles, 143
Lawis, P.A. ___, 229-30
Layman, Eric, 153
LDS (see Reorganized Church of Latter Day Saints)
League of Nations, 279
Leavitt, ___ (ONI), 311
Lee, Mary Nelson, 82
Leeper [AWZ spelled as "Liepor"] Mrs. ___, 302
Legation, American, 14, 20, 25, 61, 67
Leghorn, Italy, 42
Legion of Honor, Chevalier, 278
Legion of Merit, 61, 282, 286
Leishmaniasis, 44
Lend-Lease, 124, 249
Letty ___, 108, 112
Lewis, 1st Lt Clarence J., Jr., 2, 59, 72, 83, 94, 158-9
Liang Shih-yi, 278
Liberator – U.S. B24 bomber, 12, 27, 309
Liberty ship, 124-5, 148
Libya, 21
Lidirer, CDR ___, MC, USNR, 47

Lido, 131
Life, 313
life raft, 124-5
lime squash, 29
Limerick, Republic of Ireland, 32
Linaweaver, CAPT Paul Glenwood Jr., (ONI), 111, 316
Lincoln, Anne, 33-4, 41, 106
Lincoln, Crosby, 41, 53, 57
Lincoln, J. Freeman and "Ginny", 2, 14, 30, 41, 129 (see also Friends of Freeman [OSS])
Lincoln, Joseph Crosby "Daddy Joe" and Flo, 17, 41, 91
Lincoln county (Lincolnshire), England, 296-7
Lincoln Liberty, 55
Lintoll, Jack, Capt. (BA), 160
Liverpool, 277
llama, 23
Lohbeck, Don, 142
Long Island, 4, 55, 105-20, 129, 131, 261
Longworth, Nicholas, and Alice (née Roosevelt), 106
Lord, Ed, 100, 104
Lorenzo Marques, 14, 58
Lorre, Peter, 18
lorry / lorries, 64-5, 228, 232-6, 256
Los Angeles, California, 26
lotus flower, 23
Lowari Pass, 60-4, 78, 203-20, 256-8, 299
Lowis, ___, 227, 234
Loy, Myrna, 25
Lucky Pierre, 278
lungi (Afghan head wrap), 68
Lutheran, 277
Lyautey, Morocco, 12, 17
Ma Kabul – Enders' jeep, 199, 279
MacArthur, General Douglas, 279
Macedonia, 68
Macomber, William B., Hon., 59
Macy, Clarence E., 14, 36, 84, 107-8, 116, 126, 141-3, 158
Madame Curie, 105
Madras, 316
Mae West – life preserver, 28
Mahsud, 234-40
Maidstone Club, 4
Malakand, 62-3, 168-9, 173, 186, 191-202, 219
Malakand Field Force, 62, 198
malaria, 14, 29, 44, 71, 114-119, 129, 280
malik, 185, 189-91, 220, 230, 234, 236
Mallin, LT ___ (RN), 108
Manchuria, 281-2
Mandarin, 277-8
mangrove, 121
Manney Steel, 37

Marcian ___, 72
Marine Airport (Karachi), 98, 110
Marine Corps, U.S. (USMC), 21-2, 59, 85, 115, 159, 277-9
Mariposa, 81
Markey, CDR (later RADM) Gene, USNR, 12, 25, 103, 172, 311-3
Markey, Melinda, 103
Markley, ___ and Ruth, 67, 134, 151, 155
markor, 63, 219
Mathews, ___ (Pan Am pilot), 93
Mawhin, 90
McCarthy, Joseph (Senator), 24, 280
McClanahan, Walter, 22-7
McCown, ___, 160
McDonald, Bruce, 86
McIntosh, Elizabeth P., 81
McNeely, George and Allie, 72, 80-1
McWilliams, Julia, 82, 92, 110, 119
  (m. Paul Child; was then Julia Child)
Meadowbrook, Camp, 38
Mecca, 189, 274-5
Medal of Honor, 89
Mediterranean Sea, 28, 89, 300
Mediterranean Theater, 25
mehter (aka *mehtar*; also see Chitral) 59, 60, 62-3, 216-9, 236, 299
Mena, Egypt, 45
Menzies, MG Sir Stuart, 61, 126
Merion Cricket Club, 1-2, 151
mertensia, 96
M.E.S. (Military Engineer Services, Corps of Engineers, Indian Army), 234, 249
Mesopotamia, 29
Metropolitan Museum of Art, 59
Miami, Florida, 80
Military Observer, 280
MIS (Military Intelligence Service), see G-2
Miles, Commodore Milton "Mary", 74-5, 79-80, 87-8, 125, 157, 177-8, 307, 312-8
militia, 16, 60, 64-5, 223-5, 241-8, 297
Minora Island, 94, 108
Minter, Mary Miles, 74
Mir Ali, 222, 227, 234
Miram Shah, 59, 64, 227-35
Miranda, Carmen, 117
Mississippi, 21
Mohammed, 150
Mohammedan, 62, 64, 269, 272, 274
Mohmand Tribe, 61, 185, 220
Money, Maj. Gen. ___, 252
Mongolia, 282
monkey, 43, 88, 90, 96, 269-70
Monomoy Pointm 35, 43
monsignor, 148
monsoon, 30, 35, 43, 103, 116, 121

Montclair, New Jersey, 278
Montgomery, Marshal Bernard, 18, 27, 178
Moon to Dawn, 34
Moor / Moorish, 270-2
Morina Goa, 58
Mormon Church (LDS), see Reorganized Church of Latter Day Saints
Morocco, 12, 17, 281
Morpheus, 143
Morris, Duane, 308
mosquito, 14, 21, 29, 44, 47, 51, 104
Mother Carey, 160
Mountbatten, Admiral Lord Louis, 14, 25, 81, 98, 110, 141-2, 152, 307
movie, 2, 7, 12, 13, 17, 18, 23-4, 27, 30-1, 34, 43, 45, 48, 51-6, 78, 83-4, 99, 104-5, 110-11, 117-8, 128, 144, 153, 157, 218-9, 221, 261, 296
Mrs. Miniver, 105
Mt. Lavinia, 111
Mullargaris, 220
mullik, 60, 61, 64, 220 (also see *malik*)
Muscat, 121
Muslim / Moslem, 12, 13, 50, 54, 65, 81, 249, 254, 270, 274, 277
Mussolini, Benito Amilcare Andrea, 13, 34
Mutterings, 261
Muzaffar-ul-Mulk, Mehtar of Chitral, 63
N.Y. (see New York)
Nagasaki, 281
Nagpur, 96
Nanda Devi, 277
Nanie (see Zimmermann, Helene)
Napier Barracks, 160
Natalie ___, 100, 103
National Cathedral, 66
National Geographic, 124, 261, 302
National Hotel, Cairo, 22
National Security Agency, 61, 278
Naval Amphibious Base-Little Creek, 282
Naval censor (see censor, and Navy)
Naval District, 1, 2, 161
Naval Group, China (US), 74
Naval Liaison Office/Officer (NLO), 10-15, 25, 30, 32, 34, 38, 46, 87, 109, 132, 151, 162-4, 267, 306-9, 314-5, 310, 320 (also, implied for most correspondence in pp. 11-178)
Naval Officer in Charge (NOIC, British), 101, 102, 108, 154
Navy Day (27 October), 125-6, 129, 133-4
Navy Storage Facility, 222
Navy, Royal Indian, 94, 108-9, 125
Navy, US, 1-2, 8-12, 17-28, 32-3, 37, 45, 60, 63, 69, 70-1, 74, 81, 85, 87, 89-93, 96, 103, 108-9, 114, 121-9, 133-4, 144, 147-51, 155-

6, 159, 185, 219, 221, 261, 268, 302-16
  (also see entries, above, for Naval ___)
nawab, 60, 62-3, 68, 184, 199, 202-4, 219
Nazi, 2, 7, 18, 26, 34, 135
Nepal / Nepali, 96-7, 106, 114
Netherlands, 279
New Delhi, India (also see Delhi), 12, 81-2,
  109, 112, 142, 269-70, 307, 309
New Jersey, 278
New Year, 66, 68-9, 77, 149, 151
New York City, N.Y., 2, 10, 12, 22-3, 26, 59-
  62, 79, 80, 89, 95, 104-41, 147, 151, 155,
  158, 161, 315
New York Hospital, 278
New York State, 1, 2
New York Times, 103, 120, 141
New Yorker, 79, 131
New Zealand / New Zealander, 18, 23, 29
Nicholson, Sergeant Thomas, 225-6, 229, 245,
  256, 277, 280, 299-300
Nico, Charlie, 66
Night and Day, 42
Nile River, 27
Nixon, President Richard, 26
Nizam of Hyderabad, 56
Norfolk, Va., 282
Normandy, 312-3
Norquist, Eliot, USNR (ONI), 315
North West Frontier (aka Northwest,
  including NWF Province, India), 14, 58,
  63-4, 68, 71, 98, 184, 220-1, 228, 279, 302
North West RR, 182
Northfield, Mass., 115
Northwestern University, 281, 309
Norton-Hayes ambulance unit, 277
notre victorie est votre liberte, 19
Nowhere Else in the World, 60, 278, 286
nurse, 27, 55, 57, 69, 78, 145-8, 261, 277,
  299-300
NWFP (see North West Frontier Province)
Oahu, Hawaii, 281
O'Connor, LCDR Edward F., USNR (ONI),
  313
OBE (Order of the British Empire), 60, 279,
  302-3
Oberlin College, 66
Off, LCDR Clarence (ONI), 308, 310, 317
Office of Naval Intelligence (ONI),
  specifically at 2, 15, 121, 124, 306-7
Office of Naval Operations (OPNAV), 306
Office of Strategic Services (OSS), 2, 13, 39,
  41, 51, 55, 68, 74, 78, 81, 83, 85, 127
Office of the Naval Attaché, 12, 22, 25, 72,
  115 (also see Navy)
Officers Club, American (Karachi), 24
Okinawa, 280-1

One for My Baby, 117
One Gun Hill, 224
OP-16, 307-18
  also see Office of Naval Intelligence, ONI
OP-16-FE (ONI-Far East Division), 307-18
OPNAV (see Office of Naval Operations)
Operation Dragoon, 119
Operation Husky, 26
Operation Overlord, 89, 101
Operation Torch, 18
Oram, Mr. and Mrs. Jim and Kay, 52, 78, 94
Oran, Algeria, 12, 17, 43
Orange, Plan (Plan Orange), 279
Orpheus Club, Philadelphia (O.C.), 1, 10, 22,
  49, 78, 92, 121
OSS in China: Prelude to Cold War, 74
OSS: The Secret History, 74
ostrich, 69
Ott, John "Jack", 18, 39-41, 47, 50, 54-5, 66,
  69, 71, 73, 77, 80, 83, 96-7, 101, 107, 145,
  157, 159, 261
Oxford University, 272, 311
PA (see Political Agent)
Pachman, Major ___, 62, 197
Pacific, 2, 25, 41, 45, 81, 89, 106, 111, 142,
  154-5, 261, 268, 279, 280-1, 302, 322
Paiwar Khotal, 58, 64, 225-7
Pakistan, 12, 63-4, 98, 148, 261, 297
Palm Beach, 95
Pamir Mountains, 68
Pan Am, 92-5
Panama Canal, 282
Panchan Lama (aka Panchen), 9th, 60, 278,
  286
Paoli Local, 40
papaya, 56
Parachinar, 64, 222-32
parachute, 12, 28-30, 43, 277
parce que la guerre, 38
Parker, Dr. Alan and Janet, 24-5
Parsi, 13, 31, 47, 50, 65, 87, 249
Pashtun (see Pathan)
pastine, 67
Pathan (aka Pashtun; aka Pushto; aka Pushtu),
  62, 68, 131, 190, 216, 228, 271-4, 277
Patterson, Andy, 73
Patterson, Bud, 73
Paulo Gullino, 81,
Payne, Al (ONI, ALUSLO), 14, 56-7, 69
PBY (see Consolidated PBY), 114
Pearl Harbor, Hawaii, attack, 2, 25, 181, 268
Peepul Tree, 269
Pegler, Westbrook, 24
pehmina shawl, 58
Peiwar Khotal, 64, 223-7
Peking (see Beijing), 278

392

Pemberton, Henry, 308
Pennsylvania, 1, 3-7, 10-12, 15, 24, 26, 32, 37-8, 40-3, 47-8, 52-3, 60-1, 66, 70, 71-3, 80, 92, 101, 110, 121, 126-35, 143-61, 245, 261, 264
Pennsylvania Railroad, 53, 78
Pennsylvania, University of, 1, 4, 5, 32, 37, 43, 48, 53, 70, 132, 143
Pentagon, 281
Perry ____, 108, 132
Persia / Persian, 61, 194, 269, 277
Persian Gulf, 28, 30, 46
Peshawar, India, 56, 60, 63-5, 78, 82, 93, 135, 141, 168, 170, 174-6, 179-97, 209, 219-22, 268, 271-3, 279, 297-302
Peshawar Club, 59, 60, 175, 183, 302
Peshawar Fort, 185
Petri, 92
Pettie Institute (Peddie School), 86
Pettigrew, Col. ___ and Doris, 93, 96, 98, 100, 102, 104, 106, 108, 116, 118, 119, 121, 129, 131-2, 144, 148-51, 156, 159-60,
Petty, John, 153
Phi Gamma Delta, 143
Philadelphia, Pennsylvania, 1, 3-4, 7, 10, 12, 32, 38, 40-1, 47-8, 60-1, 66, 71-3, 110, 245
Philadelphia Country Club, 1
Philadelphia Tapestry Mills, 1
Philby, Harold "Kim", 62
Philippines, 279
Phillies, 52-3, 71, 110
Phlebotimanae (see sand fly)
Photo Album, 16, 81, 84, 90, 92, 113, 119, 124, 130, 133, 145, 149-50, 163-4
piano, 33, 39, 95-6, 102, 118
piastre, 28
Pierre, Lucky, French fighter pilot, 278
piquet, 193, 229, 242
Pike, Geoffrey, and Mrs. ___, 103, 126, 128, 148-50, 156
Pike, Harvey, 103
Pittsburgh, University of, 277
Pittsburgh, Penna., 277
Plan Orange, 279
Platt, Major ____, 252
PO (see Political Officer)
Poland, 142
poliomyelitis, or polio, 50, 53
Political Agent (PA), 60, 62-3, 169, 190, 192, 197, 199-200, 216, 226, 234-6
Political Officer (PO, British), 185
Polley, Cyrus, 151
Polo, 62, 107, 216, 218-9
Port Directors School, 147, 155, 162, 315
Port Trust, 141
Portland, Oregon, 21

Portugal / Portuguese, 13, 36, 58
Potsdam Agreement, 281
poulow, 62
pound (unit of money), 27
Pratt, Mrs. ____, 86
Presbyterian, 39, 120, 277
Preston, Consul General ____, 13, 58
Pride and Prejudice, 129
Priestley, ____, 153
Princeton University, 53, 309, 319
Proceed to Peshawar, 180-1, 209, 267, 302
Pryor, Virginia (OSS), 82
Psi Upsilon, 43
PT boats, 89
PT Squadron 102, 89
PT-41, 89
ptomaine, 118
Pu-yi, Emperor, 278
pukka (aka *pucca, puckka*), 95-6, 114, 121, 133, 144
Pullman, 69
Punjab, 56, 98, 269
Punjkara River, 201
Purdue University, 268, 271, 279
Purvis, Brig. (?) ___, 246
Pushtu, 62 (also see Pathan)
Pushto, 272 (also see Pathan)
Putman, Hal, 81
putthu, Chitrali, 276, 300
pyramids, 12, 23, 28, 45
Qantas, 115
Qatara Depression, 28
quarantine, 48
Queens Way, New Delhi, 271
Quetta, 14, 46, 64-7, 78, 98, 123, 135, 169, 174, 179-80, 219, 248-55, 267, 280, 299-301
quinine, 44
race, horse, 23, 33, 84, 93, 104, 132, 156, 159, 160
Racquet Club, 134
R.A.F. (Royal Air Force), 42, 51, 56, 69, 76, 135, 152, 228, 272
Ralli Brothers, 102-3
Ram, 88
Rangoon, Burma, 102, 308
ransom, 64-5, 227-8
Rapid Creek, Wyoming, 17
Rauch, Mr. and Mrs. ___, 58
Rawan, 88
Reader's Digest, 69, 124
Red Cross (American Red Cross), 27, 67, 76, 147-8, 277
Refbord, ____, 80
regatta, 94
Reggio Calabria, Italy, 27
Reilly, Mrs., 27

Reinitz, Dr. ____ and Mrs., 44, 58, 99
Remington Rand, 151
Reorganized Church of Latter Day Saints (LDS; Mormon Church), 1, 120
Rethbord, LT ____, 66
Reynolds, Quentin, 12, 24
Richardson, ____, 110
Rio de Janeiro, Brazil, 141
Rivett-Carnack, Mrs., 76
Roberts, Lord Frederick, VC, 16, 224
Robinson, ____, 272
Rochester, N.Y., 24, 33
Rockefeller Foundation, 148
rodeo, 145-6
Rogers, "Chub", 36
Rome, Italy, 43
Rommel, Marshal Erwin, 12, 18
Room, The (FDR's secret spy group), 2, 121
Roosevelt, Eleanor, 280
Roosevelt, Elizabeth (née Donner), wife of (1) Elliott Roosevelt and (2) Curtin Winsor, 66, 77, 308, 319
Roosevelt, Elliott, 66, 319
Roosevelt, Franklin Delano (FDR), 2, 18, 22, 50, 61, 66, 121, 141, 280
Roosevelt, Kermit "Kim" Jr., 62
Roosevelt, President Theodore, 106
Roosevelt, William Donner, 66
Rosengarten, Peter, 156
roulette, 20, 95
Royal Air Force (see RAF)
Ruatan, Frances (née Sloan), 76
ruby / rubies, 57
rupee (Indian currency), 47-8, 55, 83-4, 94, 117, 127, 129, 300
Russell, Richard, U.S. Senator, 14, 37-8, 167
Russia / Russian, 13, 31-2, 42, 61-2, 68, 80, 273, 280
Ruter, Major General ____, 91
Rye, Robert, USNR (ONI), 315
SACSEA (Supreme Allied Command Southeast Asia), 307
Sadd, CDR ____ (ONI), 317
Sadler, CDR ____ (ONI), 309-10, 317
Sahib, Khan, 302
Sahib, Mir Ali, 222
Sahib, Sheikh, 222
Saidu, Swat, 194-5
Saif-ul-Mulk Nasir, Mehtar, 63
Saif-ur-Rehman, Mehtar, 63
Sambaza, 241-2, 246
Sampan, 125
Samuels, Barney, 309
San Francisco, Calif., 18, 41
sand fleas (sand flies), 29, 44, 68
sand fly fever, 13, 44, 129, 300

sandalwood, 26
Sandeman, Fort, 58, 65, 240-52
Sandhurst. 60, 297, 306
Sandspit, 13, 76-7, 92, 98, 102, 121, 132, 168
sapphire, 92
sari, 51
Saturday Evening Post, 17, 114
Saveney, France, 278
Say, Allen, 150
Schrader, CAPT ____ (ONI), 317
Schroeder, Harry, M.D., 148
Schuirmann, RADM ___ (DNI), 312, 317
scorpion, 49-50, 53
Scottish, 274
Scouts, 62, 64, 204, 209, 211, 227, 229, 233-7, 299
sea sickness, 122
Searle, P.A. ____, 243-4
Seavens, 53
Secret Intelligence Service (MI-6), 60
Secretary of the Navy, 8, 90, 121
Sennett, Mack, 52
sepoy, 62, 113, 132
serai, 203
Services Club, Peshawar, 176, 221
Services Hotel, Peshawar, 60, 299-302
Services of Supply (SOS), 45, 142, 163
Severn, Hecker and Ellie, 70, 72, 79, 92, 131, 145, 156
Sexias, Amelie, 1, 58
Sexias, Vic, 1
Shabkadar, 135, 185-7
Shagai, 220
Shahigu, Fort, 248
Shanghai, 271, 278
Shanghai Emergency Medal, 279
Shangri-la, 96-7
Shah of Iran (Reza Shah Pahlavi), 320
Shawet, 249
shawl, 300
sheep, fat tail, 78, 250-1
Sheikh, ____, 13, 31, 304
Sheiku, ____, 222
Sheikh, Mir Ali, 13, 31, 310
Shen Shung, 278
Shingari, 220
Shirley, ____, COL (U.S.), 158
Shiva, 88
Shoemaker, Ann, 89
Shoemaker, Barbara (see Zimmemann)
Shoemaker, Dorothy "Dot", 58, 156
Shoemaker, Dr. William Toy, 4, 48, 88
Shoemaker, Jesse Warren, 89, 100, 142
Shoemaker, Mabel "Mabs", 89
Shoemaker, Ralph Warren, 89
Shoemaker, Robert Comly, 89

Shoemaker, Theodore "Ted", 100, 102, 105
Shoemaker, William, 89
Shoemaker, William Brock, 72
Shriner, 51
Sicily, 12, 26, 28
Sigma Tau, 1
Sikh, 59, 64
Silk Road, 279
Silver Prawn, 112
Simmy ____, 141
Sims, Joseph Clark, Jr. "Sandy", 92
Sims, Joseph Patterson, and Nancy, 50, 92
Sind, 13, 31, 33, 47, 56, 59, 63, 82, 85, 97, 117, 142, 182
Sind Club, 13, 51, 55, 67, 87, 96, 127, 133-4, 148, 153
Sind Observer, [Karachi], 94
Singh, Jowar, 277
Singer Sewing machine, 274
Singhalese, 57
SACO, 74
SATO, 74
Sisterhood of Spies, 81
Siva / Shiva, 58, 88
Skull and Bones, 2, 121
Sloan, Burrows, Jr. "Buzz", 76, 108, 149-50, 321
Smith College, 78, 82, 92, 94, 101, 141,
Smith, F. Howard, LCDR, USN, 13, 31-3, 37, 59, 66, 74, 87-90, 94, 157, 165, 182, 310-7
Smith, Joel, 79
Smyth, Maj. D. Montgomery and Joan, 13, 33, 38, 47-8, 58, 96, 98, 144
Snow, Edgar, 12, 26
Sora Rogha, 236-7
Sorby, Stella, 148, 150
Sousse, Tunisia, 12, 17, 44
South Waziristan (see Waziristan)
Southampton, N.Y., 120
Southern Home for Children, 261
South-East Asia Command (SEAC), 14, 81, 92, 98, 110, 141
Spain / Spanish, 29, 142, 277
Sparks, Joseph (U.S. Vice Consul), 158
Sphinx, The, 12, 28, 45
Sphinx Honorary Society, 1
Spinwam, 227-8
Spring Fever, 55
St. Albans Naval Hospital, 161, 261
Stair, ____, 76
Stalin, Marshal Josef, 61
Standard Oil, 13, 31, 43, 51, 86
Standard-Vacuum Oil Co., 67
Star Spangled Banner, 52
Stare, LCDR Pat and Mrs. ____ (RN), 80-3
Steel, CDR ___, MC, USNR, 148

Steel, Manney, 38
Steinberg at the New Yorker, 79
Steinberg, Saul, 79
Stilwell, GEN Joseph, 141-2
Stores, Lt. Com. and Mrs., 76
Stratemeyer, LTGEN George, 141-2
Strauss, ____, 66, 120
Streather, Col. Anthony, CBE, 63
Street, Ed, 48
Strong, MG George Veazey, 12, 23-4
Stroud, Ms. ____, 33
Stubbs, LCDR Daniel, USNR, 319
Strubing, Jack, 85
Stuble, Louisa, 127
Stuble, William H. "Bill," Esq., 54, 73, 85, 146, 151, 155
Sturt, Lieut. ____, 76
Stuttgart, Germany, 1, 2, 7, 33
Suez Canal, 282
Suleiman Kheyl, 249
Sunni, 227
Sunny ____, 78
Suter, Morgan, 83
Swabe, ___, Flight Lieut., RAF, 70, 86, 95
Swat, 62, 169, 192-8, 257
Swat River, 62, 192-3, 196-7
Swayne-Thomas, Geoff & April, 79, 167-70
Swedenborgian Church, 120
Sweet, Sidney (ONI), 311
Syria, 141, 279
Tai Li, Lt. General (China), 74, 178
Tiarze, 236
Taj Mahal, 112
Tale of Two Cities, 27
Tambazzi, ____, 144
tank traps, 135, 220, 224
Tannai (aka Tanai), 234, 246
Tarbaby, 54
Tari Mangal, 224
Tatarar, Mount, 271
Tata airline, 310
TATA industry, 103
Taylor, Jeanne (OSS), 82
Technicolor, 144
Teetor, Hilda, 85
Teheran, 61
Teheran Conference, 61
Tel el Kebir, 23
Tennyson, ___, 109
Tenth Air Force, U.S., 141, 222
Texas, 21
Thai / Thailand, 79, 282
Thal, 222-33, 300
Thanksgiving, 70, 144, 146, 149, 265
Thayer, Charles W., 14, 24, 60-1, 67-8, 172, 276, 280

Thayer, George, 60-1
Thayer, John W., Sr., 61
Thayer, John W., Jr., "Jack", 2, 60-1, 67
The Centerville Ghost, 143
The Glass Key, 77
Colonel Blimp, 144
The New Yorker, 79
The Room (see Room, The)
Thief of Bagdad, 273
Thiry, Eleanor (OSS), 32
Thomas, Gould, 83, 311
Thomas, Mrs. Dwane, 301
Thomas, Lowell, 60, 175
Thorne, Freddy, 128
Thornburg, ____ (Assistant P.A.), 216, 219
Thornton, CAPT Thomas A., 1, 12, 25
Tibet / Tibetan, 60, 156, 268, 277-8, 282
Tigris River, 46
Tim's Air Express Edition, 54
Timoney, Arthur, 36, 71, 73, 75, 78, 85, 114
tikala, 234, 245
Tirichmir, Moutain, 63, 215-8
Tirwaza, 237
Titanic, 2, 61
Tochi, 64, 227
Tochi Scouts, 227-34
tonga, 271-4
Tora Bora Mountains, 64, 224
Tora Gharu, 242, 245
torpedo guidance, 103
torquoise [sic], 268
Tower of Silence, 87
Town & Country, 123
Trafalgar Day, 134
Train, Rear Admiral Harold C., (DNI), 10
Treanor, Tom, 26
Treichler, F. T., 285
Tribal Territory, 60-1, 64, 123, 184-90, 220, 271
Tripoli, Libya, 12, 17-18, 21, 36, 39, 43
Trippe, Juan, 93
Tsu Hai-san, 278
Tunisia, 12, 18, 281
Turkey, 61
typhus, 148
U.A.R. (United Arab Republic), 282
ulcer, 83, 91, 95, 162, 261
Ulla, Masih, 277
un seul bout – Victoire, 19
Union Station, Washington, D.C., 68
University of California, 61
Uramur, 236
Urdu, 37, 62, 80, 134, 277
U.S. Army in Korea (see Korea)
Van Dusen, William, and Lillian, 86, 103-5, 158

Vassar College, 50, 82
Vaswani, Professor ____, 80
Venice, 131
Viceroy of India, 63-4, 82-5, 177, 219-21
Victory Ribbon, 278
Victrola, 48, 55, 76, 121, 146
Vienna / Viennese, 35, 39, 44, 61
Villanova, Pa., 67
Vogue, 85, 92, 97
Voorhees, LT Howard, USNR, 87, 94, 97, 104, 108, 120, 124-5, 129, 150, 159, 161-2, 166, 307-17
vulture, 65, 87, 249, 269
W. Colston Leigh Bureau (agency), 521 Fifth Avenue, NYC, 288
Wakhan, 61
wali, 60, 62, 184
Wali of Swat, 62, 194
waliard (aka *walihad*), 194, 199, 202, 204
Walihad of Swat, 194, 199, 202, 204
Walter J.____, to Philadelphia, 105
Wana, North-West Frontier Province, 58, 64, 236-47
War Services Exhibition, 108
War Shipping Administration (W.S.A), 39, 78, 145
Warden, BG John A., 45, 108, 133, 141-2, 158, 163, 173
Warren Brothers Co., 69, 141, 148
Warren, Frederick, 142
Warren, Herbert Marshall, 142-3
Warren, Mabel, 3
Warren, Ralph and Catherine, 57, 143
Washington, D.C., 315
Washington Navy Yard, 9
Washington Post, 106
Watt, CAPT ____ (RN), 158
Wavell, Marshal Archibald, 63, 82, 88, 142, 178, 270-1, 279, 293
Wazir, 227, 234, 236
Waziristan, 64, 100-1, 132, 135, 218-9, 235-8, 257, 272, 297, 299, 304
W.B.A.A, 269
Wellesley College, 94
Welsh, Stan, 59
West Point, 280
Weston, Tommy, 80, 84, 126
Wharton Business School, 151
Wheeler, RADM C. Julian, 125-6
Wheeler, MG Raymond A., 270, 273
Wheeler, "Skinny", 120
Whitehead, ____, 122
Wilder, Major ___ (British), 69
William I. Kip, 125
Williams, Col. and Mrs. ____, 83
Willow Grove, Pa., 52-3

Wilmer, Harry, 48
Winsor, Curtin "Curt", LT, USNR, 10, 16, 31, 39, 41, 54, 66, 70-1, 73-4, 78, 87-8, 93, 124, 129, 132, 151, 153, 155, 157, 267, 307-19
Winsor, Curtin, Jr., 66
Winsor, Elizabeth (née Donner, née Roosevelt), 66, 319
Winsor, James Davis III "Jim"; and Sally, 39, 41, 54, 58, 72-3, 76, 81, 94, 105, 120, 127, 156, 308
wog ("worthy oriental gentleman"), 27, 41
Women's Army Corps, 260
Wood, "Father" ___, and Mrs. (Quetta, and in Hanger, south of Kohat), 252, 256, 300
Woods, Dr. ___, 148, 322
Woods-Ballard, Maj. A. A., 252, 255
Wooster, Ohio, 277, 291
Word, MAJ Karl (US), 98
World War I, 277
WREN = aka Wren, 27, 111-2, 142, 158
Wright, Lucille, 25
Wu Ming-fu, 277
Wynn, "Pop", 249
Wynnewood, Pa., 52
Xray and Electrotherapy Institute, 274
Xmas (see Christmas)
Y.M.C.A., British, 75, 79, 82
Yale University, 2, 53, 74, 120, 309
Yardley's, 54
yellow fever, 14, 44, 47
yogi, 69
York (aka Avro York), 315
Yorke-Torr, Mr. ___ and Mrs. (Janah), 76, 102, 104
Young, Connie, 73
Yu, Maochun, 74
Yugoslavia, 280
Yusef Khel, 60, 185-91

Zacharias, Rear Admiral Ellis, 10
Zedong, Mao, 26
Zhob, 58, 65, 234, 241-9, 297
Ziarat, 168, 209, 214-5, 219
Ziegfield Follies, 117
Ziesing, Dick and Martha, 40, 77, 99, 101
Zimmerman, Anna Cecelia Mackinlay de, wife of John C., 301
Zimmerman, of Penna, son of John C., 301
Zimmerman, John C., friend of Nancy Bromhead's grandfather, 301
Zimmermann, Albert Walter, Jr. "Albie", 6, 15, 26, 36, 42-3, 58, 67, 71-2, 79, 82, 87-8, 94-6, 100, 102, 105, 118, 126, 128, 132, 146, 154, 156-7, 264-5
Zimmermann, Albert Walter (Al, AZ, AWZ), 1-135, 141-64, 166-8, 171-9, 181-4, 196, 208-9, 218-22, 224, 226, 234, 256-68, 275-6, 286, 297, 301-2, 306-22
Zimmermann, Barbara (née Shoemaker) "Dearest" (BSZ) 1, 4-6, 11, 15, 17-135, 141-61, 172-3, 261-6, 307-10, 314-5
Zimmermann, Barbara Warren "Babs" (m. Melvin Johnson), 6, 15, 31, 33, 35-6, 38-9, 42-3, 45, 48-50, 54, 57, 59, 67, 73, 75-9, 82, 89, 95-104, 108, 110-17, 152, 122, 126, 130, 141, 145-7, 152, 156-7, 160, 264-5
Zimmermann, Helene "Lanie" "Nanie", 6, 15, 35, 37, 48-50, 53, 66-7, 69-71, 75-8, 82, 92, 95, 101-2, 104-5, 109-10, 114, 118, 121-2, 126, 130, 146, 157-60, 264-5
Zimmermann, John, Sr, 1, 92
Zimmermann, Warren "Warr", 6, 15, 26, 28, 32-3, 35-40, 48-52, 57, 65-72, 79, 82, 89-102, 106, 109-10, 113, 115, 118, 121-2, 126, 128, 131-2, 146, 152, 154, 156, 264-5
Zug, James, 151

of making many books there is no end
Ecclesiastes 12:12 (KJV)

Other Books by the Author

Leprosy in Five Young Men
Outpatient Surgery (3 editions)
Clinical Oncology, with John Horton
Edison's Environment (2 editions)
Intimate Relationships: Church and State in the U.S. and Liberia (2 editions)
Proceed to Peshawar

©JanPressPhotomedia

ABOUT THE AUTHOR

GEORGE J. HILL, M.D., M.A., D.Litt., is Professor of Surgery Emeritus at the New Jersey Medical School, Rutgers University. A fifth-generation Iowan, he was born in Cedar Rapids and graduated from high school in Sac City, where he became an Eagle Scout and received the Bausch and Lomb Science Award, the National Thespian Award, and was a member of the Iowa State prize-winning percussion ensemble and two-mile relay team. He was placed on his high school's Honor Roll of Graduates.

He attended Yale College, where he graduated with High Honors in history, and Harvard Medical School, on scholarships, while working part time at many jobs: as a ranch hand, cub reporter, salesman, lab tech, and science teacher at a junior college. He then began a 40-year career as a surgeon, scientist, and medical school professor. He wrote prize-winning books and papers on subjects as diverse as leprosy, malaria, toxicology, public health, cancer, and surgery. He served as chairman of advisory panels for the U.S. government, many professional societies, the American Cancer Society, Boy Scouts of America, and the Y.M.C.A. He was elected as a Fellow of the Royal Society of Medicine, the American College of Surgeons, and the Explorers Club.

After retiring from surgery and oncology, he continued to work as a historian, author, genealogist, and popular lecturer. He has also travelled with his wife and family to more than 50 countries and they have trekked on all seven continents.

He has had a simultaneous career in U.S. military service, beginning in 1950, when he volunteered for the U.S. Marine Corps. He became a military parachutist, and he retired as a Navy Reserve Medical Corps captain in 1992, having been on active duty in four wars: the Korean War; the Cuban Missile Crisis of the Cold War; the Viet Nam War; and the First Gulf War. His Reserve service included duty in the Navy Command Center at the Pentagon, as Chief of the Contingency Planning Branch of U.S. Navy Medicine. His many awards and decorations include the Navy Meritorious Service Medal, the Gorgas Medal, the New Jersey Distinguished Service Medal, the Outstanding Service Medal of the Uniformed Services University; and as a civilian, the St. George Medal of the American Cancer Society and the Distinguished Eagle Scout Award.

In 2006, he was invited to give the keynote address for the annual Navy Birthday celebration at the National Reconnaissance Office. He spoke to the NRO about the fictional Doctor Stephen Maturin in Patrick O'Brian's novels about Jack Aubrey and warfare at sea at the end of the eighteenth century. The title of his lecture was "Master and Commander, Surgeon and Spy." His book, *Proceed to Peshawar*, won a Finalist's Medal at the Indie Book Awards festival in 2015.

www.ingramcontent.com/pod-product-compliance
Lightning Source LLC
Chambersburg PA
CBHW081414230426
43668CB00016B/2238